Methodological Challenges in Nature-Culture and Environmental History Research

This book examines the challenges and possibilities of conducting cultural environmental history research today. Disciplinary commitments certainly influence the questions scholars ask and the ways they seek out answers, but some methodological challenges go beyond the boundaries of any one discipline. The book examines: how to account for the fact that humans are not the only actors in history yet dominate archival records; how to attend to the non-visual senses when traditional sources offer only a two-dimensional, non-sensory version of the past; how to decolonize research in and beyond the archives; and how best to use sources and means of communication made available in the digital age.

This book will be a valuable resource for those interested in environmental history and politics, sustainable development and historical geography.

Jocelyn Thorpe is Associate Professor of the Women's and Gender Studies Program at the University of Manitoba, Canada. Her interdisciplinary research focuses on the history and legacies of, as well as challenges to, colonialism in the Canadian context, examining how past discourses and relationships of power influence the present.

Stephanie Rutherford is Associate Professor in the School of Environment at Trent University, Canada. Her research inhabits the intersections among political ecology, critical animal studies, biopolitics, and historical geography.

L. Anders Sandberg is Professor of Environmental Studies at York University, Canada. His two most recent books are the co-edited *Urban Forests, Trees, and Greenspace: A Political Ecology Perspective* (Routledge 2014) and *Post-Industrial Urban Greenspace: An Environmental Justice Perspective* (Routledge 2015).

Routledge Environmental Humanities

Series editors: Iain McCalman and Libby Robin

The *Routledge Environmental Humanities* series is an original and inspiring venture recognising that today's world agricultural and water crises, ocean pollution and resource depletion, global warming from greenhouse gases, urban sprawl, overpopulation, food insecurity and environmental justice are all *crises of culture*.

The reality of understanding and finding adaptive solutions to our present and future environmental challenges has shifted the epicenter of environmental studies away from an exclusively scientific and technological framework to one that depends on the human-focused disciplines and ideas of the humanities and allied social sciences.

We thus welcome book proposals from all humanities and social sciences disciplines for an inclusive and interdisciplinary series. We favour manuscripts aimed at an international readership and written in a lively and accessible style. The readership comprises scholars and students from the humanities and social sciences and thoughtful readers concerned about the human dimensions of environmental change.

Methodological Challenges in Nature-Culture and Environmental History Research

Edited by Jocelyn Thorpe, Stephanie Rutherford and L. Anders Sandberg

Routledge
Taylor & Francis Group
LONDON AND NEW YORK

First published 2017
by Routledge

2 Park Square, Milton Park, Abingdon, Oxfordshire OX14 4RN
711 Third Avenue, New York, NY 10017

Routledge is an imprint of the Taylor & Francis Group, an informa business

First issued in paperback 2018

British Library Cataloguing-in-Publication Data
A catalogue record for this book is available from the British Library

Library of Congress Cataloging-in-Publication Data
Names: Thorpe, Jocelyn, editor. | Rutherford, Stephanie, editor. | Sandberg, L. Anders, 1953- editor.
Title: Methodological challenges in nature-culture and environmental history research / edited by Jocelyn Thorpe, Stephanie Rutherford and L. Anders Sandberg.
Description: New York : Routledge, 2016.
Identifiers: LCCN 2016019446 | ISBN 978-1-138-95603-2 (hbk) | ISBN 978-1-315-66592-4 (ebk)
Subjects: LCSH: Human ecology—Research—Methodology. | Natural history—Research—Methodology. | Nature—Effect of human beings on—Research—Methodology.
Classification: LCC GF26 .M46 2016 | DDC 304.2072/1—dc23LC record available at https://lccn.loc.gov/2016019446

ISBN: 978-1-138-95603-2 (hbk)
ISBN: 978-1-138-61520-5 (pbk)

Typeset in Goudy
by FiSH Books Ltd, Enfield

"The book opens up a world of possibilities. Contributors invite us to rethink established modes of academic production, to decolonize our methodological inheritances, to find ways to understand and trace nonhuman actors as well as engage with the full palette of human sensory perception. Doing research requires us to take chances and to put ourselves out there – this collection does that and more." — ***Steven High, Concordia University, Canada***

"Thorpe, Rutherford and Sandberg have brought together a team of brave and gifted interdisciplinary scholars, who assiduously and judiciously scrutinize the generative intersections amongst place, body, mind and spirit. The contributors provide effective commentaries on the methodological challenges faced by scholars looking for voices beyond conventional texts and archives, and give me hope that navigating such methodological challenges will yield fresh, elegant and generative results." — ***Joy Parr, University of Western Ontario, Canada***

Contents

Figures

Contributors

Marco Armiero is the Director of the Environmental Humanities Laboratory at the Royal Institute of Technology, Stockholm. He has been post-doctoral fellow and visiting scholar at Yale University, UC Berkeley, Stanford, the Autonomous University in Barcelona, and the Center for Social Sciences at the University of Coimbra, Portugal. In English he has published *A Rugged Nation: Mountains and the Making of Modern Italy* (2011), several articles and book chapters, and he has co-edited with Marcus Hall *Nature and History in Modern Italy* (2010) and with Lise Sedrez *The History of Environmentalism* (2014). His research focuses on nation and nature, migration and the environment, and environmental justice. Marco serves on the editorial boards of *Capitalism Nature Socialism* and *Environmental Humanities*.

Bawaka Country, including **Laklak Burarrwanga, Ritjilili Ganambarr, Merrkiyawuy Ganambarr-Stubbs, Banbapuy Ganambarr, Djawundil Maymuru, Sarah Wright, Sandie Suchet-Pearson, Kate Lloyd** and **Jill Sweeney**, are an Indigenous and non-Indigenous research collective. They are four sisters, elders and caretakers for Bawaka Country in northeast Arnhem Land, and their daughter, Djawundil. They are also three non-Indigenous academics, Sarah, Sandie and Kate from Newcastle and Macquarie universities, who have been adopted into the family as granddaughter, sister and daughter. Non-Indigenous academic Jill is a human geographer who has supported the collaboration's work, contributing valuable insights into the theoretical literature. Bawaka Country refers to the diverse land, water, human and nonhuman animals (including the human authors of this paper), plants, rocks, thoughts and songs that make up their Indigenous homeland of Bawaka. Theirs is a story of lives entwined and of new places of being and belonging. It is also a collaborative narrative of unexpected transformations, embedded families and the spirituality and agency of nonhuman elements in and of the landscape. The group members have worked together as a research collective since 2006, and have written two books and several academic and popular articles together.

Stephen Bocking is Professor of Environmental Policy and History in the School of the Environment at Trent University in Peterborough, Canada. He joined

Trent in 1994, after receiving his PhD in the history of science from the University of Toronto. He teaches courses on science and environmental politics and environmental history. In his research he studies the evolving roles of science in environmental politics. Case studies now underway include the salmon aquaculture industry, biodiversity conservation, and science in northern Canada. He is also interested in local environmental history. He has published many scholarly articles, as well as a few books: *Nature's Experts: Science, Politics, and the Environment* (Rutgers University Press, 2004); *Biodiversity in Canada: Ecology, Ideas, and Action* (Broadview Press, 2000); and *Ecologists and Environmental Politics: A History of Contemporary Ecology* (Yale University Press, 1997).

Sonja Boon is Associate Professor of Gender Studies at Memorial University in Newfoundland and Labrador, Canada. She has research interests in the areas of identity, citizenship, embodiment, migration, life writing, and feminist theory. Her work appears in *Life Writing*, *Journal of Women's History*, *SubStance*, *International Journal of Communication*, and *The European Journal of Life Writing*, among others. Her second monograph, *Telling the Flesh: Life Writing, Citizenship, and the Body in the Letters to Samuel-Auguste Tissot* (McGill-Queen's University Press), appeared in September 2015.

Kimberly Coulter directs the Environment & Society Portal, a digital gateway to openly accessible resources on human-environment relationships, at the Rachel Carson Center for Environment and Society at Ludwig-Maximilians-Universität in Munich. She also teaches Digital Environmental Humanities and co-edits the blog "Ant Spider Bee." Before arriving to Munich in 2009, she worked as an architectural draftsperson, lecturer, editor, and cartographer and earned a Ph.D. in geography from the University of Wisconsin–Madison. Her dissertation, supported by DAAD fellowships at the University of Bonn, traced appeals to territorial interest and identity in European film production and distribution processes. Her ongoing research interests concern the transnational production and distribution of media and knowledge products.

Aimée Craft is an Indigenous lawyer, professor of law at the University of Manitoba and Director of Research at the National Centre for Truth and Reconciliation. Her expertise is in *Anishinaabe* and Canadian Aboriginal law. In her decade of legal practice at the Public Interest Law Centre, Professor Craft has worked with many Indigenous peoples on land, resources, consultation, human rights and governance issues. Her award-winning 2013 book, *Breathing Life Into the Stone Fort Treaty*, focuses on understanding and interpreting treaties from an *Anishinaabe inaakonigewin* (legal) perspective. Professor Craft is past chair of the Aboriginal Law Section of the Canadian Bar Association and member of the Speaker's Bureau of the Treaty Relations Commission of Manitoba. In 2011, she received the Indigenous Peoples and Governance Graduate Research Scholarship.

Wilko Graf von Hardenberg is research scholar at the Max Planck Institute for the History of Science in Berlin, where he coordinates the working group "Art of Judgement." He holds a degree in history from the University of Turin in Italy and a Ph.D. in geography from the University of Cambridge, UK. He also co-edits the blog "Ant Spider Bee" and is the curator of "Arcadia: Online Explorations in Environmental History," a collaboration of the Rachel Carson Center for Environment and Society and the European Society for Environmental History. Immediately before moving to Berlin he was DAAD Visiting Professor of Environmental History at the University of Wisconsin–Madison, USA, where, beside other topics, he taught digital history. Previously he also worked and researched at the Rachel Carson Center and the Deutsches Museum in Munich, Germany, the University of Trento, Italy, and the Scuola Normale Superiore in Pisa, Italy.

Michael J. Hathaway is Associate Professor of Anthropology at Simon Fraser University in Vancouver, Canada. His first book, *Environmental Winds: Making the Global in Southwest China* (University of California Press, 2013), explores how environmentalism was refashioned in China, not only by conservationists, but also by rural villagers and even animals. His second project examines the global commodity chain of the matsutake, one of the world's most sought after mushrooms, following it from the Tibetan Plateau to markets in urban Japan. He works with the Matsutake Worlds Research Group, looking at the social worlds this mushroom engenders in Canada, the US, China, and Japan.

Sabine Höhler is Associate Professor of Science and Technology Studies at KTH Royal Institute of Technology in Stockholm. She holds a MSc in physics and a PhD in history of science and technology. Her research addresses the earth sciences in the nineteenth and twentieth centuries: aviation and atmospheric physics; oceanography and deep-sea exploration; and space flight and ecology. Recent book publications include the co-edited volume *Civilizing Nature: National Parks in Global Historical Perspective* (Oxford/New York 2012; in paperback 2015) and *Spaceship Earth in the Environmental Age, 1960–1990* (London 2015) on the spaceship as a key metaphor in the late-twentieth century debate over the world's resources and the future of humankind. Among her ongoing projects are a study of remote sensing technologies and earth imagery titled "Views from a distance" (with Nina Wormbs) and a project on collecting, classification and archiving technologies in nature conservation.

Katherine Jones is a Research Associate on the Arts and Humanities Research Council project Towards Hydrocitizenship. She comes from a human geography and planning background, having been awarded a PhD in 2015 for her work with an eco-village in Pembrokeshire, Wales, which focused on the interactions between different understandings of human-environment relationships – particularly the ways in which more radical or alternative understandings challenge policy-based understandings of sustainable development. She maintains an interest in how humans can live in a more respectful way on this beautiful planet, and what research can contribute to uncovering alternative ways of living and being.

Owain Jones gained a PhD in Cultural Geography at the University of Bristol in 1997 and since then has researched and written about many aspects of nature-society relations, landscape, place, memory and the environmental crisis. He has conducted research projects on trees, place and landscape; floods, communities and memory; food chains and ecology; tidal landscapes; animals and society; and children, nature and place. His is currently leading a £1.5 million Arts and Humanities Research Council Connected Communities project which involves eight UK universities, community partners and artists in four case study areas across the UK. This project is seeking to creatively explore and transform connections within and between communities, and communities and nature, in relation to water issues. Owain was appointed as the first Professor on the Environmental Humanities in the UK in 2014 at Bath Spa University. He has published over 70 scholarly articles and two books – *Geography and Memory: Identity, Place and Becoming* (2012) with Jo Garde-Hansen; and *Tree Cultures: The Place of Trees, and Trees in their Place* (2002) with Paul Cloke.

Dolly Jørgensen is an environmental historian with broad research interests, ranging from medieval to modern history. Her research areas have included medieval forestry and agriculture, late medieval urban sanitation, offshore oil policy, animal reintroduction, and environmentalism in science fiction. She runs an active project blog "The Return of Native Nordic Fauna" (http://dolly.jorgensenweb.net/nordicnature) about her current project on animal reintroduction, rewilding, and deextinction. She has co-edited two volumes: *New Natures: Joining Environmental History with Science and Technology Studies* (2013) and *Northscapes: History, Technology & the Making of Northern Environments* (2013). She was a practicing environmental engineer before earning a PhD in history from the University of Virginia in 2008. She is Associate Professor of History of Environment and Technology at Luleå University of Technology, Sweden.

Finn Arne Jørgensen is Associate Professor of History of Technology and Environment at Umeå University, Sweden. Before coming to Umeå, he earned a PhD in Science and Technology Studies at Norwegian University of Science and Technology (2007). His research includes studies of waste and recycling histories in Scandinavia and the US, the history of the Norwegian leisure cabins, material culture and consumption studies, and the connections among environmental humanities, media studies, and digital humanities. He is the author of *Making a Green Machine: The Infrastructure of Beverage Container Recycling* (Rutgers University Press, 2011) and co-editor of *New Natures: Joining Environmental History with Science and Technology Studies* (University of Pittsburgh Press, 2013). His most recent article is "Why Look at Cabin Porn?" in *Public Culture* 3, 2015. He is also the Digital Content Editor for *Environmental History* and co-edits the blog "Ant Spider Bee."

Sean Kheraj is Associate Professor of Canadian and Environmental History at York University, Toronto, Canada. He is also the director and editor-in-chief of the Network in Canadian History and Environment. In 2013, he published *Inventing Stanley Park: An Environmental History*.

Lianne C. Leddy is *Anishinaabe kwe* and a citizen of the Serpent River First Nation in Ontario. She is an Assistant Professor of Indigenous Studies at Wilfrid Laurier University's Brantford campus. Leddy specializes in Indigenous history and is interested in research questions around methodology and the impacts of colonialism on Indigenous homelands, stewardship, and gender. She is author of "Poisoning the Serpent: Uranium Exploitation the Serpent River First Nation, 1953–88" which appeared in *The Natures of Empire and the Empires of Nature*, edited by Karl Hele, and "Interviewing Nookomis and Other Reflections of an Indigenous Historian" in *Oral History Forum*.

Fred Metallic is a citizen of Gespe'gewa'gi, Mi'gma'gi. In 2011, Fred graduated from York University and obtained his Ph.D. in Environmental Studies. His dissertation, written and defended in Mi'gmaw in Listuguj, was a landmark achievement. This was the first time a dissertation was written solely without translation in an Indigenous language. He has written and presented numerously to Mi'gmaq and non-Mi'gmaq communities, on topics related to Mi'gmaq history, Mi'gmaq political philosophy, culture, spirituality and governance. Fred also sits on the Mi'gmaq Grand Council as a Ge'ptin and works closely with the leadership in the advancement of Mi'gmaq rights and responsibilities as a nation. Fred is currently working with the Listuguj Mi'gmaq Government as Director of Natural Resources. Fred, his wife Amy, and their three children – Emma, Je'gopsn and Erika – live in the Mi'gmaw community of Listuguj.

Ian Mosby is an award-winning historian of food, health and colonialism in twentieth century Canada. He recently made international headlines for his work on the history of human biomedical experimentation in a number of residential schools and First Nations during the 1940s and 1950s. You can read more about his research at www.ianmosby.ca.

K. Jan Oosthoek is an environmental historian based in Brisbane, Australia. For many years he has lectured and researched at the Universities of Newcastle (UK) and Edinburgh. He has published on a wide range of topics including forest history, the history of industrial water pollution and history of the ozone problem. His latest book entitled *Conquering the Highlands: A History of the Afforestation of the Scottish Uplands* was published by ANU Press (2013). He has also served as vice-president of the European Society for Environmental History (2005–2007) and is author of the leading environmental history website "Environmental History Resources."

Roger Owen is a Lecturer in Theatre Studies at the Department of Theatre, Film and Television Studies, Aberystwyth University, Wales, UK. His teaching and research interests are varied, and include Welsh language theatre and drama, issues in Performance Studies, rural performance culture and farming. He is the author of a full-length study of the Welsh playwright Gwenlyn Parry (2013), and a history of Welsh language theatre and identity between 1979 and 1997 (*Ar Wasgar* ['Dispersed'], 2003). He has been a regular collaborator with Eddie Ladd and recently acted as dramaturg for Theatr Genedlaethol Cymru's revised production of *Dawns Ysbrydion*.

Salvatore Paolo De Rosa is an Italian political ecologist. He is PhD candidate at the Department of Human Geography of Lund University, Sweden, and a Fellow of the ENTITLE Network for Political Ecology. His research areas are political ecology and environmental anthropology. His work focuses on socio-environmental conflicts, community-based management of resources, grassroots eco-politics and cultural change. His PhD project deals with the conflicts over waste management and ecosystem contamination in the Campania region (Italy). In particular, he aims to understand why conflicts over waste erupt and how the mobilization of affected communities contributes to the emergence of alternative imaginaries and practices of socio-ecological relations. Salvatore has been on the editorial board of the independent magazine *Napoli Monitor* for seven years. His work is published in magazines, research blogs, edited books and NGO websites.

Stephanie Rutherford is Associate Professor in the School of Environment at Trent University in Peterborough, Canada. Her research inhabits the intersections among political ecology, critical animal studies, biopolitics, and historical geography. She is currently working on a history of wolves in Canada. She has published in *Progress in Human Geography*, *Geography Compass*, and *The Journal of Canadian Studies*, among others. Stephanie is also the author of *Governing the Wild: Ecotours of Power* (University of Minnesota Press, 2011).

L. Anders Sandberg is Professor in the Faculty of Environmental Studies at York University, Toronto, Canada. His two most recent books include the co-edited *Urban Forests, Trees, and Greenspace: A Political Ecology Perspective* (Routledge, 2014) and *Post-Industrial Urban Greenspace: An Environmental Justice Perspective* (Routledge, 2015).

Lisa H. Sideris directs the Consortium for the Study of Religion, Ethics, and Society and is Associate Professor in Religious Studies at Indiana University, specializing in environmental ethics, science and religion, and religion and nature. She is author of *Environmental Ethics, Ecological Theology, and Natural Selection* (Columbia University Press, 2003) and an edited interdisciplinary volume on the life and work of Rachel Carson, titled *Rachel Carson: Legacy and Challenge* (SUNY, 2008). She has published a variety of essays on religious environmental ethics, Darwinism, ecotheology, and Rachel Carson in interdisciplinary journals such as *Soundings* and in environmental journals such as *Worldviews* and the *Journal for the Study of Religion, Nature, and Culture*. Her work also appears in prominent guides to the field such as *The Blackwell Companion to Science and Christianity* and the *Oxford Handbook of Religion and Ecology*. Sideris serves as Associate Editor of the *Journal for the Study of Religion, Nature, and Culture*. Her current research interests include the emergence of Anthropocene narratives, and the role of wonder in environmental discourse and scientific discourse.

Sverker Sörlin is a Professor of Environmental History in the Division of History of Science, Technology and Environment, KTH Royal Institute of Technology, Stockholm. His core area of research centres around the roles and functions of knowledge in environmentally informed modern societies. Another major area of interest is research and innovation policy where he serves as a policy analyst and advisor. His current research projects encompass the science politics of climate change through the lenses of glaciology and sea ice; the emergence of and changes within environmental expertise; historical images of Arctic futures; and the environmental turn in the humanities and the social sciences. Among his publications is a co-edited volume on the formation of environmental expertise with Paul Warde and Libby Robin, entitled *The Future of Nature* (Yale University Press, 2013). He has published numerous articles in journals such as *BioScience, Ecological Economics, Environmental Science and Policy*, and *The Journal of Historical Geography*.

Martha Stiegman is a scholar and documentary filmmaker currently based in Toronto, where she is an Assistant Professor at York University's Faculty of Environmental Studies. Her work has screened in festivals around the world from Tunisia and New Zealand to Brazil. Her first two documentaries, *In Defense of our Treaties* (2007) and *The End of the Line* (2007) explore alliances between Mi'kmaq and non-Native fishing communities in her home province of Nova Scotia. *Honour Your Word* (2013) is an intimate, behind the blockades portrait of the Algonquins of Barriere Lake and their decades-long fight to defend their territory. *Seeking Netukulimk* (2013) is a lyrical exploration of the traditional laws that govern fishing for Mi'kmaq people. Underlying Stiegman's filmmaking and scholarship is a commitment to collaboration and social justice. Indigenous self-determination struggles and settler solidarities have been the focus of Martha's film work, community-arts practice and academic research for more than ten years.

Jesse Thistle is Métis-Cree from Saskatchewan and a master's student in history at the University of Waterloo. His work centres on trauma and memory within populations of Métis and Cree in Northern Saskatchewan, and the Algonquin of Timiskaming, Ontario. Specifically, he looks at how history can be applied to understand the effects of intergenerational trauma in contemporary Indigenous populations. His work is directed towards community healing and cultural reclamation as well as retrieval of oral history archives –

termed as *truth-telling* as defined by the Truth and Reconciliation Commission – challenging orthodox settler histories in the narrative of Turtle Island. Jesse is also interested in how the arts – film, writing, dance, painting and illustration – can be used as a mode of resistance, cultural revitalization, and land reclamation among Indigenous peoples.

Jocelyn Thorpe is Associate Professor of Women's and Gender Studies at the University of Manitoba in Winnipeg, Canada. Her interdisciplinary research focuses on the history and legacies of, as well as challenges to, colonialism in the Canadian context, examining how past discourses and relationships of power influence the present. With Innu artist Joanna Barker, she is currently working on a five-year project about the history of relationships among Indigenous and non-Indigenous Newfoundlanders and the territory they have come to share. She is the author of *Temagami's Tangled Wild: Race, Gender, and the Making of Canadian Nature* (UBC Press, 2012).

Marsha Weisiger is the Julie and Rocky Dixon Chair of U.S. Western History and Associate Professor of History and Environmental Studies at the University of Oregon. She is the author of *Dreaming of Sheep in Navajo Country* (University of Washington Press, 2009) and *Land of Plenty: Oklahomans in the Cotton Fields of Arizona, 1933–1942* (University of Oklahoma Press, 1995) and the principal author of *Buildings of Wisconsin* (University of Virginia Press, forthcoming 2016). Her work has won the Carol and Norris Hundley Award and the Hal Rothman Book Award, among others, as well as fellowships from the National Endowment for the Humanities and the American Council of Learned Societies. She is currently working on three book projects, *The River Runs Wild*, *Danger River*, and *Ecotopia Rising*, the third of which explores the significance of the counterculture, including artists, to the environmental movement.

Nina Wormbs has an MSc in Engineering Physics, a PhD in History of Technology, and is Associate Professor of History of Science and Technology in the Division of History of Science, Technology and Environment, KTH Royal Institute of Technology, Stockholm. From 2010 to 2016, she served as chair of said Division. Nina has written mainly on the media and technological change during the 20th century with a focus on infrastructure and politics. She takes special interest in ideas about what technology can achieve, not the least digital technology. Recent publications include the co-edited volume *Media and the Politics of Arctic Climate Change: When the Ice Breaks* (Palgrave 2013). Together with Sabine Höhler, she is running a research project named "Views from a distance", which investigates how remote sensing and digital data construct the environment.

Acknowledgements

The editors would like to thank the Routledge Environment and Sustainability editorial team (particularly Helen Bell, Khanam Virjee and Rebecca Brennan) who have been instrumental in bringing this book to life. Thanks also to Iain McCalman and Libby Robin, the editors of the Environmental Humanities series at Routledge, for their support of the volume. Additionally, three anonymous reviewers provided insightful comments that allowed us to rethink and refine the objectives of the book. Finally, thank you to the contributors, who have written thoughtful and engaging chapters that grapple with the methodological opportunities and challenges presented by nature-culture and environmental history research.

1 Introduction

Methodological challenges

Stephanie Rutherford, Jocelyn Thorpe and L. Anders Sandberg

This collection of essays started life as a round table on research methods at the World Congress of Environmental History in Guimarães, Portugal, in the summer of 2014. We, the editors of this volume, asked round-table participants to reflect on their research methods that take them outside of archives, as well as on how they communicate that research beyond scholarly books and articles. While archives can be valuable sources of information about the past, there is also a lot they do not tell us, particularly if we are trying to understand the past of other-than-human beings and/or of human beings whose perspectives are difficult to find in stored files. Like the archives, scholarly ways of communicating can obfuscate as well as illuminate, for example by using words such as obfuscate instead of perfectly acceptable alternatives such as confuse.

Much to our surprise, given the warm weather and stunning beauty of our mountainous surroundings, conference attendees packed the room for the round table, making possible and contributing to a rich discussion about the challenges and possibilities of conducting and communicating research on the history of relationships between humans and the rest of the world. Scholars from a diversity of disciplines and inter-disciplines attended the conference, some of them hailing from history and geography departments and conducting their research under the respective banners of environmental history and historical geography. Others work in anthropology, English, and interdisciplinary departments and programmes such as Indigenous studies, cultural studies, environmental studies, gender studies, and science and technology studies. The level of interest in the round table led us to consider the importance of making space and time at conferences to share ideas about the common elements of our work—the *how* of research and teaching—as well about our more individual research interests and projects. It also motivated us to continue and expand the conversation in book form.

Despite their varied intellectual traditions, researchers who seek to comprehend the histories of relationships between humans and the more-than-human world take as a starting point that key to understanding the past is paying attention to how humans shaped and were shaped by the world in which they lived. Disciplinary commitments certainly influence the questions scholars ask and the ways they seek out answers, but some methodological challenges go beyond the boundaries of any one discipline. In this collection, we grapple with such methodological challenges by bringing to bear the insights of a number of scholars working in different geographical and disciplinary areas on questions related to nature, culture and history.

The methodological challenges that authors in this collection explore are inspired by the conversations we had at the round table, and do not have easy solutions. Our aim in collecting these stories is to continue the conversation that began for us at the round table in Portugal, but that we know also takes place in supervisors' offices, on walks, and online as we grapple to figure out how to answer the questions we pose in our research and how best to communicate what we find out. While we know that this book represents just one part of this broader conversation and that it does not fully reflect the geographical diversity of our shared world, we hope that it offers a promising continuation of the conversation about the 'how' of our research and communication. We would like to see this conversation continue over the variety of media available to us today, including in scholarly books such as this one.

This book is about the challenges and the possibilities of conducting nature-culture-history research today. The methodological challenges that authors in the collection examine include: how to account for the fact that humans are not the only actors in history, yet they dominate in archival records; how to decolonize research when archival sources and their embedded narratives are both colonial and colonizing; how to attend to the non-visual senses, to materiality and to affect when traditional sources offer only a two-dimensional, non-sensory version of the past; and how effectively to use sources and means of communication made available in the digital age. The book is divided into sections (or 'parts') according to these main challenges, and the chapters within each section consider pieces of the larger methodological challenges addressed within each section. The themes of the sections emerged both from the round table and from abstract proposals submitted by contributors, not all of whom participated in the round table. By emphasizing the research process rather than product, we aim to challenge ourselves and our readers to think creatively and to make use of research tools we might not have considered previously. While the focus on process rather than product sets this collection apart, it is exciting also to think about how different research processes might also lead to innovative research findings and communication strategies. Below, we explain the sections and chapters in further detail.

Part I: Nonhuman actors

In the past 20 years, the 'question of the animal' and discussions about the agency of the more-than-human world have gained currency. Environmental historians, animal geographers, scholars of posthumanism, anthropologists, sociologists, and literary theorists have produced a range of studies that consider the nonhuman not simply as fodder in the human story of change over time, but as agents that shape the course of history (see, for example, Coleman 2004; Fudge 2006;

Haraway 2007; Wolfe 2003). This kind of research can offer particularly nuanced ways of apprehending the world. As geographers Stephen Hinchliffe and Nick Bingham note, 'All kinds of things become more interesting once we stop assuming that "we" are the only place to begin and end our analysis' (2008: 1541). Including the nonhuman in conversations about historical and landscape change offers the opportunity not only to understand the past in a more holistic way, but also to begin to encounter with fresh eyes the variety of environmental problems we now face.

This kind of research, however, remains fraught with methodological dilemmas. The most central among them is how to understand and trace the histories and contemporary realities of beings whose patterns of communication, movements, and life-worlds are often outside of human comprehension. How do we come to know animals, trees and glaciers as agents in historical change if the only record we use to understand them is that made by humans? The chapters in this section engage with these questions, and, in so doing, offer students and practitioners concerned with the roles in history played by nonhuman actors tools with which to conduct research.

The first two contributions in this section consider landscapes—often imagined as passive terrains to be written upon—as agents that shape relationships in the more-than-human world. This section opens with a chapter by Sverker Sörlin that considers glaciers as protagonists in global environmental change. Working against the grain of scientific knowledge, Sörlin suggests that we pay attention to the ways that glaciers act as local 'truth-spots': sites where climate change is made manifest. Drawing together cultural representations of glaciers, Sörlin contends that glaciers give us a visual and conceptual language to apprehend our changing environment. In the next chapter, Martha Weisiger takes the reader on a journey to another iconic landscape: the American West. Accessing methodological questions through the lens of art, Weisiger offers an analysis of what we can learn about environmental history by walking the land, including through landscapes shaped by Earth art. The following two chapters consider how animals might be written into environmental histories and nature-culture research, and what methodological challenges and opportunities this might pose. Stephanie Rutherford attends to how sound might be an entry point into histories of the animal, exploring how the howl of the wolf has been recognized (and misrecognized) through time in Canada. Rutherford suggests that paying attention to sound offers a way to understand the affective registers wolves have induced, especially in settler Canadians, while offering new insights into human-nonhuman relationality. Moving from Canada to China, Michael Hathaway contends that changing conceptions of agency can open up data (gained through interviews, participant observation, and oral histories) to new interpretations, and queries the uses and limits of using conventionally generated scientific information to enrich our writings about animal histories. The final chapter in this section considers neither landscapes nor animals, but weather. Roger Owen's contribution traces the history of a dance-theatre work that illuminated a political act of Welsh nationalism set within the context of, and indeed affected by, a heavy and prolonged snowfall in the UK. In his description of the event, Owen asserts that theatre practice may effectively combine and collide human and nonhuman actors within a single assembly of action, spectatorship, and objects in space.

Part II: Decolonizing research

The idea that historical documents provide a clear window into the past has long given way, at least in many circles, to the scholarly notion that objectivity is an illusion and truth is an effect of power (Brown and Vibert, 2003, p. xiv; see also Mills, 1997; Smith, 2012). Indeed, anthropologist and historian Ann Laura Stoler (2002) has encouraged students of colonialism to think of archives not simply as sources of information, but as sites of knowledge production that require their users to take seriously the implication of archives in colonial power relations (see also Burton 2005). The relationship between archives and colonialism comes as no surprise to Indigenous scholars, activists, and community members who have consistently critiqued written historical documents for their failure to include Indigenous perspectives (see, for example, Cardinal et al. 2005). The recent Truth and Reconciliation Commission on Indian Residential Schools in Canada makes powerfully clear the disconnect between the written historical record and the experiences lived by people in their communities and in residential schools. It is only through the persistence and courage of residential school Survivors that their stories of a system designed to eradicate their cultures by removing children from communities are becoming part of the mainstream historical record (Truth and Reconciliation Commission of Canada 2012). It is impossible to go to the archives and access an unmediated and unbiased past.

Yet scholars examining historical relationships between 'nature' and 'culture,' like others interested in understanding and illuminating parts of the past, often find themselves conducting research in colonial archives. Certainly archives can help researchers reveal important (though not apolitical) insights about the past, even as work such as Julie Cruikshank's (2005) demonstrates the need to conduct research beyond a colonial archival record. In *Do Glaciers Listen?*, she shows how Indigenous oral traditions can offer quite different, though no less true, readings of the past than the written colonial record, and makes clear how colonial power relationships can reappear through the telling of historical stories as though European perspectives are the only ones. Decolonizing nature-culture-history research, then, requires conducting research differently, both at the archives and beyond.

This section opens with a chapter co-written by Bawaka Country, which includes Indigenous and non-Indigenous women from Bawaka, an Indigenous homeland in the north of Australia. Together, the chapter's authors examine how their research team has grappled with the concept of time in their research, and show how understanding time within Yolŋu ontologies of co-becoming presents unique challenges and possibilities for decolonizing nature-culture-history research. Next, Lianne C. Leddy examines the possibilities of *dibaajimowinan* (stories) as a method for conducting environmental history research on *Anishinaabe* territory, demonstrating that while oral history research might be 'new' within environmental history, it is anything but new within Indigenous communities. Its centrality to Indigenous ways of knowing, however, does not mean that there exist no challenges to balancing oral histories and archival material as sources of knowledge about the past, and Leddy addresses these challenges as well. In her chapter, Aimée Craft reflects upon the process of coming to learn about *Anishinaabe* water law, describing the research as a practice of learning and growing together. She shows that ceremony and research depend on each other, much as humans depend on and indeed are made of water. In an interview with co-editor Jocelyn Thorpe, Fred Metallic reflects upon decolonizing academic work through language, describing how he conducted his PhD work in Mi'gmaw, including the defence of his dissertation. He also argues that how one does research is at least as important as the outcome of the research. Jocelyn Thorpe's chapter concludes this section with an analysis of what she has learned so far, often from Indigenous scholars and knowledge holders, on her journey toward decolonizing research.

Part III: Senses and affect

In recent years, an exploration of 'sensory histories,' through attention to sound, smell, taste, touch, vision and movement, has enlarged the possibility of using different sources to piece together an understanding of the past (see, for example, Harvey 2003; Fitzgerald and Petrick 2008; Jay 2011; Parr 2006; Picker 2003). Simultaneous to this embrace of the senses has been what's named 'the affective turn' (Ticiento, Clough and Halley 2007). Studies rooted in affect recognize that precognitive, unstructured intensities shape how humans encounter the world on a physiological level. To return to the senses, one might suggest that a foot that unconsciously taps to the beat of a song offers one expression of affective attunement to sound. Scholars of affect suggest that these intensities give feelings strength and resonance. As such, the emphasis on materiality and sensory understandings of the past offer new research horizons in the social sciences and humanities. This attention to senses is a particularly important endeavour when one's study focuses on historical engagements with nature, where so much of what we know is felt rather than seen.

The chapters in this section employ eclectic strategies to access the senses and to consider affect, demonstrating the myriad approaches scholars in nature-culture and environmental history can use to understand human/nonhuman relations through time. Owain Jones and Katherine Jones start off this section emphasizing the affective dimensions of narrative. Drawing on their relationship with tidal landscapes as places, spectacles, experiences, and as manifestations of a living, mobile planet and biosphere, they offer vignettes that attempt to mirror the ebb and flow of the tides themselves to tell the story of environmental change. Ian Mosby offers an examination of the palate, thinking with taste as a novel access point into environmental history. Mosby explores eating practices through time, using the changing experiences of food's smell, taste and presentation to offer an innovative set of tools for examining the past. From the delight of a tasty meal, we move on to the smell of garbage in Marco Armiero and Salvatore Paolo Da Rosa's exploration of environmental injustice in Naples, Italy. Armiero and Da Rosa use waste, and the smells it produces, as a source of environmental history, exploring the way that perception can lead to new kinds of alliances between activists and scientists in the matter of what gets thrown away. Sonja Boon's chapter turns to movement and embodiment as ways of knowing. Using the minuet, an eighteenth-century dance form, Boon argues that our understanding of the nonhuman environment is shaped by the way that we physically and physiologically move through space, and that such movement is fundamentally gendered and classed. Moving from the scale of the body to the cosmic, Lisa Sideris's chapter takes on Big History, investigating stories such as the 'Epic of Evolution' as the basis for a new metanarrative that inspires wonder and affective connection to Earth. She contends that these stories reveal a phenomenology of human-as-species, offering an intriguing, if problematic, means to craft an environmental cosmopolitanism. This section is rounded out with a chapter by Stephen Bocking on cycling as a novel method in environmental history scholarship. Bocking asserts that the unique vantage point of a bike provides access to sensory histories, enabling one, at least in part, to recapture how people sensed and experienced past environments.

Part IV: Digital research

The Internet has certainly changed the way people do research, and indeed many forms of work. Digital history has emerged as a sub-discipline ('Interchange' 2008), and the journal *Environmental History* (2014) recently appointed a digital content editor who edits a web-only 'field notes' section that provides insights into the practice of environmental history scholarship. (Its editor is also one of the contributors to this book.) There is a growing number of archival sources that are available on the Internet rather than in the reading rooms of remote archives. The open-access *Programming Historian* (2014) is an online, open-access, peer-reviewed suite of tutorials that assist humanists to learn a wide range of digital tools, techniques, and workflows to facilitate their research and help organize and present their data. Many digital historians are public historians, seeking to reach a wide audience with open-access publications, websites, podcasts, blogs, and tweets. They often use innovative ways to present research material through augmented reality technology, repeat photography, composite imaging, and GIS mapping (Bonnell and Fortin 2014). Yet, in the digital age, researchers are often faced with the dilemma of how to learn technological skills and how to sort through, make sense of, and present the welter of digitized sources and material made available through the click of a button. How do the demands of the digital help and hinder, reveal and obscure, and connect and disconnect the researcher to research matters (Barringer and Wallace 2014)?

Contributors to this section explore specifically the effects of new technologies on research that seeks to understand nature-culture-history relations. Sean Kheraj and Jan Oosthoek trace the history of digital communication and connection for the environmental history community. By looking at email listservs, audio and video podcasts, as well as blogging and micro-blogging, Kheraj and Oosthoek contend that digital technology opens the possibility for truly collaborative research ventures that reach audiences beyond academic ones. Dolly Jørgensen's contribution explores the role blogging can play not only in attracting non-academic audiences to academic work, but also in shaping the research process itself. Crafting her chapter as five pseudo-blog posts, Jørgensen suggests that 'doing historical research in public' has made her a better writer, one more willing to take risks on 'side' stories that have led to robust academic work. Kimberly Coulter and Wilko Graf von Hardenberg pay attention to the building of communities in the digital realm. They argue that the dynamism of environmental history makes it particularly well-suited to digital collaboration and explore examples they contend work to demonstrate a 'spirit of the commons' in digital environmental history. Sabine Höhler and Nina Wormbs look at environmental change from a distance, considering how remote sensing crafts particular kinds of environments through visualization techniques. Their chapter dwells on both the opportunities and challenges of this way of understanding environmental change, grappling with the question of what happens to environmental history if sensory data become the source and digital data management becomes the practice of making sense of the earthly environment. Like Stephen Bocking in the previous section, Finn Arne Jørgensen is concerned with understanding history through mobility. But for Jørgensen, this mobility is accompanied by the digital technology of GPS, allowing him to reflect on the ways technology shapes our encounter of the natural world. The final chapter in this volume also considers movement. L. Anders Sandberg, Martha Stiegman and Jesse Thistle recount the development and expansion of their alternative campus tour at York University in Toronto, Canada, and the way the information generated through this practice has been disseminated online. Sandberg, Stiegman and Thistle suggest that the different elements of this project—walking, telling stories, video-recording, and providing a digital record—disrupt the dominant narrative of Canada as settler nation, highlighting instead the presence and actions of Indigenous peoples.

As a whole, the collection shows that how we do our work matters fundamentally for what kind of work we do. Discussing research practices and processes as well as research products makes possible different ways of doing research, ways that might help us answer the pressing questions of today. We hope that readers' engagement with this book will allow for the continuation of conversations about how to conduct excellent and engaged nature-culture-history research.

References

Barringer, Terry and Wallace, Marion (eds). (2014). *African Studies in the Digital Age. Disconnects?* n.p.: Brill Online.

Bonnel, Jennifer and Fortin, Marcel. (2014). *Historical GIS Research in Canada.* Calgary: University of Calgary Press.

Brown, J., and Vibert, Elizabeth (eds). (2003). *Reading Beyond Words: Contexts for Native History* (2nd edn). Peterborough, ON: Broadview Press.

Burton, A. (2005). *Archive Stories: Facts, Fictions, and the Writing of History.* Durham, NC: Duke University Press.

Cardinal, T., Highway, T., Johnston, B., King, T., Maracle, B., Maracle, L., Marchessault, J., Qitsualki, R. A., and Taylor, D. H. (2004). *Our Story: Aboriginal Voices on Canada's Past.* Toronto: Doubleday Canada.

Coleman, Jon. (2004). *Vicious: Wolves and Men in America.* New Haven, CT: Yale University Press.

Cruikshank, J. (2005). *Do Glaciers Listen?: Local Knowledge, Colonial Encounters, and Social Imagination.* Vancouver: UBC Press.

Deloria, Vine. (1969). *Custer Died for Your Sins.* New York: Macmillan.

Environmental History. (2014). Field notes. http://environmentalhistory.net/field-notes (accessed 27 December 2014).

Fitzgerald, Gerard and Petrick, Gabriella. (2008). In good taste: Rethinking American history with our palates. *Journal of American History*, 95 (September): 392–404.

Fudge, Erica. (2006). *Brutal Reasoning: Animals, Rationality and Humanity in Early Modern England.* Ithaca, NY: Cornell University Press.

Haraway, Donna. (2007). *When Species Meet.* Minneapolis, MN: University of Minnesota Press.

Harvey, Elizabeth (ed.). (2003). *Sensible Flesh: On Touch in Early Modern Culture.* Philadelphia, PA: University of Pennsylvania Press.

Hinchliffe, Stephen and Bingham, Nick. (2008). Securing life: The emerging practices of biosecurity. *Environment and Planning* A, 40 (7): 1534–1551.

Interchange: The promise of digital history. (2008). *Journal of American History*, 95 (September). www.journalofamericanhistory.org/issues/952/interchange (accessed 27 December 2014).

Jay, Martin. (2011). In the realm of the senses: An introduction. *American Historical Review*, 116 (April): 307–315.

Mills, S. (1997). *Discourse* (The New Critical Idiom). London: Routledge.

Parr, Joy. (2006). Smells like? Sources of uncertainty in the history of the Great Lakes environment. *Environmental History*, 11: 269–299.

Picker, John. (2003). *Victorian Soundscapes.* Oxford, UK: Oxford University Press.

Programming Historian. (2014). http://programminghistorian.org (accessed on 27 December 2014).

Smith, L. (2012). *Decolonizing Methodologies: Research and Indigenous Peoples* (2nd edn). London: Zed Books.

Stoler, A. (2002). Colonial archives and the arts of governance. *Archival Science*, 2 (1–2): 87–109.

Ticiento Clough, Patricia with Halley, Jean (eds). (2007). *The Affective Turn: Theorizing the Social.* Durham, NC: Duke University Press.

Truth and Reconciliation Commission of Canada, and Canadian Electronic Library. (2012). *Truth and Reconciliation Commission of Canada Interim Report.* Winnipeg, MB:

Truth and Reconciliation Commission of Canada.
Wolfe, Cary. (2003). *Animal Rites: American Culture, the Discourse of Species, and Posthumanist Theory*. Chicago, IL: University of Chicago Press.

Part I

Nonhuman actors

2 Do glaciers speak?

The political aesthetics of vo/ice

Sverker Sörlin

Climate change is a reality, but one that is hard for most of us to sense on an everyday basis. On the timescales of meteorology it may be fast enough; the global average temperature has risen almost one degree Celsius over the last century, which is a drastic leap compared to ordinary Holocene variations. Only a few generations ago, even scientists had a hard time knowing for sure what direction global climate would take. Temperatures were not systematically measured in many places and systematic monitoring and collecting of data didn't exist. Even more importantly, the very concept of climate wasn't a global concept; it was precisely the prevailing atmospheric conditions, including average temperature, that characterized a local place.

That said, ice is a dynamic material and it was obvious to people in many parts of the world that multiyear ice demonstrated changes. When these changes were studied more systematically in the nineteenth century it became clear that glaciers, in most regions from which there were any kinds of records, were retreating. This observation gave impetus to the emerging Ice Age theory that was first proposed in the 1840s by Charpentier and Agassiz. In the coming decades glacial data were registered at an increasing pace all over the world, which corroborated the idea that there was some kind of warming trend affecting the atmosphere. Thus, glaciers came to be regarded as living markers, or signals, of changes that were otherwise extremely hard to anchor in data. Strangely perhaps, the theory of global warming based on increased levels of carbon dioxide in the atmosphere was known before there was any commonly acknowledged record of rising temperatures.

Glaciers thus played an important role in nineteenth-century climatology, as a marker of climate change. Through their specific properties of slow melting—with a negative mass balance—during low precipitation and warm years, and slow growth—with a positive mass balance—during cold years, glaciers had what we might call 'voice'. They spoke, albeit in a soft voice and with a language that required patience in the listener. It also required a presence in the field to hear that signal. Glaciers, even small ones, are massive formations. They move slowly, too slowly for any of our senses to capture. They melt very slowly vertically but at their end down a slope or in a valley they can lose meters or dozens of meters in one year. Still, in order to secure a trend you need many years of observation. The voice of glaciers is special.

The fact that anthropogenic climate change is now well-established must not permit us to forget that for a long time that was a claim that was controversial, or, before the 1950s, hardly even taken seriously. That glacial retreat was documented in the nineteenth century and was the object of systematic, comparative study around the world in the first half of the twentieth, of course says nothing about the causes for their retreat. But for anyone who wished to create an argument about climate change, regardless of its cause, the performance of glaciers became of critical importance. They acquired a role as indicators of climate change. By their numbers, size and sheer mass and, perhaps most importantly, their climate response on the directly observable timescale, this role was different from other indicators such as tree rings or layered clay. Here was an element, with local presence in mountains and Alps across the northern hemisphere and in certain locations in the southern hemisphere as well, that served as a global demonstration of a change that was otherwise not visible, not even with the most sophisticated instruments, until far later, when sediments or wooden data could be analyzed.

Yet another step was taken when *anthropogenic* climate change became more widely acknowledged in the final decades of the twentieth century. Then the retreating glaciers, and perhaps especially the melting Greenland ice cap, became sites to go to for tangible evidence of climate change. Almost like holy sites of the tragedy of human-nature relationships, glaciers now performed a role as 'truth-spots' (Gieryn 2002 & 2006), shedding any possibly remaining doubt on whether climate change is real. Thus glaciers have come to serve the role of major protagonists in stories about the human predicament, serving increasingly important political purposes.

Narratives, iconographies—and sites

What can a historian make of this slow performativity of a medium whose properties and behavior have, after all, mostly attracted scientists? More than you might think. Once you get a little deeper into the history of glaciers and glaciology in the wider science politics of climate change, you are made aware that this performativity—this slow, soft voice of glaciers—is always at the core of the attention. The distinctness of their retreat (or growth) is attractive and calls for measurement and long-term monitoring. But at the same time, what does the melting at the glacial ends really tell about the loss of mass? The relationship between changes in climate and the behavior of glaciers is diffuse and precision therefore hard to attain. These are obsessions in the history of glaciology: the quest for precision (how much exactly does the glacier shrink or grow?) and the quest for explanation (why does it shrink/grow?). The glacier itself is never the same object. It always moves: it grows and shrinks seasonally, it has good years and bad years. It was at an early stage described as an 'economy' with a 'budget'. The 'budgetary balance' of glaciers was a way of talking about ice as if it was a matter of money in an account. Some years were spending years, and others were saving years.

The environmental and science history of glaciers and glaciology that I have practiced over my entire career has been increasingly informed by issues such as these. When I started my research, first as a side interest in my dissertation work in the 1980s (Sörlin 1988), and more decisively as part of my work on the science and environmental politics of the Arctic in the 1990s—a work that I have continued on and off ever since—it was without much sense of the glaciers themselves. I regarded them as a physical property, the object glaciologists studied. Little did I think about their performativity or their agency as anything to pay attention to in my research.

I started visiting field sites. Since I was a teenager I had always liked to trek in the mountains, in summer as well as winter. I had walked and skied on glaciers. I knew they were dangerous; I had seen crevasses and had crossed them, jumping rope to friends on either side, and so I knew from first-hand experience that they had many sides that were not just attainable through physics and chemistry. When I went into the archives more and started getting under the skin of early glaciologists, I started to sense more than just their excitement about the triumphs of monitoring and data that clearly made it possible for them to track down the historical changes of the ice, a massive new experience that told entirely new stories of many countries, especially in the Alps and in Scandinavia. After the establishment of the ice age theory, glaciers were regarded as remnants of the big glaciation and thus kind of geophysical atavisms, which just like the layers of the crust of the Earth spoke of dramatic 'antediluvian' events and served as evidence of deep time, the depth of which could hardly be fathomed (Rudwick 2014). These countries also started to retell their pasts with the help of glaciers. Ice became an instrument of new national(ist) narratives in countries such as Sweden, Norway, and Italy—manifested in tourism, guide books and, perhaps most importantly, in geographical atlases and books for schools where glaciers featured prominently (see, for example, Nansen 1911; Sörlin 1998; Armiero 2011).[1]

A popular iconography of ice emerged, with pictorial representations of ice in painting (Gunnarsson 1998), and perhaps foremost in photography, which was inspired by the scientific monitoring style of repeat photography. In the far north of Sweden, where much of my own fieldwork has taken place, Storglaciären [Big glacier] was documented by geologist Fredrik Svenonius who had already photographed the glacier in 1886 (Svenonius 1910). A few years later, the glacier was mapped by the authority of Land Survey and in 1897 the front position was surveyed for the first time using measuring tape from two marked stones. The front was photographed again in 1904 and 1907 by the local Kiruna photographer Borg Mesch, showing, in fact, an advancing front: i.e., a growing glacier. In August 1910 the glacier was mapped using a photogrammetric theodolite, which has been the baseline to demonstrate a continuous retreat since then. Several more observations and measurements were made in the following decades and by the early 1940s the glacial retreat in the area was already a well-established fact (Schytt 1940), a half-century after the first documentation.

Figure 2.1 Field co-producers of climate change knowledge in Iceland, 1936. Jón fra Laug, altmuligman (left), and Skarphjedinn, glacial observer (right). Photo by Sigurður Þórarinsson

Source: H. W. Ahlmann Collection, Royal Swedish Academy of Sciences, KVA, Stockholm.

Figure 2.2 Maria Sarri, Nikkaluokta village. Field co-producer of glacial knowledge in the Kebnekaise area in Swedish Lapland. Sarri belonged to one of several Sami families which assisted glacial scientists in the field and in building and maintaining the Tarfala glaciological research station, started in 1946. Photograph: Valter Schytt.

Source: Schytt family, private collection.

Figure 2.3 Repeat photography was, and still is, a technology much used to get an overview of long-term changes in glacier extension. Suottasjekna glacier, Swedish Lapland, 1907 and 2014. Photograph: Valter Schytt.

Source: Schytt family, private collection.

Figure 2.4 Storglaciären [Big Glacier], one of the research glaciers in the Tarfala area in Swedish Lapland, photographed by Fredrik Enqvist 1910 and by Per Holmlund 1997

Source: Stockholm University, Bert Bolin Centre Database http://bolin.su.se/data/svenskaglaciarer/historik.php

Figure 2.5 The glaciers in the Kebnekaise massif surrounding the Tarfala glaciological station, where the first unbroken series of glacial data was started in 1946, was documented by glacial photographer Fredrik Enquist, here during a field session in 1908. Photographer unknown.

Source: Stockholm University, Bert Bolin Centre Database http://bolin.su.se/data/svenskaglaciarer/historik.php

Ice also became the object of heroic storytelling, an element that required conquering and harsh (male) fieldwork. Ice is an almost omnipresent element in a broad genre of polar explorer narratives that flourished broadly in the century from ca 1850 (after the demise of Franklin's Northwest Passage expedition [Riffenburgh 1993]) until the middle of the twentieth century when a more professional and less heroic research style had taken over (Roberts 2011). A common trope is that of the young explorer who ventures into the ice and returns after a long time as a secular version of the prodigal son, to receive knighthood and immortality (for example, Nansen, Shackleton, Mawson), if he returns at all (for example, Franklin, Scott, Amundsen). The original template is literary, where the smallness of the human genius faces the overwhelming powers of nature, as represented by the ice. In Mary Shelley's *Frankenstein* (1818), the experimental scientist meets his own creation, 'the Creature', for a final show-down on the sea ice, the place where Enlightenment Man has to finally face his fate near the end of the world which, as it turns out, he can't subdue. Ice is a probing element, in which civilization is put to the test. Crisis is a core element of the narratives of ice, in recent decades reinforced by projections of climate change science.

I have become increasingly aware of the importance of these extra-scientific properties of ice and have found them useful in understanding the building of narratives where the ice is enrolled. I have also found it rewarding, almost necessary, to visit at least some of the sites where historical fieldwork took place. Insofar as the glaciology also involved local co-producers of knowledge—in the form of field assistants, seasonal observers, providers of food, housing and shelter, builders of local infrastructure, etcetera—I have tried to interview them or at least track them down. To move in the landscapes and places where the work took place has offered a way to re-enact the experience, at least partly, and also to get a better sense of the sometimes excessive reporting on weather and circumstance in the sources.

'Field-linked' research

In the middle of the 1990s I was part of the planning of a field trip with humanist and social science scholars to Svalbard. The field trip itself took place in 1997 (Sörlin 1997). It was part of a conscious widening of the Swedish polar science program but tellingly it was the first organized trip by the Swedish polar science agency for these fields of knowledge. There was also some doubt among the scholars whether it was a worthwhile trip. After all, there were no written sources out there. Why go? Historians need no 'field'? The experience proved not only worthwhile, but formative for the growth of an expanding and vital humanities research that I would like to call 'field linked'. It was not field research per se; in my work I have not collected data on glaciers or in the field the way glaciologists or indeed anthropologists do, at least not to any major degree. But I have learned to make the field sites of scientists into a part of shaping my own research.

On the Svalbard trip in 1997 I was part of identifying and investigating the precise field site of the Swedish-Norwegian 1931 glaciology expedition to North East Land. We combed the place and with the help of photographs from published sources and archives we could—down to the square meter—tell where instruments and buildings were located. We also found nails and other remnants from the buildings. They had no scientific value, but they contributed to a certain excitement. We also walked in the area, and around the nearby, still-standing Kinnvika station from the International Geophysical Year 1957–58 that the Swedes put together with Finland and Switzerland, three small non-NATO states. This 'geography of science' (Meusberger et al. 2010) told me things that I would never have been able to acquire through the written sources. It has aided me ever since when I have come across photographs of scientists sitting inside icy grottos under the light of kerosene lamps, or when I have seen images of the assemblage of what seems just to be 'stuff' in a field site. Even Bruno Latour's idea of the relationship between the fragility of an observation—a piece of data in some distant field location—and the social construction of what becomes first Fact, then Knowledge (Latour 1987), became a much more tangible and interesting analytical tool to use after I had seen the vulnerable arrangements of the instruments on their shaky stones or planks. Science comes out as also 'work', this too casually used word.

These early field experiences are now almost twenty years old. I didn't think much of the posthumanities then; it was before the concept. Nor did I use terms such as nonhuman agency. But at least some of us on that week-long cruise to sites of cold science around Svalbard islands—including graves of huntsmen and memorials to scientific tragedy and disaster—who had a background in environmental history and STS (science, technology and society) studies were sensitized to the fact that there was probably good reason to look more carefully into what the field encounter did. That nature had agency was often said by environmental historians, ever since Roderick Nash's classic on *Wilderness and the American Mind* (1967); nature had agency enough to affect the self-perception of the most powerful nation in the world. The idea was formulated also by other leaders in the field (Worster, 1988, Merchant, 1993). But it was more often gestured toward, as something that we knew was true and wished we could, someday, corroborate through empirical work. Strangely, here on the outskirts of land under the North Pole, I could sense the insight of this thought. If you stripped the site of almost everything that there usually is in terms of human paraphernalia and are left only with the bare minimum, you come closer to what that agency can be.

Through Actor Network Theory we had just learned that objects of various kinds could be enrolled in the interests of actors and then become actants (Latour 1987). In the following years an entire, massive dissertation was crafted that built heavily on that conviction (Avango 2005). The supervisor, architectural and heritage historian Marie Nisser, was on the Svalbard trip and must have also seen the light. In this minimalist landscape, agency was much more distributed. Where humans were present only days and weeks, with years in between, interests needed ombudsmen that were not only humans. They could be shacks signaling presence, claim signs for prospected mine sites, and various other leftovers.

In 2008 I spent several weeks in the field and in archives in Iceland. Again it was the interaction between the different kinds of sources that provided the energy to both of them. In the written sources—mostly letters between glaciologists and meteorologists in Stockholm and in Iceland—I found evidence of several inhabitants of the Vatnajökull area who served as fieldworkers for the glaciologists. I had heard stories in the field that told of these people, and had also seen the massive forces of the glacial agency in the built environment of the villages around the glacier with traces of floods and barriers for protection.

Visiting the sites and finding out where these collaborators had lived, where they had gone for their fieldwork and how they had set up their internal relations proved necessary for the reconstruction of how the field experiments were carried out. I could again see the micro-geographies with my own eyes and, with the help of photographs, reconstruct the knowledge-gathering process over what turned out to be several years, beyond what was always called the '1936' glaciological expedition to Iceland. The longer time required, and the local labour involved, as well as the economic transactions that this labor demanded, was in fact necessary for nature's agency to become described. Only over those years could a clear trend be established and their reduction be mapped. The Vatnajökull studies were crucial in the establishment of the theory of 'polar warming', which was launched in a period when anthropogenic global warming had no purchase as an idea. It turned out to be a blind alley in the history of climate science, but an interesting one, and the data collected by these Icelanders were essential for the Stockholm-based geographers to draw it together (Sörlin 2011; Ahlmann et al. 1937, 1938, 1939, 1940, 1941, 1943).

A wider reach of agency

This field linked environmental science history and has provided for me a kind of integrated understanding of nonhuman agency that might be worthy of some articulation. It may seem a distant cry from the fields of glaciology and geography that my site visits represent and the role of glaciers when they serve as truth-spots of global warming. First, they are individual, site-specific pieces of evidence in fragmented observational landscapes with fragile relationships to overarching truth claims, and sometimes used in theories that evidently failed (partly because the theory of how to link the data was weak in early to mid-twentieth century glaciology). Second, they have been organized into a cryospheric choir of voices singing consistently according to a chord arranged by the coalescing scientific orthodoxy and the IPCC (Intergovernmental Panel on Climate Change).

Nonetheless, I would like to claim that these seemingly contrasting stories of agency in the field belong very close together. The voice-making of glaciers is present in both instances and plays a crucial role for both styles of articulation of messages from the ice. The major difference is theory. Glaciology had no theory that could comprehensively explain glacial activity. There were several attempts, drawing on existing theories in climatology or physical geography. Indeed there were even early (circa 1900) individual expressions of sympathy for human-induced changes in climate, for example by the German geologist Eduard Brückner. He had started to edit the first journal of the field, *Zeitschrift für Gletscherkunde*, in 1906 under the auspices of the International Glaciological Commission, in a situation when knowledge was still not coherent or assembled anywhere. 'So zerstreut ist die Literatur, dass der einzelne sie nicht zu überblicken vermag' (Brückner 1906, quoted in Stehr and von Storch 2000), as he noted in the first issue. Brückner was also very interested in communicating with a wider public, holding that everyday knowledge was important for the understanding of complex issues which couldn't be tackled with observation alone (Stehr and von Storch 2000, p. 12).

The situation didn't improve much in the interwar years. With the theory of anthropogenic climate change, glaciology has come under a new frame altogether, which happened more slowly and cautiously in this discipline than in fields such as atmospheric science and theoretical, computer-based meteorology, which were leading the transformation to the new orthodoxy in the post-World War II decades. The voice of glaciers thus changed, and its agency too, from something essentially local to something global, if not planetary. It is worth observing that the effects of glacial agency in the interwar period rarely reached beyond the local or national. The new narratives that the melting produced were about valleys, plains and the post-glacial fertility of soils and productivity in single countries. When the climate change, and later Anthropocene, narratives are brought to bear on glaciers they become enrolled in a storytelling on a much grander scale. Effects are now felt as rising sea levels and these effects are deemed catastrophic, on a par with what Wally Broecker (1987), drawing on the Vostok ice core, called 'surprises in the Green House'. Glaciologists never spearheaded this discourse, but gradually aligned with it from the 1990s onwards. What happened then was essentially a reinterpretation of existing data and a re-configuration of the mission of existing field sites.

A case in point is the station that was organized within the same team of Stockholm-based geographers that had previously collected data in the North Atlantic and across Scandinavia. Still with the polar warming theory uncontested, this group established the first permanent collection site for glacial data anywhere with a time series dating from 1946, two years older than the Juneau series from Alaska. The site chosen was Tarfala in the highest part of the Scandinavian mountain range in the far north of Sweden. The climate change framing of the glaciology undertaken here was obviously colored by the skeptical stance that had been taken in the 1930s and 1940s by its senior scientists vis à vis the very few attempts to suggest anthropogenic climate change, mainly the papers by British steam engineer G. S. Callendar (1938, 1949; see also Fleming 2007).

The original narrative of Tarfala as a site that demonstrated an older observational paradigm, and what we might call essentially a 'national' agency of the melting glacier, could over some years (involving some internal turbulence and a shift of generations) be retold as a narrative of climate change. Under this new narrative the national significance of the glacier and the melting has moved into the background. Instead the larger, planetary story provides the framing. Another interesting development is that collaboration with local Sami, the Indigenous population, has moved into cognitive and content issues. For several years the Sami and scientists have tried to make use of the climate change data—which they now can be called, instead of simply 'glacial data'—and link that to the movement of animals, changes in grazing conditions of reindeer and other information vital for the Sami community. In turn the Sami contribute their local knowledge from the field. Thus, the glacial site has acquired an entirely new and wider geography and the data provide a much more integrated knowledge.

The agency of the glacier gets a wider *reach*. In my own field linked research in Tarfala, in collaboration with the current generation of glaciologists there, I have been able to see how the nonhuman agency of ice has undergone this transformation. Ice is in that respect a very useful material. Its dynamic materiality and plasticity lends itself to flexible interpretations and ever new and suggestive narratives. It is also a material, part of a 'sphere'—the cryosphere—which has a fascinating conceptual and intellectual history (Dobrowolski 1923; Barry et al. 2011), and works well as an illustration of the themes that have appeared in recent years under new integrative rubrics such as the environmental humanities, sometimes including the social sciences (Castree et al. 2014). This reflects a sense of urgency in trying to understand the role of ice in the dramatically increased interest in the rates and directions of change in the natural world. It is a broad cultural, aesthetic and moral engagement with ice, including sea ice (Christensen, Nilsson and Wormbs 2013). There is now a geopolitics of ice, a thinking, sensing, and seeing about, with ice in the past, present and future, and it is changing the way we write history and conceive of the humanities (Sörlin 2013, 2015, 2016).

When ice becomes an interest of the humanities and social sciences it can be described as an extension of what is considered *relevant knowledge* of ice, in order to understand it as, at the same time, a natural, cultural and social phenomenon. It can also be regarded as part of an ongoing expansion, or molding, of concepts. Just as, for a longer period of time, concepts such as 'environment' and 'climate' have already proven to be (first) primarily scientific and then expanded in significance into other realms of knowledge, 'ice' now seems to be undergoing a similar process. Glaciers are taking on the same kind of agency that has been attributed to other features of nature and environment, which is why we now are seriously pondering an anthropological, literary, visual, and historical glaciology, and indeed also a feminist and an Indigenous glaciology (Carey et al. 2016). This has not come without controversy.[2] Indeed, to suggest that certain elements of nature rightly belong in the knowledge enterprise that is conducted by the humanities and social sciences, and could thus be approached by new methodologies, is disputed. This is worth some reflection. Why is it hard for some to see what used to be a domain exclusively for science shared by other disciplines and regarded with other lenses?

Telecouplings and the production of the planetary

The current narrative includes what we might call the production of the planetary (Masco 2012, 2015). Growing significantly only in the last twenty to thirty years, with international societies, major research programs and committees and research centers, this broad and comprehensive process includes what ecocritical scholar Ursula Heise identified as a new sensibility for the many dimensions that reach far beyond the local, national or even the continental scales. Indeed a 'sense of planet' are words that also appear in the title of her book, *Sense of Place and Sense of Planet* (2008). The two—place and planet—actually require each other and through the scaling they are connected in multiple ways. There are, with a concept that takes its roots from the geological sciences, *telecouplings* between the local place where impacts are seen and identities formed, and the planetary level where the elements of knowledge accumulate into an earth systems science understanding, and where impacts accumulate also.[3]

There is a momentum for the study of the cryosphere as a comprehensive factor in the fragile balance of the earth as a whole and in the life balance of communities around the world—in mountain regions, in vast drainage areas in India and other parts of Asia and in Europe and in North and South America (Carey 2007, 2010). There are billions of people who on a daily basis are dependent on freshwater supplies from glaciers. Indeed, the entire world population depends on glaciers for their role in climate and future sea levels. And then, more directly, people in particular communities such as the people in Qaanaaq in Greenland and the Sami in Scandinavia and Russia, and the Nenets, and the Inupiat and Inuit are among many others who hunt on snow and ice and use it for transportation.[4] It thus also became part of an emerging global change understanding. The cryosphere was important, but the world was changing on a more comprehensive scale through global change science and Earth-system science, both of which took off in the 1980s.

Livelihoods of people in the Arctic tell a story of 'Arctic amplification', which means that the warming, that occurs in most places around the world, happens on a much steeper scale in the Arctic. Average winter temperatures have in some places risen by 7–8 degrees. The result is ice that covers smaller areas which gives lesser hunting opportunities. The ice is also wetter, thinner, weaker, less reliable, and even dangerous. Kirsten Hastrup (2012, 2015), who has worked among the Inuit in Qaanaaq for many years, has given vivid descriptions of how situations can occur that put even experienced hunters to the test.

A range of scholarly and artistic and community projects over the last decades have been mapping this. During the International Polar Year 2007–2009, there was a project called SIKU (Sea Ice Knowledge and Use). The beauty of that project was that scholars and scientists worked with local residents and did fieldwork together. It was a project built on shared knowledge and also shared practice. Shari Gearheard, who led this project, lives in Baffin Island and is employed by NOAA (National Oceanic and Atmospheric Administration) in Boulder, Colorado. Their synthesis coffee table book (Gearheard et al. 2013) is a remarkable achievement. It describes a situation where everyday spatialities and temporalities not only change but also undergo processes of what we might call inadvertent scaling and synchrony.

To make this point we need to use a different optic. Kirsten Hastrup has emphasized in her work how these so-called remote places are in essence not much different in many of their current parameters from anywhere else in the world. Globalization and modernity have brought a common frame. The realities of Earth-system science and climate change are as well known here as there as anywhere.

Things happen in an Arctic community in close relationship with things happening elsewhere—the telecouplings again. Synchronicities seem to be built into the ontology of this object. The performance of the ice is a result of processes elsewhere and they are no longer only natural. The planetary scale is directly hitting down in the shape of a local manifestation. The waning ice links the excesses of modern consumer society and industrialism to disastrous impacts on the innocent original populations whose vulnerability, despite a long history of successful adaptation, is aggravated when the ice melts (Krupnik et al. 2010; Hastrup 2012). It also reflects a complex relationship between Arctic residents and modern mainstream science, which on the one hand provides the best available knowledge of ice but still does not seem to be able to capture the full magnitude of the changes; Mother Nature does, as she certainly must, hold surprises, nearly two centuries after the ill-fated Frankenstein, creating what has been called 'the politics of uneven time' (Jordheim 2014).

Under the glacier—a global predicament

I started researching glaciers back in the 1980s, 35 years ago.[5] It was then mostly science, with little openness to other knowledge traditions and basically without a global or planetary framework to locate the interest in ice within. I didn't think about it in any other way, really, and few others did either. We were less insightful then.

But then we did not have works such as Julie Cruikshank's on the role of ice in the cultural encounters between the Tlingit in the Pacific Northwest and the French Lapérouse expedition in the 1780s. *Do Glaciers Listen?* (2005), the remarkable book by Cruikshank on that encounter and the colliding world views and belief systems of European scientific culture and that of the Tlingit, circled around the magic of ice, largely unknown to the French, who had no terribly sophisticated theory with which to read the glaciers, and the Tlingit who, on the contrary, had a deep respect, if not fear and trembling, for the mighty element towering thousands of feet into the air.

According to an old notion, much rehearsed in the geographical literature in the late nineteenth century, there had once been warmer periods, several thousand years ago, when there was open Polar Sea (Wright 1953). Only in the last decade or so has it dawned on me that we can read this completely differently, as part of a long-standing reflexivity on the variability of Arctic natures and their causes and consequences. We are pretty much in flux. Today these kinds of ideas are returning, not just as a chronicle of past lore, but as part of a broadening pattern of understanding.

In a recent book, *Media and Arctic Climate Change: When the Ice Breaks* (Christensen, Nilsson and Wormbs 2013), there is another dimension of ice presented: namely, that when the Arctic sea ice reached its then summer minimum in 2007 (it has since seen a new minimum in 2012) it went well beyond what models had predicted and even beyond what most models deemed possible. This caused concern in the scientific and policy communities, and it of course confirmed worries among Indigenous communities that something was seriously wrong with the basic conditions of their livelihoods. But it also linked to another significant development, which is indicative of our time and day: the fact that nature now acts in spurts, that it displays drastic developments, on timescales we have rarely seen before, and in much higher frequencies. Ice has agency and voice.

To summarize: we have, during the past couple of generations, added conceptual layers to the formation of the planetary, in what seems more and more as a history of scaling the human predicament from the local to the global, increasingly circumventing previously core boundaries such as those of the nation and instead focusing on other kinds of boundaries: critical boundaries for survival of species, impacts of pollution, limits of resources. It is no coincidence that there has appeared a language for all this, that of *planetary boundaries*, so entitled in an influential, and much discussed, *Nature* article (Rockström et al. 2009).

It is interesting to observe the humanities and social sciences now taking stock of this new understanding, and introducing new dimensions to concepts such as ice and cryosphere. Ásdis Jónsdottír writes about the southern Iceland glaciers, which she reads as constructed natures, and the materialities of major imaginaries. "It examines how 'local' and 'global' glaciers come into being in practices of knowledge making, storytelling and taking photographs" (Jónsdottír 2013). These glaciers are enrolled in narratives of Iceland, its history, its future, its identity, but also in global narratives on climate change, vanishing life forms and cultures. That is also glaciology.

My examples build towards a core insight which is that when we start thinking about the grand challenges that we need to face, we have new knowledge in store to add to the policy work. We understand the emergence of cryospheric science, we understand its science politics much better than only a few years ago. But we can *also* provide on-the-ground evidence of the significance of ice for communities of people. And we realize that certain understandings on the local level are only possible with the new and wider frame of earth systems science, or the Anthropocene.

Perhaps most importantly, we can contextualize this work into new narratives of the planetary, some of which are truly scary and involve policies and technologies that go beyond the ethically and environmentally responsible. But the local stories become parts of the global narrative in ways that it falls on people like ourselves in the human and social sciences to provide. There are narratives of gloom, but also of hope that we can tell, emerging stories of protecting ice and conceiving of it along the lines of the environmental humanities.

I often think, when I reflect on these matters, about Haldór Laxness's 1968 novel *Under the Glacier*. Apart from the fact the title is a good illustration of the way humanity has developed—a world that through an absurd religious zeal for material wealth sanctioned by a misunderstood divine mandate—it is also a fabulous story of what it means to live on the edge, and what it means to be able to read and interpret with imagination what is going on. The bishop has picked up rumours about a congregation in a faraway village in Iceland where strange things happen. The local practice of Christianity has fallen into disrepair. The church itself is boarded up and the steps leading to the entrance are gone. He sends a young priest to investigate. The young priest arrives in a world of magical realism which couldn't have been better described by Gabriel García Márquez, another Nobel laureate who incidentally also wrote about the same element in one of the most remembered, some would also say one of the best, opening sentences in a novel of the twentieth century: 'Many years later, as he faced the firing squad, Colonel Aureliano Buendía was to remember that distant afternoon when his father took him to discover ice' (García Márquez 1967/70, p. 1).

In this small Icelandic village there appear many completely unlikely characters. The woman who is baking fine little cookies for the young priest, but there is no bread to be had. The bizarre local priest, a bricoleur. The story of a body appearing from underneath the ice, macabre evidence of the landslide changes in the mental climate, not to speak of the Andean orchestra playing their flutes and guitars, or the eccentric millionaire. Under the glacier not much seems to be left of puritan religiosity. But the young priest is wise enough to see the human comedy that is going on and to realize that there is a limit to what kind of intervention in human life it is reasonable to make. In God's house there are many mansions, we finally learn, and Christianity under the glacier isn't precisely the same as Christianity elsewhere. Maybe, in the presence of very mighty forces, you need to calibrate your own ethics, find an appropriate *genre de vie*. Things change. Perhaps the church, and the way of life on earth it has prescribed for its members, has to change, too.

Notes

1 See also the work on the formation of national identities of (a few) northern nations, e.g. Bomann-Larsen 1993 on snow and Norway; Sörlin 1988 on the role of north, including glaciers, in Sweden; or on ice as a fascinating element to be conquered in the quest for the poles, e.g. Lopez 1986; Spufford 1997.

2 A massive slandering campaign was directed towards Carey and his team of authors in media and by bloggers that traditionally have been identified as republican and denialist, e.g. Fox News, The Wall Street Journal, the Heartland Institute, reason.com, and others. Carey's response came in the form of an interview in *Science*, 14 March 2016. 'A feminist glaciology framework for global environmental change research'.

3 Paglia, E. "The Teleconnected Arctic", is planned for publication in Nina Wormbs, ed., *The Continuous Production of Competing Artic Futures: Voices, Resources, Governance*. New York: Routledge (in preparation), but is already available in his dissertation, *The Northward Course of the Anthropocene: Transformation, Temporality and Telecoupling in a Time of Environmental Crisis* (2016). Diss. Division of History, KTH Royal Institute of Technology. Stockholm.

4 There has been a tremendous growth in knowledge about ice in this respect in only the last decade, including important work during the IPY. See e.g. Krupnik et al. (2010); Gearheard et al. (2013).

5 See Sörlin (1988). The best summary version in English is Sörlin (1989).

References

Ahlmann, H.W. (1938). *Land of Ice and Fire: A Journey to the Great Iceland Glacier* (trans. K. and H. Lewes). London: Routledge & Kegan Paul.

Ahlmann, H.W. (1941). Le régime des glaciers. *Revue de Géographie Alpine*, 29 (4), 537–566.

Ahlmann, H.W. et al. (1937). Vatnajökull: Scientific results of the Swedish-Icelandic investigations. *Geografiska Annaler*, 19, 146–231.

Ahlmann, H.W. et al. (1938). Vatnajökull: Scientific results of the Swedish-Icelandic investigations. *Geografiska Annaler*, 20, 171–233.

Ahlmann, H.W. et al. (1939). Vatnajökull: Scientific results of the Swedish-Icelandic investigations. *Geografiska Annaler*, 21, 39–66, 171–242.

Ahlmann, H.W. et al. (1940). Vatnajökull: Scientific results of the Swedish-Icelandic investigations. *Geografiska Annaler*, 22, 188–205.

Ahlmann, H.W. et al. (1943). Vatnajökull: Scientific results of the Swedish-Icelandic investigations. *Geografiska Annaler*, 25, 1–54.

Armiero, M. (2011). *A Rugged Nation: Mountains and the Making of Italy: Nineteenth and Twentieth Centuries*. Isle of Harris, UK: White Horse Press.

Avango. D. (2005). The Svea Mine: Swedish mining between industry, diplomacy and geoscience. (Dissertation in Swedish with a summary in English). Department of Historical Studies, KTH Royal Institute of Technology, Stockholm.

Barry, R.G., Jania, J., and Birkenmajer, K. (2011). A.B. Dobrowolski—the first cryospheric scientist—and the subsequent development of cryospheric science. *History of Geo- and Space Sciences*, 2, 75–79.

Broecker, W.S. (1987). Unpleasant surprises in the greenhouse? *Nature*, 328, 123–126.

Callendar G.S. (1938). The artificial production of carbon dioxide and its influence on climate. *Quarterly Journal of Royal Meteorological Society*, 64, 223–240.

Callendar, G.S. (1949). Can carbon dioxide influence climate? *Weather*, 4, 310–14.

Carey, M. (2007). The history of ice: How glaciers became an endangered species. *Environmental History*, 12(3), 497–527.

Carey, M. (2010). *In the Shadow of Melting Glaciers: Climate Change and Andean Societies*. Oxford and New York: Oxford University Press.

Carey, M., Jackson, M., Antonello, A., and Rushing, J. (2016). Glaciers, gender, and science: A feminist glaciology framework for global environmental change research. *Progress in Human Geography*, 40. Online.

Castree, N., Adams, W.M., Barry, J., Brockington, D., Buscher, B., Corbera, E., . . . Wynne, B. (2014). Changing the intellectual climate. *Nature Climate Change*, 4(9), 763–768.

Christensen, M., Nilsson, A.E., and Wormbs, N. (eds). (2013). *Media and the Politics of Arctic Climate Change: When the Ice Breaks*. New York: Palgrave MacMillan.

Cruikshank, J. (2005). *Do Glaciers Listen? Local Knowledge, Colonial Encounters, and Social Imagination*. Vancouver: University of British Columbia Press.

Dobrowolski, A.B. (1923). *Historia naturalna lodu [The natural history of ice]*. Warsaw: Kasa Pomocy im. Mianowskiego.

Fleming J.R. (2007). The Callendar Effect: The life and times of Guy Stewart Callendar (1898–1964). Paper presented at the American Meteorological Society, Boston, MA.

García Márquez, G. (1970). *A Hundred Years of Solitude* [orig. in Spanish 1967]. New York:

Harper & Row.
Gearheard, S.F., Holm, L.K., Huntington, H., Leavitt, J.M., Mahoney, A.R., and Opie, M. (2013). *The Meaning of Ice: People and Ice in Three Arctic Communities*. Hanover, NH: International Polar Institute Press.
Gieryn, T.F. (2002). Three truth-spots. *Journal of the History of the Behavioral Sciences*, 38, 113–132.
Gieryn, T.F. (2006). The city as truth-spot: Laboratories and field-sites in urban studies. *Social Studies of Science*, 36, 5–38.
Gunnarsson, T. (1998). *Nordic Landscape Painting in the Nineteenth Century*. New Haven, CT: Yale University Press.
Hastrup K. B. (2012). The icy breath: Modalities of climate knowledge in the Arctic. *Current Anthropology*, 53(2), 226–244.
Hastrup, K.B. (2015). *Thule: På tidens rand*. Copenhagen: Lindhardt og Ringhof.
Heise, U. (2008). *Sense of Place and Sense of Planet: The Environmental Imagination of the Global*. Oxford: Oxford University Press.
Jónsdottír, Á. (2013). Scaling climate: The politics of anticipation. In Hastrup, K. and Skrydstrup, M. (eds) *The Social Life of Climate Change Models: Anticipating Nature* (128–43). New York: Routledge.
Jordheim, H. (2014). Introduction: Multiple times and the work of synchronization. *History and Theory*, 53(4), 498–518.
Krupnik, I., Aporta, C., and Gearheard, S.F. (eds) (2010). *SIKU: Knowing Our Ice*. Heidelberg, London and New York: Springer.
Latour, B. (1987). *Science in Context: How to Follow Scientists and Engineers Through Society*. Cambridge, MA: Harvard University Press.
Laxness, H. (2005 [1968]). *Kristnihald undir Jökli* (*Christianity at the Glacier*) new edn *Under the Glacier*, trans. Magnus Magnusson. New York: Vintage.
Lopez, B. (1986). *Arctic Dreams*. New York: Vintage.
Masco, J. (2010). Bad weather: On planetary crisis. *Social Studies of Science*, 40(1), 1–31.
Masco, J. (2015). The age of fallout. *History of the Present*, 5(2), 137–168.
Merchant, C. (1993). *Major Problems in American Environmental History*. Boston, MA: D.C. Heath/Houghton Mills.
Meusburger, P., Livingstone D. N., and Jöns H. (eds) (2010). Geographies of Science. *Knowledge and Space* 3, Dordrecht: Springer.
Nansen, F. (1911). *In Northern Mists: Arctic Exploration in Early Times, Vols. 1–2* (trans. A. G. Chater). New York: Frederick A. Stokes.
Nash, R. (1967). *Wilderness and the American mind*. New Haven, CT: Yale University Press.
Paglia, E. (2016). The northward course of the anthropocene: Transformation, temporality and telecoupling in a time of environmental crisis. Diss. Division of History, KTH Royal Institute of Technology. Stockholm.
Riffenburgh, B. (1993). *The Myth of the Explorer: The Press, Sensationalism, and Geographical Discovery*. London: Belhaven Press.
Roberts P. (2011) *The European Antarctic: Science and Strategy in Scandinavia and the British Empire*. Basingstoke: Macmillan.
Rockström, J. Steffen, W., Noone, K., Persson, Å., Chapin III, S., Lambin, E.F., ... Foley, J.A. (2009). Exploring the safe operating space for humanity. *Nature*, 461, 472–475.
Rudwick, M. (2014). *Earth's Deep History: How it was Discovered and Why it Matters*. Chicago, IL: The University of Chicago Press.
Schytt, V. (1940). Glaciologiska arbeten i Kebnekajse. *Ymer*, 67, 18–42.

Sörlin, S. (1988). *Framtidslandet: Debatten om Norrland och naturresurserna under det industriella genombrottet* [with a Summary in English: *Land of the Future: The Debate on Norrland and its Natural Resources at the Time of the Industrial Breakthrough*]. Stockholm: Carlsson bokförlag.

Sörlin, S. (1989). *Land of the Future: Norrland and the North in Sweden and European Consciousness*. Center for Arctic Cultural Research, Umeå University, Miscellaneous Publications 8.

Sörlin, S. (1997). Resa i spåren av det förflutna: Rapport från Svalbard, augusti 1997, *Forskning & Framsteg*, 32(8), 38–45.

Sörlin, S. (1998). Monument and memory: Landscape imagery and the articulation of territory. *Worldviews: Environment, Culture, Religion*, 2, 269–279.

Sörlin S. (2011) The anxieties of a science diplomat: Field co-production of climate knowledge and the rise and fall of Hans Ahlmann's "polar warming" (66–88). In Fleming J.R., Jankovich V. (eds) *Osiris 26: Revisiting Klima*. Chicago, IL: University of Chicago Press.

Sörlin, S. (2013). Reconfiguring environmental expertise. *Environmental Science and Policy*, 28, 14–24.

Sörlin, S. (2015). The emerging Arctic humanities: A forward-looking post-script. *Journal of Northern Studies*, 9(1), 93–98.

Sörlin, S. (2016). Scaling the planetary humanities: Environmental globalization and the Arctic. In Heise, U., Christensen, J., and Niemann, M. (eds) *The Routledge Companion to the Environmental Humanities*. London: Routledge.

Spufford, F. (1997). *I May Be Some Time: Ice and the English Imagination*. London: Palgrave.

Stehr, N. and von Storch, H. (eds) (2000). *Eduard Brückner—the Sources and Consequences of Climate Change and Climate Variability in Historic Times*. Dordrecht: Springer Science-Business Media.

Svenonius, F. (1910). Die Gletscher Schwedens im Jahre 1908. *Sveriges Geologiska Undersökningar* (SGU) serie Ca, Vol. 5, Part I, 1–54.

Worster, D. (1988). *The Ends of the Earth: Perspectives on Modern Environmental History*. London: Cambridge University Press.

Wright, J.K. (1953). The open polar sea. *Geographical Review*, 43, 338–365.

3 Experiencing Earth art

Or, lessons from reading the landscape

Marsha Weisiger

At high noon, we arrive at the Cowboy Bar and Café in Montello, Nevada, a godforsaken town on the alkali flats of the prehistoric Lake Bonneville near the Utah state line. Three companions and I are making a pilgrimage to *Sun Tunnels*, artist Nancy Holt's monumental earthwork, completed in 1976, where we will observe the setting sun of summer solstice. We will not be alone. Fellow pilgrims—latter-day hippies and artsy college kids from Salt Lake City and Reno—trickle into the bar for burgers and beers before venturing back into the white heat of the Great Basin Desert. I strike up a conversation with a couple of locals, who confess that they never have felt sufficiently curious to make the half-hour drive to this major work of art, even though they admit that there is little to do here but play the slot machines and fish at a nearby reservoir. That's too bad, because *Sun Tunnels* offers the chance to see their familiar landscape anew.

As evening draws near, we head out into the desert, eager to get to our destination well before sundown. Following sketchy directions, we pass active sand dunes, bump down a series of washboard graveled roads, and miss a turn before at last we see what look like sewer pipes splayed in the distance. Just as we arrive, a hellacious sandstorm whips across denuded wastes. At first we huddle by our cars, but soon we realize the wind and sand may never let up. Reluctantly, we emerge and inspect the four concrete tubes, each eighteen feet long with eight-foot-diameter openings, configured in an open cross and aligned with the rising and setting sun of the summer and winter solstices (Holt 2011). We scurry inside one of the tunnels, where it is cool and calm. Along each of the cylinders' walls, perforations admit dappled sunlight designed to represent one of four constellations—Draco, Perseus, Columba, and Capricorn—while also serving as oculi, eyes opening to the sky. The large openings at each end align so that pairs of tunnels appear to form two rings, one encircling the other, creating an aperture that frames distant peaks. *Sun Tunnels* is thus a kind of desert observatory, connecting land and sky.

Nancy Holt designed *Sun Tunnels* so that visitors would experience both the expansiveness of the Great Basin landscape and the intimacy of specific landforms located in the circular frame (Lippard 2011). "I wanted to bring the vast space of the desert back to human scale," the artist explained in an essay in *Artforum* shortly after completing the work. "The panoramic view of the landscape is too overwhelming to take in without visual reference points. The view blurs out rather than sharpens. Through the tunnels, parts of the landscape are framed and come into focus." *Sun Tunnels* encourages visitors to consider place, scale, the relationship of earth to cosmos, and "a sense of being on this planet, rotating in space, in universal time" (Holt 1977, p. 35). It has the power to alter the viewer's perspective.

At the moment, though, all I perceive is the wind, whipping up the ancient remnants of Lake Bonneville, coating my hair, my nostrils, my lungs, my entire being with silt.

Before long, more than fifty pilgrims have arrived, mostly twenty-somethings, who seem oblivious to the wind and sand. A number of them scale the cylinders, using the perforations as hand- and toeholds, and once on top, some dance while others extend their arms and legs in yoga poses. It is clear that for these millennials, *Sun Tunnels* is a tactile, playful construction. A small group congregates within one of the tubes and gyrates to recorded music with an African beat.

"It's like a night club in there," one of my friends remarks. The dancers seem not to notice the framed peaks in the distance or the clouds moving across the oculi. They engage in a full body experience, erotic and trippy in the swirling sand.

As the hour approaches, I circumnavigate the earthwork and then stand directly in front of the two tunnels that will line up with the setting sun, seeking the best vantage. Soon my friends join me. We wait. At some unspoken signal, the revelers stop the music, slide off the tubes, and assemble behind us in silence. We wait. And wait. Losing patience, three women walk back to the cylinders, hoist themselves on top, and trill ethereal melodies on flutes. Two men enter the paired tunnels and perform synchronized handstands. And then at long last, the moment comes. The orange orb arcs across the aligned orifices, which glow with golden light, like a pair of wedding bands. Three minutes later, it's over. The sun sinks below the horizon. The dust continues to blow.

One friend wonders aloud whether all this sand is the result of erosion created by the artist herself. Had Earth Art created Earth Damage? Constructing *Sun Tunnels* had involved road graders, ditch diggers, dump trucks, a concrete mixing truck, a crane, and even a helicopter (Holt 1977). Clearly, the ground immediately around the cylinders had been scraped bare, which undoubtedly allows the wind to lift sediment aloft. But was this earthwork ultimately to blame? I had to consider the fact that this was the former bed of a Pleistocene lake, and we had passed active sand dunes miles away. What we were witnessing was surely geological erosion at work, though no doubt accelerated by the artist and her pilgrims.

To see a work of Earth Art (also commonly called Land Art), the visitor must generally make a pilgrimage—to a desert, a forest, a quarry, a mine—and experience the work bodily or performatively by walking above it, across it, around it, through it (see, for example, Hogan 2008). In the process, the visitor engages in a direct encounter with the natural world. Earth Art transforms the visitor, as the art critic Jeffrey Kastner has written, "from observer of nature to participant in it" (Kastner 1998, p. 16). Earth Art also illuminates what can be learned about the environmental history of a place by walking the land.

One of the research methods that sometimes distinguishes environmental history from most other historical fields is an approach, borrowed from historical and cultural geographers, that involves examining the land as a primary source. We call it "reading the landscape." Geographer Alyda C. Hanson (1922) coined the phrase, stating that it involved "seeing, recognizing, and interpreting ... the natural and cultural phases" of the land, "especially their interdependence" (p. 217). It was akin to reading a book, she added, in that it could be skimmed or studied intensively. A generation later, the naturalist May Theilgaard Watts (1957) adopted the phrase and pioneered the methods we use to discern the human imprint in the mosaic of grasses, shrubs, trees, rivers, ponds, bogs, and dunes. Even today, her guidebook remains the best introduction for detecting the hidden human presence in what otherwise seem like "natural" landscapes. Geographers, geomorphologists, naturalists, and urban scholars have built on the foundation she laid, creating specialized methods for their particular disciplines. (Lynch 1960; Brown 1969; Marks and Gardescu 2001; Duncan and Duncan 2009; Brierley et al. 2013; Brierley and Fryirs 2014). It is, in fact, possible to "read" landscapes—both rural and urban—much as you would read a somewhat challenging text, so long as you have some knowledge of a landscape's ecological and physical processes, the cultures of those who worked in the landscape, changes in aesthetic fashion over time, and what the geographer D. W. Meinig called "the beholding eye" (Clay 1973; Lewis 1979; Meinig 1979; Francaviglia 1991; Cronon 2008).

Geographer William Wyckoff (2014) offers the most comprehensive field guide to reading landscapes in the United States, although its focus is on the American West. He advises us to pay attention to both geological and historical time and to be aware of the scale of a given landscape, keeping in mind that the extent of the area you can observe from a given vantage point may be smaller than the landscape unit you are studying. When you stand on the ground and look around you, what do you see? If you look down from a hill or a tall building, how does the change in scale alter your view and understanding of the landscape? If you fly over the area, how does the scale of the landscape change, and what new patterns and connections do you see? Now, observe the physical evidence of the ways that people have both transformed and adapted to the natural environment. This may be obvious or require considerable detective work, using maps, historical photographs and documents, field guides of various sorts, and an eye for anomalies on the ground. Follow the flow of water and note how it is exploited and managed. Examine the buildings, structures, and infrastructure and consider what they communicate about fashion, culture, economics, power, and connections. Notice the density of development and the transition zones between neighborhoods, between cities and suburbs, between grasslands or sagebrush and forest. Pay attention to the ways in which government or private ownership, local regulations, power, wealth, poverty, labor, ethnicity, race, gender, and cultural diffusion are manifest in the landscape. Finally, look for symbolic expressions of nationalism, ideology, cultural identity, and nostalgia. Each landscape has different combinations of these elements, and some are more readily apparent than others.

Because landscapes are culturally and historically produced, you must be attuned to the different types of features you might encounter in a particular place. The geographer Richard Muir's guide to reading the landscapes of Britain (2000), for example, explains how to identify and interpret standing stones, Roman roads, dykes, moats, and medieval wood-management systems. Trees with numerous trunks rising from the same point and looking much like inverted octopus tentacles, for example, indicate coppicing, a medieval method of harvesting woodlands that encouraged trees to produce new shoots, which could be harvested on a regular cycle. Earthen embankments, crowned by hedges or pollards—coppiced trees with the branches high above the reach of livestock—embraced these woodlands, reflecting the trend away from commons and toward enclosures. Different landscapes offer different possibilities. And yet, as Wyckoff writes, "every landscape has a human imprint" (2014, p. 23). It is up to us to figure out how to recognize, reveal, and render its meaning (Ingold 2000).

One might think of the land as a palimpsest. Remnants of past developments form layers, one atop the other, or side-by-side. If we were archaeologists, we might dig down through the strata to develop a history of the *longue durée*. But it is also possible to examine the surface and see layers of history over time. Novices might take along an archaeologist, an ecologist, a geomorphologist, or an architectural historian, someone specially trained to read the stories inscribed, embedded, or built in the environment. The assistance of an elder from a local Indigenous community might offer especially valuable insights passed down from one generation to the next (see, for example, Breen 1989; Basso 1996). To be sure, observing landscapes is insufficient. One must look, read, go to the archives, and look again (Lewis 1979). Documentary research and the examination of historical maps and photographs are necessary to corroborate, correct, contextualize, and comprehend what you see on the ground. And yet the landscape holds many clues to the past that likely cannot be found in the written record, and walking the land can raise research questions that one would otherwise never know to ask.

I have long taken students in my environmental history classes on field trips to learn the value of reading landscapes. A recent trip near our campus in Eugene, Oregon, took us by a duck pond, which seems to emerge from underneath a shuttered Chinese restaurant. As we walked, we noted that the pond narrows into a creek, overgrown with weeds, and an old boathouse sits on one side of the water, across from the university's art studios. The creek continues for a little more than five hundred meters and then disappears among clumps of trees and a cluster of apartment and commercial buildings. We made our way toward the Willamette River, beyond a park and a strangely stagnant stream, perhaps backwater. A path carried us through a tunnel under a railroad crossing, and when we emerged, there was the river, where a sign proclaiming CAUTION, PUMP INTAKE SCREEN drew our attention. We looked through the murky Willamette and saw a rusting pipe with a screen over its open end. Further down the path, an old graffiti-covered concrete embankment lay below us, near the stream. Here the path comes to an end near a tangle of roadways, railway tracks, and a pedestrian bridge. We took the bridge across the Willamette. But about midway, we stopped and looked back, peering through binoculars. To our right, a broken slab of concrete juts from the shore. To our left, in the middle distance, four concrete piers rise in a line across the river, and from the pier closest to the south bank, a concrete beam extends to a sand island. Several of the students mentioned that they had floated by those structures while kayaking down the river but were not sure what to make of them. Our investigation of the land raised questions but did not give us clear answers.

Fortunately, I had brought with me a collection of historical maps, which—along with investigations of photographs and the written record—revealed that the duck pond was once a millpond, and what looked to us like a creek or a stagnant stream was once a millrace. First excavated in 1851, five years after Eugene Skinner founded the town, the race once powered a cluster of mills that produced flour, lumber, furniture, doors and sashes, excelsior, and wool. The traces of these mills lie buried beneath a modern commercial district and a highway interchange, about a half-mile from the millpond. Following a series of floods between 1874 and 1890, which scoured the Willamette's channel, the owner of the millrace built a small dam to divert water from the river and increase flow into the raceway. The embankment and the concrete slab, piers, and beam are the remnants of that diversion structure. The pipe lying in the river was added a half century later as part of a pumping system to bring fresh water into the millrace, which by the 1920s—after electricity made water power obsolete—had become a popular recreational site for college students on romantic canoe rides. (Sanborn Map Co. 1888, 1895, 1902, 1925; "The 1890 Flood," *Eugene City Guard*, 8 Feb. 1890; Gill, et al. 1979; Kramer, Minor, and Toepel 2007).

Across the bridge, we found more puzzles. "Pre's Trail," named for track star Steve Prefontaine, a local legend, follows what's known as the "Canoe Canal." We assumed that this was a continuation of the millrace. It flows past some expensive-looking houses, passes through a large pond, and eventually empties into the Willamette River, across from the old mill district. Was the pond here another millpond? As it turned out, the answer is no. Aerial photographs reveal that the pond was created sometime in the mid-1970s. Moreover, the Canoe Canal is not a canal at all. It is a slough, a remnant of the braided Willamette River, which changed course in the disastrous flood of 1890 and left what modern maps label "Patterson's Slough" as a side channel (U.S. Army Corps of Engineers 1936; City of Eugene 1974; U.S.G.S. 1975, 1978). Maps and photographs helped solve some of the mysteries of this landscape, but it was through walking the land itself that we discovered our first clues.

Reading the landscape has been important not only to my pedagogy, but also to my scholarship and that of other environmental historians. Historian Mark Fiege (1999), for example, gained crucial insights regarding hybrid landscapes as he walked the irrigation canals and ditches of southern Idaho's Snake River valley. While hiking along the Blitzen River, historian Nancy Langston (2003) stumbled on five slaughtered coyotes, which signified the vicious conflicts over land management that have shaped Oregon's high desert, for good and for ill. As for myself, reading landscapes proved crucial to my study of the environmental history of Diné (Navajo) pastoralism in the southwestern United States (Weisiger 2009). That study examined the history of livestock grazing on the Navajo Reservation and a New Deal program to cut the numbers of sheep, goats, and horses by half, in an unsuccessful effort to conserve rapidly eroding rangelands. At the beginning of my research, I was mystified as to how soil conservationists could distinguish overgrazing from the natural effects of aridity, drought, wind, and water, and how those men could produce a detailed evaluation of a landscape stretching across 25,000 square miles within only a few years. Walking through desert grasslands with range ecologists, I learned to recognize hummocky ground, snakeweed, and other plants as markers of overgrazing. Afterwards, I could not travel through New Mexico without seeing the evidence of hammered land everywhere I looked. Hiking through canyons with archaeologists, I acquired a spatial understanding of early Diné settlement and the ecological challenges of raising livestock on stingy ground. Flying in chartered airplanes over the Navajo Nation—just as New Deal conservationists had done in one of the earliest examples of aerial photographic surveys—I gained a landscape view of large-scale erosion resulting from a combination of overgrazing and climate change. But Earth Art taught me something new about reading the landscape.

Earth Art was the vanguard of an environmental art movement. It arose in the mid-1960s among a group of New York artists who shared an anti-Establishment, anti-authoritarian ethos and chafed at the commodification of art by investors (Boettger 2002). Many used remote western landscapes as their canvas so that, as sculptor Michael Heizer wrote, "the artist has no sense of the commercial or the utilitarian," which galleries and collectors favored (quoted in Wallis 1998, p. 30). The environmental art movement encompasses a wide array of approaches, ranging from Christo and Jeanne-Claude's temporary installations, such as the nylon *Running Fence* shimmering through a rural landscape in northern California, to Michael Heizer's *The City*, monumental altars in the Nevada desert that evoke the ruins of Chichen Itza, to reclamation projects, such as Nancy Holt's *Up and Under* at a sand quarry near Tampere, Finland, or Betty Beaumont's underwater *Ocean Landmark*, three miles off the coast of New York's Fire Island National Seashore (Boetzkes 2010). Some of this work helps us to see nature more clearly. Andy Goldsworthy's ephemeral arrangements of leaves articulate the turbulent current of a river, while Walter De Maria's one-mile by one-kilometer grid of stainless steel poles in the New Mexico desert makes visible subtle shifts in light from one moment to the next. What most of these artists have in common is that they facilitate a conversation about our relationships with the natural world.

Earth art emerged in the cataclysmic year of 1968, both anticipating and reflecting the rise of modern environmentalism. Robert Smithson—who would become the movement's philosophical voice—published a kind of manifesto in *Artforum*, titled "A Sedimentation of the Mind: Earth Projects" (Smithson 1968/1969), and organized an exhibition of ten artists at the Dwan Gallery in New York City under the simple title "Earthworks." Many visitors didn't quite know what to make of that exhibit, which featured Smithson's trapezoidal arrangement of boxes of rocks and Robert Morris's formless mound of dirt, peat, grease, and felt, pierced by aluminum, zinc, copper, and brass rods, bringing to mind the messy rubbish of a construction site. None of this was the kind of art that buyers sought to acquire. That was the central narrative, but there was a subtext, too. Smithson had named the exhibit after Brian Aldiss's dystopian novel, in which a burgeoning population has so exceeded Earth's resource base that even soil is a precious commodity and the only solution to overpopulation appears to be apocalyptic global war (Wallis 1998). Within this context, then, the "Earthworks" exhibition—though not political in any conventional sense—offered a caustic commentary on American materialism and the commodification of both art and nature.

The most celebrated extant earthwork is Smithson's *Spiral Jetty*, created in 1970 near Rozel Point on the edge of the Great Salt Lake in northwestern Utah. Made of six thousand tons of basalt rock and mud, the fifteen hundred-foot-long coil seems to unwind from the shore into the lake. The lake's hypersalinity was crucial to the work. Pigmented bacteria (*Halobacterium*) and algae (*Dunaliella salina*) thrived in an environment that was transformed from highly saline to hypersaline in the 1950s, when the construction of an earthen railroad causeway isolated the lake's northern arm from fresh water inlets to the south (Post 1977). Those microorganisms gave the water a deep red hue, which contrasted with the black basalt. For Smithson, the red water evoked images of blood coursing through veins and arteries. The lake thus brought to mind, he wrote, the essence of life—not only blood, but also the "masses of cells consisting largely of water, proteins, lipoids, carbohydrates, and inorganic salts" (Smithson 1972, p. 228). At the same time, *Spiral Jetty* expressed Smithson's fascination with spirals, with geological processes, with the physical laws of entropy, and with the dialectic between nature and industry (Smithson 1971).

Smithson explicitly rejected the romanticism often associated with environmentalism. While he appreciated the sublime qualities of the Great Salt Lake, what most interested him was the juxtaposition of the sublime with signs of industrial ruin—old abandoned piers, fragments of oil rigs, a sticky black tar pool, all remnants of intermittent oil drilling operations between the 1920s and the 1980s. The site near Rozel Point, he wrote, "gave evidence of a succession of man-made systems mired in abandoned hopes" (Smithson 1972, p. 223). When people came to view his work, he hoped they would see the detritus of abandoned oil drilling operations along the way. (Unfortunately, in 2005, the state of Utah, which owns the land, hauled off the rusting junk relating to these drilling operations (Sinor 2008).) This interest in drawing attention to industrial sites later led him to create or propose earthworks at disused quarries and exhausted strip mines. Speaking of proposed projects for strip mines, he mused in an interview with the art historian Stella Russell that it seemed that "on the one side you have ... the extreme preservationist ecology groups operating, and then on the other side you have the rather blind profiteering, and there seems to be no connection, no dialogue between the two areas. So I hope to create some consciousness in that area" (Smithson 1973, p. 2). Smithson sought to create art that would mediate the uneasy relationship between humans, industry, and nature (Jones 2005).

With *Spiral Jetty*, he also sought to reveal the nature of the Great Salt Lake. The lake changes from season to season, and from year to year. Smithson anticipated and applauded these fluctuations. In summer, he wrote, "the water evaporates and leaves the salt crystals, and the water gets redder" because of the intensified salinity (Smithson 1973, p. 1). This metamorphosis, as the *Spiral Jetty* interacted with the salt, the water, and water's erosional power, was integral to Smithson's plan. By juxtaposing black rock against red water, by making clear the changing water levels, by accumulating and then shedding salt crystals, the *Jetty* told visitors something about the material reality of the lake.

I first visited *Spiral Jetty* on a cold, blustery day in May 2008. Storm clouds darkened the sky. After first stopping at the Golden Spike National Historic Site, which commemorates the completion of the first transcontinental railroad in 1869, three girlfriends and I bumped down a rutted and rocky road for fifteen miles until it became impassable. In the far distance, we could see Smithson's earthwork uncoiling into the lake. Our first impression was disappointment. *Spiral Jetty* seemed small and insignificant. We walked along a trail for about a quarter of a mile, past an abandoned wooden jetty. As we neared, we discovered that our initial perception offered a lesson in scale. This was part of Smithson's intention. "The scale of the Spiral Jetty tends to fluctuate depending on where the viewer happens to be," he wrote. "Scale depends on one's capacity to be conscious of the actualities of perception" (Smithson 1972, p. 225). Indeed, *Spiral Jetty* brings a human scale to an expansive waterscape that seems almost limitless, despite the mountains that bound it. The rocks and the mud between them were encrusted with salt crystals, and water lapped up between the coils. "Each drop that splashed onto the Spiral Jetty coagulated into a crystal," Smithson had observed. "Undulating waters spread millions upon millions of crystals over the basalt" (Smithson 1972, p. 228). *Spiral Jetty* expressed the saline nature of the lake, exactly as the artist had envisioned. He wanted visitors to walk the length of the spiral, and so we made our way down a long basalt promenade, followed the rocky path as it wound toward the center, and then retraced our steps to the point of beginning. *Spiral Jetty* is not monumental but meditative, reminiscent of a labyrinth in a watery Zen garden.

Most importantly, it makes us mindful of the lake. The Great Salt Lake is a barometer of local climatic conditions. A closed body of water with no outlets, it is both shallow and expansive and thus fluctuates according to long-term evaporation and precipitation. Over the course of the historical record, it has varied in height by about twenty feet, from a historic low point in 1963 to a high point in 1986. (Hassibe and Keck 1991; Lall and Mann 1995; DeRose et al. 2014).

Within two years after Smithson completed his masterpiece, lake levels rose and *Spiral Jetty* disappeared. Hidden from view for two decades, it was known only through photographs. Then in 2002, the Great Salt Lake receded, and the earthwork emerged (Merrell 2002). It disappeared again in 2005 and reappeared the following year, as the lake advanced and retreated. From one year to the next, oscillating between deluge and drought, the lake submerged and exposed the work (Loe 2016; Palmer Drought Severity Index 2002–2015). *Spiral Jetty* became a marker for visualizing a volatile climate.

And yet knowing that did not prepare me for my return trip in June 2015.

After the solstice bacchanal at *Sun Tunnels*, my companions and I drove eastward for three hours to see *Spiral Jetty*. This time, the view stunned me. The earthwork had not disappeared. The lake had. Where once waters had lapped up against my shoes as I traced the arc of the *Jetty*, now the Great Salt Lake shimmered in the distance, like a mirage, a quarter of a mile or more away. White salt flats, not red water, surrounded the coil of black rocks. Might this be the new normal in an era of climate change? It is too soon to tell, but this work of Earth Art will surely let us know.

Earth Art, to be sure, provides an unconventional path toward reading landscapes. And yet it illustrates the many ways that one can learn about the nature of the land through careful observation, combined with more familiar archival research. The landscape itself can reveal the consequences of land management policies that defy understanding when we rely solely on the documentary record. Indeed, walking the land—and returning again and again, from one season to the next, one year to the next—is essential to understanding the dynamic landscapes environmental historians often write about. And best of all, it gets you outdoors and reminds you why you chose to study environment history in the first place.

Acknowledgments

I would like to thank Kate Bonansinga, Tim Cockrum, Elizabeth Horodowich, Jarma Jones, Steven Leone, Patricia McDowell, John Nelson, Colleen O'Neil, and Rachel Stevens for assisting me with this work in various ways. I also thank Julie and Rocky Dixon, the University of Oregon, the Redd Center for Western Studies at Brigham Young University, and the Southwest and Border Cultures Institute at New Mexico State University for their financial support.

References

Basso, K. (1996). *Wisdom Sits in Places: Landscape and Language Among the Western Apache*. Albuquerque, NM: University of New Mexico Press.

Boettger, S. (2002). *Earthworks: Art and the Landscape of the Sixties*. Berkeley, CA: University of California Press.

Boetzkes, A. (2010). *The Ethics of Earth Art*. Minneapolis, MN: University of Minnesota Press.

Breen, T. H. (1989). *Imagining the Past: East Hampton Stories*. Reading, MA: Addison-Wesley.

Brierley, G., and K. Fryirs. (2014). Reading the landscape in field-based fluvial geomorphology. In M. J. Thornbush, C. D. Allen, F. A. Fitzpatrick (eds) *Geomorphological Fieldwork* (pp. 231–257). Amsterdam: Elsevier.

Brierley, G., K. Fryirs, C. Cullum, M. Tadaki, H. Huang, B. Blue. (2013). Reading the landscape: Integrating the theory and practice of geomorphology to develop place-based understandings of river systems. *Progress in Physical Geography*, 37(5), 601–21.

Brown, V. (1969). *Reading the Woods: Seeing More in Nature's Familiar Faces*. Harrisburg, PA: Stackpole Books.

City of Eugene. (1974). C-ECE-74. B17-9-10. Scale 1:6,000. [Aerial photograph.] Map and aerial photography library, Knight Library, University of Oregon.

Clay, G. (1973). *Close-up: How to Read the American City*. Chicago, IL: University of Chicago Press.

Cronon, W. (2008). *How to Read a Landscape*. www.williamcronon.net/researching/landscapes.htm. Web.

DeRose, R. J., S. Wang, B. M. Buckley, M. F. Bekker. (2014). Tree-ring reconstruction of the level of Great Salt Lake, USA. *The Holocene*, 24(7), 805–13.

Duncan, N., and J. Duncan. (2009). Doing landscape interpretation. In D. DeLyser, S. Herbert, S. Aitken, M. Crang, L. McDowell (eds), *The SAGE Handbook of Qualitative Geography* (pp. 225–247). London: SAGE Publications.

Fiege, M. (1999). *Irrigated Eden: The Making of an Agricultural Landscape in the American West*. Seattle, WA: University of Washington Press.

Francaviglia, R. V. (1991). *Hard Places: Reading the Landscape of America's Historic Mining Districts*. Iowa City, IA: University of Iowa Press.

Gill, J., S. Otremva, J. Rees, G. Wingfield. (1979). *Eugene Millrace: A History*. Eugene, OR: City of Eugene Housing and Community Conservation Department.

Hanson, A. C. (1922). Reading the landscape. In *Proceedings of the High School Conference of November 17, 18, and 19, 1921* (pp. 217–20). Urbana, IL: University of Illinois.

Hassibe, W. R., and W. G. Keck. (1991). *The Great Salt Lake*. U.S. Geological Survey. Washington, D.C.: Government Printing Office, rev. printing.

Hogan, E. (2008). *Spiral Jetta: A Road Trip Through the Land Art of the American West*. Chicago, IL: University of Chicago Press.

Holt, N. (1977). Sun tunnels. *Artforum*, 15(8), 32–37.

Holt, N. (2011). Photographs and writings on selected site-specific sculptures. In A. J. Williams (ed.), *Nancy Holt: Sightlines* (pp. 73–162). Berkeley, CA: University of California Press.

Ingold, T. (2000). The temporality of the landscape. In T. Ingold, *The Perception of the Environment: Essays on Livelihood, Dwelling and Skill* (pp. 189–208). London: Routledge.

Jones, J. R. (2005). *Breaking Ground: Art, Urbanization and Industry's Role in American Concepts of Nature*. MA thesis, New Mexico State University.

Kastner, J. (ed.) (1998). *Land and Environmental Art*. London: Phaidon Press.

Kramer, G., R. Minor, K. Toepel. (2007). *An Archaeological Assessment of the Eugene Millrace Diversion Dam and Intake for the I-5 Willamette River Bridge (#08329) Project, Lane County, Oregon*. Heritage Research Associates Report No. 314. Eugene, OR: Heritage Research Associates, Inc.

Lall, U. and M. Mann. (1995). The Great Salt Lake: A barometer of low-frequency climatic variability. *Water Resources Research*, 31(10), 2503–15.

Langston, N. (2003). *Where Land and Water Meet: A Landscape Transformed*. Seattle, WA: University of Washington Press.

Lewis, P. K. (1979). Axioms for reading the landscape: Some guides to the American scene. In D. W. Meinig (ed.), *The Interpretation of Ordinary Landscapes: Geographical Essays* (pp. 11–32). Oxford: Oxford University Press.

Lippard, L. (2011). Tunnel visions: Nancy Holt's art in the public eye. In A. J. Williams (ed.), *Nancy Holt: Sightlines* (pp. 59–72). Berkeley, CA: University of California Press.

Loe, H. S. (2016). *The Spiral Jetty: Strata of Water*. www.hikmetsidneyloe.com/the-spiral-jetty.html. Web.

Lynch, K. (1960). *The Image of the City*. Cambridge, MA: MIT Press.

Marks, P. L., and S. Gardescu. (2001). Inferring forest stand history from observational field evidence. In D. Egan and E. A. Howell (eds), *The Historical Ecology Handbook: A Restorationist's Guide to Reference Ecosystems* (pp. 177–198). Washington, DC: Island Press.

Meinig, D. W. (1979). The beholding eye: Ten versions of the same scene. In D. W. Meinig (ed.), *The Interpretation of Ordinary Landscapes: Geographical Essays* (pp. 33–48). Oxford: Oxford University Press.

Merrell, L. (2002). A briny time capsule. *High Country News*, November 25.

Muir, R. (2000). *The New Reading the Landscape: Fieldwork in Landscape History*. Exeter: University of Exeter Press.

Post, F. J. (1977). The microbial ecology of the Great Salt Lake. *Microbial Ecology*, 3(2), 143–65.

Sanborn Map Co. (1888, 1895, 1902, and 1925). *Insurance Maps of Eugene, Oregon*. New York: Sanborn Map Co.

Sinor, J. (2008). Spiral Jetty: A mirror in the west desert. *Wasatch Journal*, Early winter, 39–42.

Smithson, R. (1969). A sedimentation of the mind: Earth projects. In N. Holt (ed.), *The Writings of Robert Smithson: Essays with Illustrations* (pp. 82–91). New York: New York University Press. (Original work published September 1968).

Smithson, R. (ca. 1971). Question and answer session at the National Institute of Chicago, pt. 2. Box 1, Robert Smithson and Nancy Holt Papers. Smithsonian Archives of American Art. Washington, DC

Smithson, R. (1972). Spiral Jetty. In G. Kepes (ed.), *Arts of the Environment* (pp. 222–232). New York: George Braziller.

Smithson, R. (1973). Transcript of a radio interview by Stella Russell. Box 1, Smithson and Holt Papers, Smithsonian Archives of American Art.

Smithson, R. and N. Holt. Papers, Smithsonian Archives of American Art. Microfilm.

U.S. Army Corps of Engineers. (1936). 3883. Willamette Valley Project. Scale 1:15,000. [Aerial photograph.] Map and aerial photography library, Knight Library, University of Oregon.

U.S. Geological Survey. (1975). Topographic map of Eugene East, Oregon, Quadrangle. Washington, DC: U.S. Geological Survey.

U.S. Geological Survey. (1978). 1–315. USGS GS-VEQT. Scale 1:16,000. [Aerial photograph.] Map and aerial photography library, Knight Library, University of Oregon.

Wallis, B. (1998). Survey. In J. Kastner (ed.), *Land and Environmental Art*. London: Phaidon Press.

Watt, M. T. (1957). *Reading the Landscape: An Adventure in Ecology*. New York: Macmillan. Expanded edition published as *Reading the Landscapes of America*, New York: Macmillan, 1975. Reprint edn, Rochester, NY: Nature Study Guild.

Weisiger, M. (2009). *Dreaming of Sheep in Navajo Country*. Seattle, WA: University of Washington Press.

Wyckoff, William (2014). *How to Read the American West: A Field Guide*. Seattle, WA: University of Washington Press.

4 A resounding success?

Howling as a source of environmental history

Stephanie Rutherford

The word howl has a particular sensuousness to it. In some ways, it is almost onomatopoeic, invoking particular affective registers in its enunciation. For most of the wolf's history in post-contact North America, its howl generated a sense of dread. While the Indigenous people in this place that came to be called Canada had a different and more ambivalent relationship to wolves, when settlers heard these vocalizations, they understood them as spectral and unsettling reminders that the wilderness remained unvanquished. Settlers' fears were often compounded by the elusiveness of wolves in the landscape, heard but not often seen. Received as a mournful lamentation or the epitome of ravening devilry, the wolf's howl was resolutely understood as a sound out of place, one that needed to be silenced through the civilization of a newly emerging nation.

A different way to approach the howl of a wolf is to focus less on the human apprehension of the sound and more on why the wolves do it in the first place. Wolves howl for a range of reasons, none of which is related to human interpretations of wickedness and depravity. Generally used as a means of communicating over distances, wolf howls can achieve a variety of aims, from chasing away rival packs to locating one another. In the end, the wolf's howl is a tool for social bonding in a species marked by its capacity to build enduring relationships through time. As such, the howl is a kind of hailing, potentially understood as a gesture of friendship (Mazzini et al. 2013). Instead of giving voice to aggression—as settlers interpreted it—wolves may howl in joy at reunion, or to show love.

What are the implications of settlers' misrecognition? And how has the wolf howl shaped how Canadians have chosen to encounter wolves? This chapter considers the role of sound as a significant—though little studied—source of environmental history and awareness. Building on the work of environmental historian Peter Coates (2005), I suggest that howling has long been important to how humans understand wolves and their place in nature, particularly in Canada where their numbers remained relatively robust into the twentieth century. What is perhaps most interesting about the howl, however, is its contingent nature: its sonorousness has not been heard the same way through time. Indeed, once heard as a sound in need of erasure, the howling of wolves has now become something of a tourist attraction. This chapter seeks to think with this shift, and in so doing, provide some reflections on how sound might provide important and relevant insights into the affective dimensions of environmental attunement, something that is difficult to access in a traditional archive.

The methodological possibilities of sound

As many of the authors in this volume have noted, historical sources have traditionally been textual, with vision as the primary sense to understand them. The image of the historian rooted in a dusty archive in search of a long-forgotten document seems the paradigmatic representation of this effort. As the child of a historian, this was how I understood the historical endeavour, replete with mildew, strained eyes and plenty of footnotes. And perhaps this was why I shied away from historical analysis at the outset of my academic career; my research interests lay in the relational experiences that make up the more-than-human world, which I imagined would not be found in a textual archive. This sense of mismatch was compounded by the fact that I was seeking subjects whose lifeworlds were not recorded in ways that made it easy to access them in the historical record. Animals, and in my case, wolves, were only written through the lens of human interpretation. What might historical analysis provide in my quest to understand actual wolves, to think of them as less to be written upon and more as the authors of their own stories, inextricably bound to, but not solely reliant, on the way humans perceived them? If, following Donna Haraway (2003, 2008), we understand human/nonhuman relationships as co-constitutive, then it would seem that the archive would privilege only one side of that assemblage.

As it turns out, the archive provided more insight than I predicted, offering the potential to think about affect as a historically contingent element of human/nonhuman relations. Indeed, archival accounts of wolves have proven important to my larger project of charting the histories of wolves in Canada. But I also needed more than the archive to trace the shifting terrain of this relationship through time. As a result, I have turned to alternative sources from the sciences and the social sciences to support my investigation. In the case of howling, sound studies and anthropology offered interesting ways to think about other sources of data.

In the 1970s, the World Soundscape Project was founded in Vancouver, Canada. At its forefront was R. Murray Schafer, a composer and professor at Simon Fraser University. In his 1973 book *The Soundscape: Our Sonic Environment and the Tuning of the World*, Schafer coined the term "acoustic ecology" to refer to methodological possibilities of listening for sound in the environment (1994, Chapter 14, location 4172, paragraph 3). From his and Westerkamp's (2001, 2006) work comes the idea of 'soundwalking': engaging in the world through movement and openness to the sounds we tend to filter out. These pioneers in sound studies framed their efforts within a normative and declensionist narrative; modernity, characterized by mechanized sound, blocked people from hearing the quieter sounds of the natural world. In line with the environmentalism of the time, Schafer sought to mediate the noise pollution he saw as fundamental to industrial society and encourage people to reconnect with the natural world, acting as an "earwitness" to "keynote sounds" from nature, like wind, water, trees and animals (ibid.). This suspicion of modernity, and the sense that it led down the path of deeper disconnection, is clear in the writings of Hildegard Westerkamp (2001), who discusses the importance of attunement to the natural world:

> Listening in that way can be a painful, exhausting or a rather depressing experience, as our ears are exposed often to too many, too loud or too meaningless sounds. Trying to ignore them, however, makes even less sense … Unless we listen with attention, there is a danger that some of the more delicate and quiet sounds may pass unnoticed by numbed ears and among the many mechanized voices of modern soundscapes and may eventually disappear entirely.

And so, the earliest work on soundscapes was driven by an environmentalist imperative that seemed to put nature and culture into opposition, valorizing and lamenting the loss of the former while demonizing the latter.

Anthropologists have been more circumspect in their derision of modernity, perhaps because, in large part, their scholarship is defined by the acknowledgement that nature and culture are inseparable. Stephen Feld has been particularly central to rendering a more nuanced analysis of sound studies. Eschewing Schafer's acoustic ecology, Feld favours the notion of "acoustemology" which "conjoins 'acoustics' and 'epistemology' to theorize sound as a way of knowing" (2015, p. 12). Feld is careful to emphasize relationality as ontological, in which both human and nonhuman actors co-evolve as subjects together in listening histories. He makes these assertions more concrete through the case study of the Bosavi people of Papua New Guinea, and their relationship to the rainforest they inhabit. Feld came to understand that bird song was more than simply a nonhuman vocalization encountered by humans. Instead, birds and people are companion species:

> … bird sounds are understood not just as audible communications that tell time, season, environmental conditions, forest height and depth but also as communications from dead to living, as materializations reflecting absence in and through reverberation. Bird sounds are the voice of memory and the resonance of ancestry.
>
> (Feld 2015, p. 16)

This articulation leads Feld to understand the rainforest as "a co-inhabited world of plural soundings and knowing presences" (ibid. p. 17). Following Donna Haraway, then, it's co-presence all the way down.

Historians have also been integral in refining how soundscapes are contingent on place and time. Although sound is central to the human experience, it has received scant attention in historical scholarship until recently. As Peter

Coates (2005) remarks in his article of the same title, there has been a "strange stillness of the past". This is, in part, because of the problem of the archive; as Rath (2003, p. 3) points out, "not all sounds can be reduced to print." It is hard to transcribe, for example, the affective registers one experiences when hearing a thunderclap or a gentle rain. These sounds could not be recorded until the late 1800s and, as such, all that remains is human reception of them. Perhaps more challenging for the environmental historian is the fact that pre-1880 textual materials we do have access to were not produced with the human/nonhuman relationship in mind. So environmental historians like Coates have been particularly ingenious in piecing together evidence from archival records about how natural sounds were heard in the past and to what effect.

How we have heard wolves

The notions of co-presence and historically contingent listening are important ones for the history of wolves in Canada because, for so long, newcomers maintained a view of nature that was based on dissimilarity rather than kinship with the wolf. Put more succinctly, it was them or us. Its soundings, then, served as an aural sign that the project of disciplining the landscape—rendering it productive to fulfill the aims of the colonial state—remained incomplete. Indeed, as Rath suggests, the howl came to stand in for all that was wrong with the lands they encountered: the wilderness howled its intention to destroy settlers, though "what howled most often in colonists' perceptions of their wilderness were not winds or wolves, but Indians" (2003, p. 147). In frontier legend, wolves were, much like Indigenous people, named anachronistic and hopelessly out of place. Of course, this conflation of animals with Indigenous people is well rehearsed, and a potent discursive strategy in emptying the land of some people so that it could be filled with others (Braun 2002; Thorpe 2012). The wolf, then, unthought of as a subject, became the cipher through which the colonial imperative, with its human and nonhuman dimensions, could be read. The way that settlers heard wolves' forms of communication, as haunting and fiendish threats on the landscape, necessitated the demise of wolves as part of the broader 'civilizing' mission of colonialism.

Taken from the perspective of sonic histories, the colonial endeavour in North America might be thought of as the sustained displacement of some sounds with others: the rustle of leaves replaced by the whack of an axe, the whinny of a horse drowned out by the whistle of a railcar, the sound of open treaty negotiations eclipsed by that of a gunshot. The howling of the wolf, following this line of thought, was replaced with its absence, its complete erasure from the landscape, for it had no analogous comparator, save for the barking of a farm dog. So Coates contends, "[t]he silencing of the wolf epitomized the taming of the frontier" (2005, p. 646).

While the efforts to silence wolves were largely successful in the United States, results were more mixed in Canada. Provinces set up bounties early on;

for example, Ontario (then Upper Canada) established a bounty on wolves in 1793, paying six dollars for proof of a dead wolf. The amount was steadily ratcheted up over the course of 150 years, to $10 in 1892, $15 in 1900, $20 in 1918, $25 in 1931, and $25 in 1941 (Omand 1950, p. 426). However, the geography of Ontario proved challenging for attempts at canid annihilation. The wolves that remained just had more places to go than their U.S. counterparts, and so people heard wolves well into the twentieth century. The way they heard them was often read though the registers of panic, fear and loathing.

One source for how wolves were heard is *Rod and Gun in Canada*, the Canadian equivalent to *Field and Stream* in the United States, a magazine dedicated to outdoor pursuits such as fishing and hunting. Wolf encounters were featured frequently in the magazine, in part because of the panic hunters felt around the competition wolves offered for game like deer. What's interesting for my purposes is how the authors of these pieces invoked the sonic elements of their encounter. For example, in Harold Raymond's (1907) account of a moose hunt gone wrong, an injured but not felled moose makes its way through the bush, only to encounter a pack of hungry wolves. Before recounting a vicious kill of the wounded moose—replete with great gushes of spurting blood and seemingly vampiric wolves—Raymond takes pains to describe the vocalizations: "Eager howls broke from their slavering throats as they saw their quarry ahead" (Raymond, p. 644). In Raymond's story, the wolves' depravity was heralded by their howls. Raymond also describes the death bellow of the moose, which gave the wolves pause but did not deter them from their goal. In the end, Raymond's story pits a valiant and brave moose against the harsh and abject wolves, and he renders this battle comprehensible through a description of its soundscape.

Other accounts focus less on the hunt than the fear induced when the wolf's howl is heard. Dorothy Patrick Dyar (1910) describes her experiences with a "monster" timber wolf in "Experiences of a woman homesteader". Being left alone while her husband searched for a lost cow, Dyar recounts being stalked by a wolf, first coming face-to-face with it outside her house, and, after fleeing, then being tormented by its baying. "From the depths of the ravine I now heard a long drawn, weird cry that curdled my blood. I had become accustomed to the howling of the coyotes but this sound was very different—far more savage and wild" (1910, pp. 228 and 230). Bates (1930, p. 748) invokes similar dread in his poem "Some wolf history of earlier days":

> Hark to that minstrelsy, ringing and
> clear!
> This the chorus of Death on the trail
> of the deer;
> The fiercest forest bloodhounds are gathering in might,
> Their echoing yells wake the silence of
> night

A few years later, Barager similarly describes utter panic when hunted by a pack of wolves he can hear, but not see, his experience made even more fearsome by the desolation of the environment he navigated:

> Fear gripped me for the moment and I felt unsure of my way back to camp. My body broke out in a clammy sweat, and I suddenly felt cold all over. Getting control of my nerves I started once more. But now the gloom of the woods was thicker as I plowed through the soft snow.
>
> (Barager 1944, p. 6)

He presses on, but is once again beset by a haunting cry from the wilderness:

> From behind me and across the lake came the full-throated hunting cry of the timber wolf. Rooted to the spot, I stood and listened while an answering call came from the southeast. And still, about a quarter of a mile on either side of me other blood-thirsty voices took up the challenge. One of the pack had picked up my scent and was calling the others.
>
> (ibid. p. 7)

While panicked, Barager is still a hunter. His story offers us the reassertion of control over wolves, a return to humans as predators and animals as prey. Interestingly, he describes this as an act of silencing: "But jamming the barrel against the rough bark of the tree I gripped the trigger and fired. Almost as suddenly as the cry had rent the stillness it was cut short as the bullet found its mark" (ibid. p. 7).

What do these stories tell us of human-wolf encounters? And how is sound central to their telling? What I suggest is that they are more than tall tales of wolf behaviour. Indeed, they tell us much more about the people telling the stories than about the wolves themselves. More precisely, these archival records signal a particular affective experience of the wild, as read through the body of a wolf. Affect, as I am deploying it here, refers to the kind of precognitive experience of intensity. Put another way, it is the precursor of emotion, and an attempt to bring the body back in to social theory. This turn to affect is important to the study of human-animal relationships for a number of reasons. First, it emphasizes the viscerality of experience, suggesting that cognition is not the only way we come to know things. By focusing on sensation and resonance, affect theory allows for the possibility that the body and its capacities matter as much as the brain in how we navigate the world. Second, and aligned with the previous point, because affect theory unseats reason and suggests that the notion of an autonomous subject is a fiction, it points us to the fact that people are animals too, not imbued with qualities that make us different or allow us to rise above nature. Finally, given that affect occurs without language—along the philosophical dividing line that policed the boundaries between human and nonhuman—affect theory lets us think with the possibilities of multiple forms of agency, contingent on their interaction with each other. So, animals and humans are affected by one another.

Or as Linda Birke and Jo Hockenhull (2015, p. 134, location 3550) put it, "[a]ffect, in other words, is produced and experienced relationally, through the experience of, and engagement with, each other, whether that be within or between species."

Turning back to the archival record, the affective registers invoked in many of the stories told hinge on terror, rage, fear, disgust and anxiety. And sound was the sense that generated these intensities. So, for Raymond, the howl gave rise to terror and then disgust at lupine moral corruption. For Dyar, her blood curdled at the sound of the wolf, inducing anxiety and flight. Bates names the howls a 'chorus of death', while Barager describes the bodily sensations produced via fearful affect: clamminess, sweat and chills. In each of these cases, the fear of animal power—the very clear and unavoidable fact that humans are not impervious to death—generated sensory reactions that worked through the body before they became reasoned through the mind. Each story tells of the power of sound to produce a reaction to one's imbrication with the natural world.

While promising, methodological efforts which read sound through text remain quite fraught. The fact that I even have the opportunity to access these stories means that a number of filters have already been applied before they reach my hands. Each author, in the process of writing his or her story, worked it through conscious thought, perhaps emphasizing some aspects of the encounter over others, and certainly adhering to the narrative conventions of the time, that more often than not were likely to see wolves as demonic.[1] But, of course, writing a story doesn't guarantee its publication. Each was then curated by *Rod and Gun in Canada* itself, perhaps with the less fantastic, inspiring or affective stories weeded out, and certainly omitted would be the narratives that did not fit with the intended audience of the magazine. Tales that venerated rather than defamed the wolf were unlikely to find a home in a magazine dedicated to hunting. Similarly, stories in which wolves played an agential role, such as those we might find in Indigenous thought, were not present in *Rod and Gun in Canada*. The stakes of these omissions are high. It means my hearing of wolf stories and the ways that sound played into their telling is quite limited. It also means that I find myself in what Peter Coates (2005) calls "the ostensibly paradoxical position of listening for sound in non-aural sources" (Coates, p. 656). Others in this volume have commented on the limits of archival sources (see Thorpe, Chapter 10 and Mosby, Chapter 13) and so I will not rehearse their critiques here. I will say, though, that, however limited, archival research still has merit. By examining the historical accounts of human-wolf relations and thinking about their affective dimensions, I can open up a space to chart the rather dramatic shift that has taken place in said relations in the last 70 or so years.

How we hear wolves today

Silencing wolves was one of the central aims of the North American bounty system. However, as mentioned above, the attempt to replace one sound with its absence has proven unsuccessful in many parts of Canada; although the various bounties removed wolves from the vast majority of their range, many thousands survived in more remote reaches. This resilience meant that by the 1950s and 1960s wolves remained to witness (perhaps …) a change in thought about their importance to ecosystems, led in part by biologists like Ian McTaggart Cowan and Douglas Pimlott, but also bolstered by an emerging environmental movement that celebrated wilderness protection. But even as modern ideas about wilderness and their inhabitants have changed, sound has remained an important feature in the experience of nature. As demonstrated by the World Soundscape Project described above, what we hear in nature matters very much to how we understand it, and what we choose to protect.

The importance of sound in environmental awareness can be demonstrated through Algonquin Provincial Park's public wolf howl. As part of Douglas Pimlott's research on wolves in the park in the late 1950s and early 1960s, he discovered that wolves would answer the howls of humans. Park staff saw an opportunity for nature education and translated Pimlott's efforts at wolf location into a public event (Strickland 2004). Every Thursday in August since 1965, Algonquin Park naturalists have given a public lecture to approximately 2,000 people about the biology and behaviour of the eastern wolf, the species found in the park area. If naturalists have been able to track wolf packs on the Tuesday and Wednesday before the event, all who listened to the talk get into their cars and assemble along the shoulders of Highway 60, the main artery running through the park. Everyone exits their cars and remains silent, as one of the park employees begins to howl. Will there be a response? If not, another howl is sounded. Then another. And those 2,000 people wait without speaking to get a chance to hear something that scared Raymond, Dyar, Bates and Barager so badly that they set about destroying its source completely. The sound still generates intensities of feeling, but instead of panic, it induces surprise, admiration, longing and maybe even love. The yips of pups, the throaty bellows of adults, the downward trajectory of the cadence, a cacophonous assemblage or single howl all point to the range of possibilities in wolf communication. Rather than making animals silent, the public wolf howl celebrates the biophonic soundscape that wolves produce, insisting that the humans remain mum. The result is palpable. As Chief Park Naturalist Rick Stronks noted in a 2013 interview with me, the public wolf howl changes people forever: it allows for visceral connection to nature that remains long after the sound has dissipated. While some may feel kinship and others alterity, the wolf howl reminds its audience that nature is always more than we can know of it, and perhaps that possibility alone makes it worth protecting.

In my view, the wolf howl offers interesting possibilities to think about the potential for more affirmative relationships between humans and nonhumans, and ways for researchers to get at these relationships. First, following Karen Dalke and Harry Wels (2015, p. 183, location 4748), I suggest that the public wolf howl offers the possibility of a transspecies "methodology without words". Dalke and Wels, taking up the work of scholars like Donna Haraway and Barbara Smuts, contend that ethology (the science of animal behaviour) offers a way into the kind of empathetic engagement needed to understand animals as agents.

Interacting in close quarters for extended periods of time, watching and mimicking movement, exchanging looks and noting responses are a means of embodied communication that take place without the need for language. Indeed, Henry Buller (2015, p. 378) argues that "we may not share language with nonhumans but we do share embodied life and movement and, in doing so, different—yet both biologically and socially related—ways of inhabiting the world." The kind of multispecies ethnography (Kirksey 2014) implied by Buller is appealing for researchers interested in coming to know animal lifeworlds. The trouble for my purposes, however, is that this kind of sustained engagement isn't possible in the case of the wolf howl, or more generally, with animals that tend to eschew human contact. That said, I assert that Algonquin's public wolf howl is still a valuable site for study. For me, the wolf howl is a gesture, a kind of opening or an interstitial space, where humans and nonhumans can reconfigure their ways of relating. And it is precisely because it relies on sound rather than language that it evokes affective registers like joy, excitement, and the desire to come to know the animal in an altogether different way. Sound, I contend, is the entry point for this to happen.

Second, the public wolf howl offers a shared space rather than one dominated by humans. Certainly, the space of provincial parks is asymmetrical—wolves would have no protection there if we did not afford it to them. However, the wolves also move freely through the park and the surrounding forty townships. They also demonstrate a particular kind of agency that is hard to deny. In tracking the wolves on the Tuesday and Wednesday before the howl, park staff try to ensure that they will be in close enough proximity to achieve their desired end. But it does not always go according to plan. For example, the wolf howl has been cancelled for the last two years.[2] It appears that this is not because the wolves have disappeared, but rather may have simply moved away from Highway 60 and deeper into the park (Chung 2015). In this way, the wolves of Algonquin choose their enmeshment with these humans who hail them. Sometimes they answer, but often they don't. The contingency of the wolf howl suggests, as Donna Haraway might, that there is a range of actors at play, and not all of them are human. Of course, not all people hear wolves in this way. For some, howling is considered anachronistic and those who seek it thought of as fanatical tree huggers, as exemplified by the reinstitution of wolf culls in British Columbia (Genovali 2015). But what Algonquin's public wolf howl does is present the possibility of thinking about sound as a catalyst for new kinds of environmental attunement. This is an interesting place to start in recasting relationships between humans and the more-than-human world.

Conclusion

> Sound, in my view, is neither mental nor material, but a phenomenon of experience—that is, of our immersion in, and commingling with, the world in which we find ourselves.
>
> (Ingold 2011, p. 137)

This chapter has argued that wolves are affective animals, generating a range of registers from disgust to desire. Indeed, Edward Hoagland (in Elhard 1996, p. xvi) has rather evocatively asserted that wolves "are an aperture through which wounds drain". As I've argued here and elsewhere (Rutherford 2013), wolves often act as containers for all sorts of human desire and dread. The sonic capacity of wolves and the way their sounds are apprehended by humans generates these affective registers. This emphasis on sound, on embodiment, and on encounter matters not just for wolves, but more generally for the more-than-human world that settlers to this country have hushed as we colonized people and landscapes. How we experience the cry of a hawk, the pounding of a wave, the chirp of a cricket, the growl of a bear, the rustle of a tree, or the howl of a wolf is important. If how we hear is historically contingent, then we can think about better ways of hearing, and strategies to foster particular sounds and specific kinds of listening which have the potential to lead to mutual flourishing. Sound can be a means through which we understand the past, present and future of human/nonhuman relationality.

Acknowledgements

I would like to thanks the Social Science and Humanities Research Council of Canada that provided an Insight Grant to support the larger project from which this chapter stems.

Notes

1 This is, of course, not as straightforward as I suggest. Authors like Ernest Thompson Seton and R.D. Lawrence penned 'realistic animal stories', fictional tales that attempted to get the reader to see animals as individuals with specificity, agency and personality (Dunlap 1992; Jones 2011). For instance, Seton's *Lobo the King of Currumpaw* (1898) told the tale of an uncatchable wolf, one that demonstrated keen intelligence in its evasion of human traps. The animal is only caught when its mate is slaughtered, and Lobo's desire to follow her leads to his own death. As such, there was more than one narrative of the wolf circulating in the early-20th century.

2 Rather unfortunately, 2013 is when I began trying to attend Algonquin Park's public wolf howl. Four research trips have resulted in disappointment – the agency of the nonhuman, indeed.

References

Barager, C.P. (1944). An experience with wolves. *Rod and Gun in Canada*, 46(8), 6–7.

Bates, M.U. (1930). Some wolf history of earlier days. *Rod and Gun in Canada*, *31*(9), 748–749.

Birke, L. and J. Hockenhull. (2015). Moving (with)in affect: Horses, people, and tolerance. In Nyman, J. and N. Schuurman (eds). *Affect, Space and Animals* (123–139). Kindle edition, retrieved from Amazon.ca.

Braun, B. (2002). *The Intemperate Rainforest: Nature, Culture, and Power on Canada's West Coast*. Minneapolis, MN: University of Minnesota Press.

Buller, H. (2015). Animal geographies II: Methods. *Progress in Human Geography*, 39(3), 374–384.
Chung, E. (2015). Algonquin Park public wolf howls cancelled 2 years in a row. CBC News, September 14, www.cbc.ca/news/technology/algonquin-wolf-howl-1.3227035.
Coates, P. (2005). The strange stillness of the past: Toward an environmental history of sound and noise. *Environmental History*, *10*(4), 636–665.
Coleman, J.T. (2004). *Vicious: Wolves and Men in America*. New Haven, CT: Yale University Press.
Dalke, K. and H. Wels. (2015). Ethnographic research in a changing cultural landscape. In Nyman, J. and N. Schuurman (eds) *Affect, Space and Animals* (181–195). Kindle edition, retrieved from Amazon.ca.
Dunlap, Thomas R. (1992). The realistic animal story: Ernest Thompson Seton, Charles Roberts, and Darwinism. *Forest & Conservation History*, 36(2): 56–62.
Dyar, D.P. (1910). "Experiences of a woman homesteader: An adventure with a timber wolf." *Rod and Gun in Canada*, *12*(2): 228, 230.
Elhard, J.R. (1996). *Wolf Tourist: One Summer in the West*. Logan, UT: Utah State University Press.
Feld, S. (2015). Acoustemology. In D. Novak and M. Sakakeeny (eds) *Keywords in Sound* (12–21). Durham and London: Duke University Press.
Genovali, C. (2015). B.C.'s wolf cull is unscientific, unethical and unwarranted. *Huffington Post*, May 15, www.huffingtonpost.ca/chris-genovali/bc-wolf-cull_b_7261320.html.
Haraway, D. (2003). *The Companion Species Manifesto: Dogs, People, and Significant Otherness*. Chicago, IL: Prickly Paradigm Press.
Haraway, D. (2008). *When Species Meet*. Minneapolis, MN: University of Minnesota Press.
Ingold, T. (2011). Four objections to the concept of soundscape. In *Being Alive: Essays on Movement, Knowledge and Description* (136–139). London and New York: Routledge.
Jones, K. (2011). Writing the wolf: Canine tales and North American environmental-literary tradition. *Environment and History*, *17*: 201–228.
Kirksey, E. (2014). *The Multispecies Salon*. Durham, NC: Duke University Press.
Mazzini, F, S.W. Townsend, Z. Viranyi, and F. Range. (2013). Wolf howling is mediated by relationship quality rather than underlying emotional stress. *Current Biology*, *17*(9), 1677–1680.
Nyman, J. and N. Schuurman. (2015). Introduction. In Nyman, J. and N. Schuurman (eds) *Affect, Space and Animals* (1–10). Kindle edition, retrieved from Amazon.ca.
Omand, D.N. (1950). The bounty system in Ontario. *The Journal of Wildlife Management*, *14*(4), 425–434.
Rath, R.C. (2003). *How Early America Sounded*. Ithaca and London: Cornell University Press.
Raymond, H. (1907). A tragedy in the wilds. *Rod and Gun in Canada*, 8(8), 642–644.
Rutherford, S. (2013). The biopolitical animal in Canadian and environmental studies. *Journal of Canadian Studies*, *47*(3): 123–144.
Schafer, R.M. (1994 [1973]). *The Soundscape: Our Sonic Environment and the Tuning of the World*. Kindle edition, retrieved from Amazon.ca.
Strickland, D. (2004). Wolf howling in Algonquin Provincial Park. *Algonquin Park Technical Bulletin No*. 3: 1–8.
Thorpe, J. (2012). *Temagami's Tangled Wild: Race, Gender, and the Making of Canadian Nature*. Vancouver: UBC Press.

Westerkamp, H. (2001). Soundwalking. www.sfu.ca/~westerka/writings%20page/articles%20pages/soundwalking.html. Web.
Westerkamp, H. (2006). Soundwalking as ecological practice. www.sfu.ca/~westerka/writings%20page/articles%20pages/soundasecology2.html. Web.

5 Animals as historical actors?

Southwest China's wild elephants and coming to know the worlds they shape

Michael J. Hathaway

What does it mean to understand animals as historical subjects or historical actors? In this essay, I explore ways of considering China's wild elephants as historical actors. For centuries, historical accounts gave little notice to animals, and if animals were included, they tended to exist as a side note, not as a central concern. In histories of North American colonization, for example, we learn of the beaver's critical importance for international trade or the extermination of buffalo. We tend to learn little about animals as beings other than their near extinction.

During the 1980s, however, environmental historians began to seriously include animals in their accounts, with path-breaking books including William Cronon's *Changes in the Land: Indians, Colonists, and the Ecology of New England* (1983); Alfred Crosby's *Ecological Imperialism: The Biological Expansion of Europe, 900–1900* (1986); and Harriet Ritvo's *Animal Estate: The English and Other Creatures in the Victorian Age* (1987). Since that time, scholars in different fields, including literature, geography, sociology, and my own field of anthropology, have shown interest in the lives of animals (Calarco 2008; Lorimer and Srinivasan 2013; Wilkie 2013). We are asking new questions, including about the issue of nonhuman agency, yet there is more contention over this term than some scholars realize. At a 2014 conference at the University of California at Santa Cruz, entitled "Anthropocene: Arts of living on a damaged planet," William Cronon suggested that while animals could be actors in history, they could not be agents. For many members of the audience, this was a surprising statement, for Cronon's own work seemed to contradict this claim. This leads to larger questions about exploring our understandings of nonhuman agency. Do our differing notions of agency shape the kinds of environmental historical projects we undertake and craft?

My training in anthropology taught me that agency was only applicable to human beings; in a discipline interested in human lives, nonhuman animals were often viewed through a materialist or symbolic perspective. For the materialists, animals often mattered only to the degree they affected human lives, and symbolic anthropologists primarily looked at the roles animals played in language and myth (Leach 1964; Lévi-Strauss 1963). My training spanned both materialist and humanist perspectives. During my fieldwork in Southwest China, however, I found myself increasingly curious about elephants, including their material effect in the larger natural landscape, their symbolic role in generating and shaping international conservation efforts, and the ways their presence shaped the lives of those who shared the same land. Yet, this interest in elephants was neither part of my original research design, nor present in my initial papers. It was only years later, after telling stories about the co-existence of elephants and villagers, and in noticing my own excitement and that of my peers in talking about this, that I began to consider how elephants could be central characters in research narratives.

Villagers introduced me to the notion of elephants as historical actors. By this term, I mean something more than just the fact that elephants play a role in shaping human history. My extension is twofold: first I assert that elephants shape the larger landscape in ways that humans don't necessarily understand or notice. Recognizing this can help us imagine a history in which humans are not always at the center, where change is not only understood in relationship to human lives.[1] Although we have some non-anthropocentric accounts of history, these tend to be in epochal terms, over tens of thousands of years, if not more. Second, and this is a more radical claim, I assert that elephants themselves, in their ways of being and in their behaviors, are changing, not only through the *long durée*, but also in historical time. We know, for example, by watching movies or reading books written fifty years ago, that people's sensibilities are changing, but it is much less widely appreciated that other animals, too, might also be changing. Even those animals most intimately associated with humans, such as dogs or cats, are not often seen as dynamic. Though new breeds may come into being or new sports such as agility training demonstrate novel physical capacities, few people seem to talk of the German shepherd of the 1950s compared to the German shepherd of the 1980s. During my fieldwork in China, however, this is precisely what I encountered: villagers talked about how elephants behaved during the time of Mao Zedong (leader from 1949–1976), Deng Xiaoping (1978–1992), and Jiang Zemin (1993–2003). Such statements were surprising to me, as I expected elephants, like other nonhumans, to be mainly instinctual. I assumed that changes in their behavior would only happen over the long term, beyond humans' capacity to observe or comprehend. Villagers didn't expect all Asian elephants to change in the ways they describe, but saw elephants as historically influential and historically influenced in ways that were new to me. In this chapter, I explore what it might mean to view elephants as historical actors and to write accounts of their lives.

A brief history of elephants during China's past century

This region, known as Xishuangbanna or Banna, was ruled by the Dai kingdom, a Buddhist group speaking a language closely related to neighboring Thai, one of several elephant-centric polities in the region. The Dai gave tribute to the Chinese to the north, but also kingdoms to the south, such as in present-day Thailand and Laos (Sethakul 2000). In Banna, the Dai ruled the lush valleys, grew abundant rice and taxed a vibrant trade network, including silk cloth and black tea shipped along long-distance international routes. In the mountains surrounding the valleys, other ethnic groups spoke over a dozen languages, paid tribute to the more numerous and politically powerful Dai, but seem to have largely been left to their own ways. For the Dai, the elephants had a special value: for centuries, they kept a stable of royal elephants. Such deep interest in pachyderms was also true for neighboring kingdoms, such as in Laos (meaning "land of a million elephants"), where massive expeditions were carried out in pursuit of wild elephants, in part for their great military importance. Unlike horses, elephants could ford powerful rivers and travel across terrain where there were no roads. White elephants, in particular, were deemed to have special powers, and were in great demand by rulers across Southeast Asia. Ancient Dai manuscripts reveal that royalty also had a kind of responsibility for elephants' actions: there was an obligation to protect villages when elephant herds became too brazen in farmers' fields.

These Southeast Asian polities were fading by the late 1800s, as the British expanded into Burma and the French into Indochina. Banna was the middle ground between these colonial territories and China. As villagers reminded me, elephants can live as long as humans, and some likely remember the time before Mao Zedong's rule in 1949. His People's Liberation Army (PLA) appeared in Banna a few years later and they gained control with little resistance. By the time the PLA arrived, in fact, the Dai kingdom was down to its last royal elephant. The soldiers shipped it to a zoo in Shanghai. This fit with older patterns of Chinese emperors going back over a millennia, where rulers did not actively seek out elephants, but welcomed tribute of ivory and live elephants, which they struggled to keep in stables in their northern capitals. After submitting to the PLA, the Dai leader was provided with a lifelong stipend, and the Buddhist government was gradually replaced with a socialist one. Surprisingly, whereas tigers were deemed a "pest" and wiped out through organized hunts and a bounty system, elephants were not killed off. Somehow, elephants managed to co-exist under a regime where the human population soared, and much of the landscape was cleared of their food and cover, replacing forest with vast plantations of rubber and tea. Narratives of animal retreat are typical in environmental histories, and indeed the title of one of the few environmental histories of China is *The Retreat of the Elephant* (Elvin 2004). But elephants are doing more than retreating these days: although they face increasing challenges, they are expanding their territory. Their presence affects multiple levels of governance and social life. As physical beings they shape the daily lives of thousands of humans around them, and they are also having sometimes surprising effects on vegetation patterns and human infrastructure, such as the building of new highways through the region. At national and international levels, their presence fosters a strong sense of interest, stimulating a range of conservation projects and regulations. The elephants, in other words, became powerful conservation subjects and actors.

Ways of knowing

There are several ways of knowing, which are often as deeply connected to epistemology (how do we know what we think we know?) as with actual methods. My own studies have mainly combined the hallmarks of anthropology (long-term participant observation, open-ended interviews and oral histories) with anthropologically inflected readings of wildlife ecology.

Using anthropological methods

As a cultural anthropologist, my main method is participant observation: spending long periods of time living with a group of people. Although I bring my own research agenda, I also try to understand what is important to the group itself, and the topic of elephants showed itself quite meaningful. If I had only come for a short visit to conduct formal interviews about international conservation projects, it is unlikely that I would have learned about the critical role of elephants. I began to read what I could about these elephants, but in the year 2000 (when I began work) there were only a handful of articles. Much of what villagers knew about elephants was not of primary concern to the biologists: their views might be seen by scientists as overly anthropomorphic, since they cared most about elephants' interest in their own crops and how elephants reacted to their attempts to protect crops. Villagers told many stories, including those of violent and intelligent elephants, the medicinal qualities of their skin, how the state now seemed to care more about elephants than about their own welfare, and how elephants interacted with them, their dogs, and water buffalo. Unlike some of the biologists, villagers tended to know little about elephants' migratory routes, life cycles and mating patterns, and overall population trends. Combined, however, these different bodies of knowledge about elephants help build a richer picture of elephant life.

In the village everyone knows something about elephants, but few hold specialized knowledge as some hunters have about wild pigs or muntjac, a small barking deer. This knowledge, of course, is centred on the hunter's interest, focused on figuring out when and where animals will appear and how to approach them undetected (how to disguise scent, wind patterns and so forth), but it is nonetheless rich, full of more knowledge than is actually necessary for the hunt itself. Game animals have not only been a critical source of food for people when famine has struck, but hunting has also been a major source of adventure, and a powerful engine of masculine companionship for the performance and cultivation of prowess, ability, and verve. Hunting helps forge friendships between age-mates and reinforces age hierarchies, as older hunters are often more successful.

By and large, elephants have not been an intentional target for hunters. Here, hunters use muzzle-loading guns, with limited power, and shots are rarely lethal for elephants. Balls of lead and, more recently, ball bearings acquired from guns, can sometimes penetrate the thick skin of the elephants, but are not powerful enough to penetrate the layers of fat, muscle and bones protecting the animals' vital organs. Indeed, I was told of villagers coming across a large pile of elephant bones, and finding forty or fifty musket balls in the pile, the metallic evidence of a long history of crop-raiding. If these balls were carbon-dated, what kind of accounts could we craft, of elephant desires for rice and corn, of human aggression and defense?

When elephants move through the forest they create opportunities and dangers. In terms of opportunities, humans and other animals sometimes employ elephant-made paths. During the monsoon season, when the soil turns soft with rain, elephant feet sink into the soil. In the dry season, these tracks harden and people use them as convenient stepping-stones when walking off the main path.

In terms of dangers, I learned a lot from one elderly woman, Lao Na. One morning, while looking at some elephant tracks in the village, I asked her how fast elephants could run in comparison to a horse. She said she didn't know how fast a horse could run, but elephants' speed was dependent on whether they were running uphill or downhill. The surrounding terrain was undulating—there were few flat stretches. Uphill, elephants are quite slow, she said, mimicking how their feet would be fairly close together, but downhill they move at tremendous speeds. Lao Na said that elephants' fast speed caused problems for water buffalo. Much of the year, when not plowing rice paddies, buffalo forage for themselves in the forest. Unlike oxen, which generally sleep standing up, buffalo might lie down, especially in a muddy wallow. She related how several members of her family, when searching for buffalo, came across their bodies. One or more elephants, while running downhill, had crushed the buffalo's ribcage or leg, moving too fast to skirt out of the way.

Lao Na explained that the elephants' presence, in part, encouraged residents to buy tractors. "For a long time, we had no choice. We had to use water buffalo—you can't plow a paddy field by hand. When the small tractors came, some of us wanted one to avoid the risk of getting our buffalos crushed—the [forestry] officials do not give us compensation for that." To explain, Lao Na said that in the 1980s, the local forestry bureau gave villagers compensation when their crops were raided by certain wild animals, mainly elephants. "It had to be an animal they were in charge of protecting," she said, "so for a while monkeys didn't really count. The process was difficult, and it really helped to have connections among the forestry staff."

When I lived at the village, I became part of this elephant compensation process, at the nexus between villagers, elephants and the state. The state took responsibility for elephant actions in a way it did not for other wild animals species. I was in demand to document the damage elephants did to crops for several reasons, but mainly because I was the only person who owned a camera. I was also viewed as less biased and had more sway with state agents. Villagers would write up a report and attach my photographs to the forestry bureau form for compensation at a very low rate. Being part of this process allowed me to understand that villagers had much respect for elephant capacities, even in the face of great personal hardship. For residents of this village, their grain crops were their main source of sustenance. One morning, surveying the damage of his ruined corn field, a young man picked up an ear of corn: "See this, He Wei? They don't just eat the husk, they peel it back. No hands—amazing. See this … (pointing to an area of flattened stalks, a rough circle about twenty feet in diameter), here they didn't even eat. They just rolled around in my corn. They are like a child, a naughty child." Papers written by conservation agents portrayed this as "HEC:" human-elephant conflict, but did not mention the potential for farmers' appreciation of elephant's abilities.

I found another common theme in the conservation literature at odds with what villagers told me about elephants and food. Elephants mainly eat massive amounts of low-calorie roughage, like bamboo, so farmers' grain is a feast that can rarely be found in the surrounding forest. Conservationist accounts of elephants often declared that raiding farmers' crops was a recent phenomenon, especially replacing tropical rainforest (containing a wealth of food) with tea and rubber plantations (a food desert for elephants). Yet, villagers said that elephants had hungered after human crops for thousands of years. This idea that elephants had for millennia sought out and negotiated human settlements, thus creating long-term dynamics, is seldom found in the conservation texts.

I learned that over the last century elephants have changed their habits in several ways, including that they are more likely now to travel at night when seeking food. When elephants come into central rice paddy fields, local people come in groups, banging pots and pans, throwing rocks, lighting firecrackers and even lobbing dynamite. Many families had a gun, and would shoot towards the herd, and sometimes hit an elephant. However, as mentioned earlier, it was rare to kill an elephant with a muzzleloader, and people knew that if they did, they would face serious punishment, since elephants are state-protected animals. As one person said, "Elephant bodies are hard to hide: many animals are attracted to the body and the smell carries a long way." The tables of human-elephant encounters were turned after a regional hunting ban was imposed and guns were confiscated in a large-scale campaign. Villagers said that this campaign inadvertently began what some called the "revenge of the elephants."

Dozens of people were killed by elephants after this time, and many more have been injured. The increased fear of elephants is shaping patterns of agricultural plantings and labor, albeit in gendered ways. For instance, there is a long tradition of making small swidden fields far from the village where the soil is especially rich. These are being abandoned for necessitating too much walking through the forests, especially in the mornings and early evening—the times of greatest elephant movement—but this is especially true for women. People are ceasing to grow patches of sugar cane, especially desired by elephants. Elephants greatly enjoy grain, which is always planted en masse, thus making it vulnerable. Elephant interest in eating the inner pith of banana stalks, however, has stimulated farmers to disperse these plantings: elephants like them so much that they will seek out scattered trees, and these can be tucked right next to dense housing clusters where elephants are reluctant to go.

The presence of elephants has both increased the amount of international interest in this region and created new challenges. In the 1990s, representatives from international conservation organizations arrived at my field site, drawn by wild elephants. As one of their staff said, "China is often seen as a lost cause, as environmentally destroyed. The fact that there are still elephants left gives people hope." He and several others worked in villages, promoting forms of "ecoforestry" and "agroforestry." They encouraged farmers to plant grain under trees, but in several cases farmers were frustrated when elephants, in seeking out the grain, destroyed the young tree seedlings. The elephants were both the recipients of conservation actions and a force to be reckoned with.

Using research in wildlife ecology

One of the great benefits of wildlife ecology is that unlike knowledge from farmers, hunters, or bureaucrats, often motivated by a narrow sense of human interest in a particular aspect of animal existence, ecologists can potentially take a much wider view. They can pursue knowledge about animals' life worlds and their relations to other beings. Field biologists might spend substantial periods of time observing and trying to learn about species, cultivating an intense connection with them (Lorimer 2015).

Although such studies can be quite insightful, I suggest that when anthropologists or other social scientists draw on wildlife studies in their own disciplines, their work can be enhanced by critical engagement. Anthropologists interested in animals can gain much knowledge from biologists, but must do so with an awareness that all knowledge is situated (Haraway 1989), and try to uncover hidden assumptions that biologists might make about notions of gender or heteronormativity (Bagemihl 1999; Haraway 1989; Mortimer-Sandilands and Erickson 2010; Roughgarden 2013); what is wild and domesticated (Cassidy and Mullin 2007); the role of humans in managing landscapes (Tsing and Satsuka 2008); or a lack of historical understanding about how contemporary behaviors and population distributions are shaped by long-term human-entangled legacies (Ogden 2011; Trautmann 2015; Whatmore and Thorne 1998).

Although wildlife research always emerges from studies of specific animals in a particular place, researchers' written work typically frames knowledge as disembodied, about the species writ large. Thus, their aim is not to explore the historical lives of a population in a particular place, which is what environmental historians typically care most about. Especially when a species is dispersed around a large area, it is unlikely that members of a species will share a common history. As Alexandra Horowitz suggests "scientists consider animals as representatives of their species first, as individuals second" (2010: 8). Even paying attention to animals' sex was uncommon before scientists such as Jane Goodall, who, after recognizing all individuals in a chimpanzee troop in 1960, was one of the first to use gendered pronouns in scientific articles. For some time, she was rebuffed as this was considered "anthropomorphic and sentimental" (Thompson 2010). Currently, we recognize that we need to take sex into account when observing animals, that it matters if a chimpanzee, for example, is male or female, and how old he or she is: animals are no longer regarded only as representatives of their species.

In wildlife ecology, history is often regarded as either cyclical or slow. In cyclical history, scientists are interested in life stages, which are seen as fixed and repetitive. I use the term *slow history* to refer to typical understandings of evolutionary change, which are often based on speculation, and quite difficult to pin down to a particular time. Although it is theoretically possible to see in the fossil record how a species may evolve morphologically, it is especially difficult to track behavioral change with fossil evidence. It is even harder to discover the reasons why it came about. There are a few alternatives to cyclical or slow history, but these are rare. Occasionally, wildlife biologists can discover an important innovation, such as a new form of tool-use in primates or dolphins (where they use sponges in their mouth to protect themselves from sharp rocks when hunting fish in crevices). They can also study how these innovations spread. We also see how animals adapt to finding new foods when their old ones disappear. For example, when herring, a major food source for humpback whales, in one area began to decline, the whales switched to a more common fish, the sand lance. Their old technique of making a bubble net (swimming under a school of fish and blowing bubbles that capture fish in a "net" of bubbles) was not effective with lances because, unlike herring, they did not swim in tight groups. However, the whales discovered that making a loud tail slap caused lances to form tight groups and then the whales could swim underneath and catch them up in their bubble net. This behavior spread rapidly (Allen et al. 2013). It should be noted that witnessing a novel behavior is an extremely fortuitous event, but such observations are likely to increase as more scientists become interested in how animals learn socially, a category often previously regarded as unscientific.

In summary, it will be difficult to know elephants on their own terms. Their senses, with their calls and ways of knowing the world (which include long-distance subsonic calls, and not only their extraordinary ears but the use of sonic receptors, including a layer of sonic fat in their feet), may largely be outside of human comprehension. We know far more about African than Asian elephant biology. They communicate over vast differences. In one reserve, scientists discovered that they could recognize the call of a known individual nearly ten kilometers away.

Gaining a better understanding of how elephants know the world provides new insights into elephant histories. In India, Dr. Sukumar is interested in elephant-human dynamics, including how elephants learn to negotiate human crops and actions (1993, 2003) and gender differences in feeding behavior (Sukumar and Gadgil 1988). In China, we do not yet have a researcher like Sukumar, but interest in elephant behavior is growing in a country that did not cultivate an elephant-centric polity, unlike in India.

African elephants are far more easily and intensively studied. Most African elephants live on the savannah, where they are easily seen by airplane and Range Rover. Dozens of full-time researchers, often well-funded, use GPS trackers to keep continuous watch over herds that face serious threats from poaching. In some locations, researchers know the animals as individuals and keep track of their biographies. In other locations, where paramilitary conservation officers successfully deter poachers, populations soar and contract hunters are called in to shoot females with contraceptives such as Depo-Provera, or kill herds outright (Pimm and van Aarde 2001).

China's elephants, on the other hand, are not outfitted with GPS trackers, and we are only just beginning to know their population numbers and where herds travel, using DNA analysis based on their scat (Yang and Zhang 2012; Zhang et al. 2015). We are starting to know what kinds of plants they like, learning about 106 species by direct observation and eighty-three species by examining the dung (as the plants are often not fully digested). Their diet is quite different from Asian elephants in India, suggesting that they learn about new plants as they move over space. Elephants change their diet seasonally, not only in terms of what plants are available, but may seek different plants if the season is getting hotter and they need to lose weight or if water may be more or less available, and so on. Thus, these studies are beginning to explore in detail the larger worlds of plants and animals that elephants are connected with in more-than-human worlds, in ways that are not always centred around human lives. Gaining these understandings, and trying to work out how elephants engage in the world is critical to creating new kinds of environmental histories. What might be more challenging is trying to grasp how elephants' diets and behaviours are changing through historical time: critical to this kind of research is the notion that elephants are historical actors. Seeing them as actors, as agents playing a role in these larger landscapes, is important and such changes are just beginning to expand.

Acknowledgements

I would like to thank generous funding from the Social Science Research Council, American Council of Learned Societies, National Science Foundation, University of Michigan, and Simon Fraser University. This chapter has greatly benefited by conversations with Thom van Dorren, Fiona Probyn-Rapsey, Mark Werner, Ursula Münster, Michael Dove, and Leslie Walker Williams.

Note

1 It is quite challenging to create interspecies accounts without humans at their center. Most work in multispecies ethnography, for example, are accounts of humans and one other species (Kirksey 2014). Even this is often challenging, especially exploring the lives of particular animals in specific places.

References

Allen, J., Weinrich, M., Hoppitt, W., and Rendell, L. (2013). Network-based diffusion analysis reveals cultural transmission of lobtail feeding in humpback whales. *Science*, 340(6131), 485–488.

Bagemihl, B. (1999). *Biological Exuberance: Animal Homosexuality and Natural Diversity*. New York: Macmillan.

Calarco, M. (2008). *Zoographies: The Question of the Animal from Heidegger to Derrida*. New York: Columbia University Press.
Chen, J., Deng, X., Zhang, L., and Bai, Z. (2006). Diet composition and foraging ecology of Asian elephants in Shangyong, Xishuangbanna, China. *Acta Ecologica Sinica, 26*(2), 309–316.
Cronon, W. (1983). *Changes in the Land: Indians, Colonists, and the Ecology of New England*. New York: Hill and Wang.
Elvin, M. (2004). *The Retreat of the Elephants: An Environmental History of China*. New Haven, CT: Yale University Press.
Haraway, D. (1988). Situated knowledges: The science question in feminism and the privilege of partial perspective. *Feminist Studies, 14*(3), 575–599.
Haraway, D. (1989). *Primate Visions: Gender, Race, and Nature in the World of Modern Science*. New York: Routledge.
Haraway, D. J. (2003). *The Companion Species Manifesto: Dogs, People, and Significant Otherness*. Chicago, IL: Prickly Paradigm Press.
Haraway, D. J. (2008). *When Species Meet*. Minneapolis, MN: University of Minnesota Press.
Hardin, R., and Remis, M. J. (2006). Biological and cultural anthropology of a changing tropical forest: A fruitful collaboration across subfields. *American Anthropologist, 108*(2), 273–285.
He, Q., Wu, Z., Xu, F., Guo, X., and Zhu, Z. (2014). Human-elephant relations becoming crisis in Xishuangbanna, Southwest China. In Sun-Kee Hong, Jan Bogaert, Qingwen Min (eds). *Biocultural Landscapes* (pp. 69–80). New York: Springer.
He, Q., Wu, Z., Zhou, W., and Dong, R. (2011). Perception and attitudes of local communities towards wild elephant-related problems and conservation in Xishuangbanna, southwestern China. *Chinese Geographical Science, 21*(5), 629–636.
Horowitz, A. (2010). *Inside of a Dog: What Dogs See, Smell, and Know*. New York: Simon and Schuster.
Jones, C. G., Lawton, J. H., and Shachak, M. (1997). Positive and negative effects of organisms as physical ecosystem engineers. *Ecology, 78*(7), 1946–1957.
Kirksey, E. (2014). *The Multispecies Salon*. Durham, NC: Duke University Press.
Leach, E. R. (1964). Anthropological aspects of language: Animal categories and verbal abuse. In E. H. Lenneberg (ed.), *New Directions in the Study of Language* (23–63). Cambridge, MA: MIT Press.
Lévi-Strauss, C. (1963). *Totemism*. Boston, MA: Beacon Press.
Lorimer, J. (2015). *Wildlife in the Anthropocene*. Minneapolis, MN: University of Minnesota Press.
Lorimer, J., and Srinivasan, K. (2013). Animal geographies. In N. C. Johnson, R. H. Schein, and J. Winders (eds), *The Wiley-Blackwell Companion to Cultural Geography* (332–342). Chichester, UK: Wiley-Blackwell.
Mortimer-Sandilands, C., and Erickson, B. (2010). *Queer Ecologies: Sex, Nature, Politics, Desire*. Bloomington, IN: Indiana University Press.
Ogden, L. A. (2011). *Swamplife: People, Gators, and Mangroves Entangled in the Everglades*. Minneapolis: University of Minnesota Press.
Pan, W., Lin, L., Luo, A., and Zhang, L. (2009). Corridor use by Asian elephants. *Integrative Zoology*, 4(2), 220–231.
Pimm, S. L., and van Aarde, R. J. (2001). African elephants and contraception. *Nature, 411*(6839), 766–766.
Roughgarden, J. (2013). *Evolution's Rainbow: Diversity, Gender, and Sexuality in Nature and People*. Berkeley, CA: University of California Press.

Santiapillai, C., Dong, Y. H., and Li, Y. (1991). Management of elephants in the Xishuangbanna Nature Reserve, PR China. *Tiger Paper, 18*, 1–5.

Sethakul, R. (2000). Tai Lue of Sipsongpanna and Miiang Nan in the nineteenth-century. In A. Turton (ed.), *Civility and Savagery: Social Identity in Tai States* (pp. 319–329). London: Routledge.

Shou, Z. H., Gao, Y. T., and Lu, C. K. (1959). Yunnan nanbu de xiang. (云南南部的象). The elephants of Southern Yunnan Province. *(动物学杂志) Journal of Zoology, 5*, 204–209.

Sukumar, R. (1993). *The Asian Elephant: Ecology and Management*. Cambridge: Cambridge University Press.

Sukumar, R. (2003). *The Living Elephants: Evolutionary Ecology, Behaviour, and Conservation*. Oxford: Oxford University Press.

Sukumar, R., and Gadgil, M. (1988). Male-female differences in foraging on crops by Asian elephants. *Animal Behaviour, 36*(4), 1233–1235.

Thompson, S. (2010). *The Intimate Ape: Orangutans and the Secret Life of a Vanishing Species*. New York: Kensington Publishing Corporation.

Trautmann, T. R. (2015). *Elephants and Kings: An Environmental History*. Chicago, IL: University of Chicago Press.

Tsing, A. (2013). More-than-human sociality. *Anthropology and Nature, 14*, 27–42.

Tsing, A., and Satsuka, S. (2008). Diverging understandings of forest management in matsutake science. *Economic Botany, 62*(3), 244–253.

Whatmore, S., and Thorne, L. (1998). Wild(er)ness: reconfiguring the geographies of wildlife. *Transactions of the Institute of British Geographers, 23*(4), 435–454.

Whitehead, H., and Rendell, L. (2014). *The Cultural Lives of Whales and Dolphins*. Chicago, IL: University of Chicago Press.

Whiten, A. (2011). The scope of culture in chimpanzees, humans and ancestral apes. *Philosophical Transactions of the Royal Society B: Biological Sciences, 366*(1567), 997–1007.

Wilkie, R. (2013). Multispecies scholarship and encounters: Changing assumptions at the human-animal nexus. *Sociology, 49*(2), 323–339.

Wu, J. L., Jiang, W. G., Hu, J. S., Li, Z. Q., and Li, C. H. (1999). Changes of distribution range of Asian elephant in Xishuangbanna in last 40 years. *Chinese Wildlife, 20*, 8–9.

Yang, F., and Zhang, L. (2012). Population genetic structure and population genetic diversity analysis based on mitochondrial DNA of Asian elephant (Elephas maximus) in China. *Acta Theriologica Sinica, 32*(2), 90–100.

Zhang, L. (2007). Current conservation status and research progress on Asian elephants in China. *Gajah, 27*, 35–41.

Zhang, L., Ma, L., and Feng, L. (2006). New challenges facing traditional nature reserves: Asian elephant (Elephas maximus) conservation in China. *Integrative Zoology, 1*(4), 179–187.

6 *Dawns Ysbrydion 09.02.63/ Ghost Dance 09.02.63*

Performance as the instantaneous precipitation of traces

Roger Owen

This chapter will discuss performance, and the way in which performance in a theatrical context might highlight and add complexity to perceptions of the environment, along with associated questions about dwelling and the relationship between individuals and environmental objects. Its central point of enquiry will concern people and weather, with particular regard to the way in which performance may act as a form of mediation or negotiation between human and nonhuman actors.

The key work bridging the discussion of performance and weather will be a physical theatre piece, *Dawns Ysbrydion 09.02.63/Ghost Dance 09.02.63* (hereafter referred to as *Dawns Ysbrydion*), devised by the Welsh performer Eddie Ladd, with music and sound by Rhodri Davies and Lee Patterson, and poetic texts by the author. The show was staged in June 2013 at Aberystwyth University and centred on a significant event from the recent political history of Wales, namely the bombing of an electrical transformer at the building site for the Tryweryn dam in 1963 (on 9 February, as the full title of the performance suggests). *Dawns Ysbrydion* linked this human event to the nonhuman action of the weather by virtue of the fact that the bombing took place during a remarkably harsh winter—by UK standards at least—which saw unusually widespread and long-lasting snowfall and freezing temperatures across the country between late December 1962 and early March 1963. The show tried to highlight the interrelatedness of the bombing and the harsh weather: a central feature of its form and subject matter. Moreover, it was linked to a discussion of weather in its association with a series of outputs in an extended project titled 'The Snows of Yesteryear: Narrating Extreme Weather', which took a broad multidisciplinary view of the ways in which extreme weather events had been remembered, documented and mediated in the UK since the eighteenth century. In terms of both its narrative sources and its academic provenance, therefore, it operated as a speculative statement about the possible status and efficacy of performance as a method of enquiry into the relationship between human and nonhuman actors.

Performance has a (very considerable) number of features which may affect its salience as a method of enquiry, and it would be far beyond the scope of the current chapter to try to account for these in any kind of definitive way. For the sake of the present discussion, however, we may note the importance of the following: performance is specific to a place and time, and its exposition is either directly or residually embodied; it is also allusive in character, and implicitly posits situations, experiences and events other than those being embodied; similarly, while it is either directly or residually sensuous or affective in character, it also involves the deployment of conceptual materials and comprises a very particular and ongoing negotiation between these affective and conceptual dimensions; and finally, the sense of reality which it creates is always contingent and speculative. These characteristics, while they may not be exclusively the preserve of performance, define it as a process which dynamically and continuously reconstitutes itself in its audience's awareness. These features of performance will become key to the way in which I discuss its relationship with issues of environmental perception and ecological awareness below. I will relate much of this discussion to the work of the anthropologist Tim Ingold. Meanwhile, there is one other related feature of performance which is worth noting here: that its various specificities create some equally specific challenges in relation to describing and documenting its action. Like a number of the other kinds of cultural situations and phenomena which yield the methodological studies in this volume, accounting for performance through the 'suspension' of writing raises many questions and problems in terms of its relation to the reality which it describes and the ways in which it relates to its broader environmental, human/nonhuman context. I will also examine this point below, with particular reference to the theatre/archaeology of Mike Pearson and Michael Shanks.

Before going on to discuss the challenges associated with formulating a description of the nonhuman agency of the weather, it is worth noting quickly some of the ways in which *Dawns Ysbrydion* dealt with its implicitly *human* material, particularly its political-historical and cultural narrative. The story of the drowning of the Tryweryn Valley is a very well-known and celebrated political cause in Wales, and has often been cited as a key moment in the progress of Welsh nationalism; but here it was reframed by looking at it as a parallel to the Native American Ghost Dance of the late 1880s, as adopted and enacted by the Lakota Sioux prior to the massacre at Wounded Knee in 1890. *Dawns Ysbrydion* proposed a common feature between the two events, namely that both were, to apply Rebecca Solnit's (2004) phrase, "a technology … a device for altering the world or the experience of the world" (p. 114). As such, it argued that the bombing, like the dance, was carried out in response to a situation which was already largely beyond repair and where catastrophic loss had already occurred; like the Ghost Dance, it was seen as performance *in extremis* – as a bold, almost absurd, assertion of hope at the prospect of transformation, acted out against the material degeneration of the host's world and dramatically altering their sense of their own continuity and the meaning of their persistence.

For those who may be unfamiliar with these two historical acts, I will provide a brief working overview of both the Ghost Dance and Tryweryn here. In both cases, readers are referred to some of the more specialised literature on the subject, particularly, in the case of the Ghost Dance, Hittman (1997), Coleman

(2000) and Smith (1981); and with regard to the bombing of the Tryweryn dam, Thomas (2013).[1]

Essentially a Messianic end-of-days cult, the origins of the 'Ghost Dance craze' of the late 1880s lay in the revelations and the reputation of Wovoka, a Paiute prophet whose visions of a new world for the Native American people was communicated to the Lakota Sioux in 1889. Wovoka's description of this new world was subject to a good deal of adjustment and embellishment as it was told and retold by those who heard it, but in essence it predicted a cataclysm which would sweep away the old world, along with the White Man, and would reinstate the plains and the plentiful wildlife, including the buffalo, upon which the Indigenous people's way of life had depended for generations. In order to achieve this blessed state of restoration, Wovoka noted, they should perform the Ghost Dance "for four successive nights, and the last night keep up the dance until the morning of the fifth day, when all must bathe in the river, and then go to their homes" (Coleman 2000, p. 9). Whilst this was intended as a purely peaceful expression of the people's desperate desire to reimagine and retain the integrity of their society (which was disintegrating as their lands were taken from them), the Ghost Dance became the subject of fevered speculation and suspicion amongst the settler community, whose fears of an Indigenous resurgence, violently driving out the White Man, were fuelled by a hysterical and profit-hungry press. This suspicion is said to be one of the contributing factors which led to the massacre at Wounded Knee in 1890, which effectively brought the Messianic cult associated with the dance to an end.

The story of Tryweryn is also concerned with a loss of land. In December 1955, the valley of the small river Tryweryn in northwest Wales, near the town of Bala, was earmarked as the site of a reservoir to serve the English city of Liverpool. The creation of the reservoir, which was to contain 16 billion gallons of water, would necessitate the flooding of the village of Capel Celyn along with about 800 acres (320 hectares) of farmland, and the displacing of 48 of the 67 people who lived in the valley. The scheme was deemed to be necessary to ensure that Liverpool could be confident of securing a sufficient supply of water to meet not only the immediate needs of its own population (which at the time stood at around 780,000) but also to provide the basis for post-war urban expansion. While that expansion took place,[2] Liverpool Corporation would sell the water which it obtained from its increased supply to neighbouring local authorities.

Opposition to the plan was voiced widely (although not universally) within Wales. Liverpool's land- and resource-grabbing opportunism was contrasted with the plight of the population of the valley itself, for whom the effects would be devastating. Most of them were farmers and farm labourers, and Capel Celyn was their *bro*,[3] marked by complex patterns of habituation, family ties, religious affiliation, work and a culture rooted in the Welsh language: it was not, they argued, to be reduced to a geological convenience. The arguments raged, formed around different ideas of dwelling, different notions of scale and ways of life, and different conceptions of human need. In spite of much objection, the bill to dam the Tryweryn received Parliamentary approval and Royal Assent on 1 August 1957.

For nationalists in Wales, the impending flooding and destruction demonstrated the powerlessness and subordination of the Welsh experience to the interests of the urban English, and many of the acts involved in vacating the village and closing down its various amenities (houses, the railway, the school and particularly the chapel) prior to their demolition acquired a powerful symbolism and became significant national events.

Although the village's cause was already lost, direct opposition to the scheme began in the autumn of 1962, when an electrical transformer on the site was attacked; later, a fitter's shed was burnt in December. Then, on the night of 9 February 1963, the transformer was attacked again, this time blown up by a bomb. The three men found to be responsible for the bomb attack were Emyr Llywelyn Jones, a student at Aberystwyth University; Owain Williams, a café owner from Pwllheli; and John Albert Jones, a former member of the RAF, also from the Pwllheli area. Following Emyr Llywelyn Jones' arrest and imprisonment for nine months, the two others continued their campaign by planting another bomb under an electricity pylon in nearby Gellilydan, but they too were eventually arrested, with Williams being jailed for a year and Jones receiving a three year suspended sentence. Work on the site resumed very soon after these attacks, and the floodgates of the dam were eventually closed on 1 September 1964 in order to allow the reservoir to gradually fill during the subsequent 12 months. Llyn Celyn (Celyn Lake), as it became known, was formally opened by representatives of Liverpool City Council on 21 October 1965 at a ceremony which itself became the occasion for heated protest.

Reframing the narrative of Tryweryn to consider it side-by-side with Ghost Dance was a bold step, not only because it implicitly posed the loss of Capel Celyn against the horrendous massacre at Wounded Knee, but also because it overlooked the role which the Welsh may have played as settlers and colonists in the United States during the nineteenth century. This possible (and serious) historical oversight notwithstanding, the reframing of the Tryweryn narrative was a significant act in itself, in that it argued for recontextualization and reappraisal of an important and increasingly mythologised human event. Making it the subject of theatre performance was not just a matter of narrating the events which occurred ("worrying the carcase of an old song" in the words of the poet R. S. Thomas (1993, p. 37)): it was a matter of *transforming* those events by subjecting them to the specific features of performance as listed above. Thus, the bombing of Tryweryn was to be seen not only as historical, but also theatrical—and thus subject to the assertive instability of embodiment, and the implicit speculation and contingencies of spectatorship.

For the performer Eddie Ladd, the moment at which the historical and theatrical became critically engaged with each other—and, for us, the point at which the human aspect of the story of the bombing encounters the nonhuman world of weather—was in the latter stages of the bombers' approach to their target, which, according to the personal account by Williams (1979), they had to negotiate by crawling on their bellies through the snow. It was here, at this shift "from the optic to the haptic" (Pearson 2011), that the political act and the

Figure 6.1 Eddie Ladd in *Dawns Ysbrydion*

Source: Photograph: Warren Orchard.

Figure 6.2 Eddie Ladd in *Dawns Ysbrydion*. The author in the background, left, with sound artists. Lee Patterson and Rhodri Davies, right

Source: Photograph: Warren Orchard.

condition of the climate became enmeshed in an immediate physical experience, with the human and nonhuman actors exerting a significant effect on each other.

This movement from optic (or scopic) to haptic is key to the present discussion, since it denotes how performance might be associated with a basic change in the terms under which one perceives or experiences the environment. Pearson (2010), in his discussion of site-specific performance, applies it first to the playing of a durational Lincolnshire game, the Haxey Hood. In the midst of the game, he notes, the 'sway'—a huge male scrum—is pushed towards one of two neighbouring villages and offers its participant-performers a '*haptic* understanding of site-specificity' (p. 50). "In the sway", Pearson says, "there is nothing *scapic* or *scopic*. It's underfoot and then not ... The swaymen know landscape through feel rather than perspective. Any contours are felt not measured. Landscape is *somatic space* as much as visual construct" (p. 49). Pearson's swaymen do not perceptually locate themselves in relation to their task by adopting an externalized perspective of it: there is nothing to be gained for them in imagining the scene of the sway in a broader visually or pictorially configured landscape. Similarly, observers of the Haxey Hood cannot fully comprehend or follow the action from a traditionally configured audience position: as Pearson notes in a different context, "You can't tell by looking ..." at it.[4] As I will argue below, this haptic understanding of space underscores all kinds of theatrical performance.

The perceptual 'location' of experience is a matter which is implicitly taken up by the anthropologist Tim Ingold (2000), when he argues that perceptions of the environment are either 'global' or 'spherical' in character. The 'global' perspective is associated with visual perception and stresses "the externality of the perceiver," and a world which, "far from coming into being as and through a life process, ... figures as an entity that is, as it were, presented to or confronted by life" (p. 210). Conversely, he argues, a 'spherical' conception of the environment is more closely related to *aural* perception, and tends to be perceived 'from within' (ibid.). Thus, it locates the perceiver at the centre of a lifeworld. Ingold's distinction between these two sense-worlds is useful because it allows us to position the haptic somewhere between the visual and the aural. The haptic certainly requires a physical engagement with one's surroundings (and, to echo Pearson, with the ground under one's feet) and thus denies the objectifying and distanced visual consumption of spectacle. On the other hand, it defines the world as an external body to a greater extent than does the aural sense: the haptic explicitly marks the encounter between the surface of the body and the surface of the world, though it does not share the visual's propensities for projection, generalization and totalization.

As a performance event, *Dawns Ysbrydion* constituted itself through an array of different sense-worlds. In terms of its audience's perception of the action, the visual (via Ladd's dance movement work, for example, which was presented in a low-standing 'vat' filled with two and a half tons of flour) and aural (via Davies and Patterson's electronically modified harp music and derived noise) were amply catered for. However, it also strongly asserted the haptic sense through these features too, and not only in its performer but also in its spectators. The physical nature of the action and the resonant properties of the sound created an affective, implicitly haptic, environment which—to paraphrase Shakespeare—was to be experienced 'feelingly'. While this ascription of the haptic sense to actions which may appear to be primarily visually or aurally orientated is a commonplace in Theatre and Performance Studies, recent developments in what Thrift usefully summarizes as "subconscious processes of neurophysiological imitation" (2009, p. 7)—including work on mirror neurons—strongly suggest that such affective identification is rooted not (only) in cultural convention but in the innate activity of the brain, "so that we are able to intuit what the other is thinking … not only because we see others' emotions but because we share them" (ibid. p. 8). There is a powerful imperative towards imitation on a neurological level within human beings which is modelled and exploited by performance: as Thrift notes, "to see others perform is to perform ourselves," a remark which he qualifies with reference to Stafford (2008): "we reciprocally paint our affective lives inside one another" (p. 76).

This kind of affective imitation is a crucial element within what Ingold (after Heidegger) describes as the 'dwelling perspective' offered by performance. This notion of dwelling is crucial to the relationship between *Dawns Ysbrydion*, the Tryweryn bombing and our understanding of the interaction between human and nonhuman worlds. Ingold notes how the dwelling perspective attempts to surmount the distinction between human and nonhuman worlds by stressing how deeply implicated into its nonhuman environment the 'human' actually is. Ingold argues that both the human and nonhuman experience (here he refers specifically to the animal) are produced by immersion in a world which has already been "decisively modified by the labours of its forebears" (2000, p. 186), and the interaction within and between the human and nonhuman is to be understood as a practical, ongoing engagement. In that sense, the experience of creating and watching a performance, although it may in terms of its application of creative thought and action consist of "the imposition of an arbitrary framework of symbolic meaning upon reality" (Geertz 1964, p. 39, cited in Ingold 2000, p. 178), derives that creativity from the embodied experience of dwelling rather than from an original act of disembodied speculation. This has a twofold importance for *Dawns Ysbrydion*. First, like any other performance, the enactment of the show involved the identification and negotiation of an ephemeral lifeworld, constituted by and as a provisional sense of dwelling which underscored the action and was a consistently articulated part of its dramaturgical *unfolding*. Second, the show's historical material itself noted the disenfranchisement of two communities, the Welsh and the Lakota, whose sense of dwelling was to be overwritten by the imposition of remotely conceived and built structures. The creation of the Tryweryn dam and Native American reservations proposed to undo the link between language, culture and land, upon which the coherence of the inhabitants' sense of dwelling had been negotiated; and so the rage at the drowning and at the theft and enclosure of traditional lands was not so much at the loss of that particular patch of ground, but at the implicit debasing of the skills required to fulfil the requirements of dwelling there (and, by extension, elsewhere).

There remains the question of the relationship between people and weather as a nonhuman agency, on the part of the performers in *Dawns Ysbrydion* and of the 1963 bombers themselves. Discussions about the relationship between theatre as a human activity and the nonhuman (or more-than-human) world have been intensifying in recent years, with the emergence of debates about environmental responsibility and ecocriticism within the fields of Theatre and Performance Studies and an intensified interest in ecological ethics on the part of theatre artists and companies. One of the most direct manifestations of this concern is the notion of 'ecodramaturgy', posited by Theresa J. May (2012), which aims to discover and articulate means of stressing the continuum between perceiver and environment (a project which she shares with several examples of recent scholarship, not least Hans-Thies Lehmann's *Postdramatic Theatre* (2006)). This ecodramaturgy, in its historical dimension, aims at "excavating the often-occluded environmental history against which many dramas unfold, and at revealing the dramatic structures that continue to foreground human conflict against a background (rather than in reciprocity with) a natural environment" (Arons and May 2012, p. 6). Crucially, it also assesses the ecological impact of theatrical production by stressing the reciprocity between story told and the means of production: "For theatre to matter at all", May argues (with Wendy Arons), "we must think of it as an ecological actor, and cease producing work that privileges the metaphorical over the material to such an extent that the plight of the material gets lost in the spectacle itself" (2012, p. 6).

Arons and May's ecodramaturgical approach to performance seeks to realise an (scientific) aspiration to "recognise our fundamental *nonseparateness* from the more-than-human world" (2012, p. 5). They find a basis for this in material culture, which may be defined according to Daniel Miller's (2005) assertion "that much of what we are … exists not through our consciousness or body, but as an exterior environment that habituates and prompts us" (ibid. p. 5). It is here that Ingold, and his interpretation of the 'nonhuman' world of weather would demur. As we have seen, his avowal of a dwelling perspective rejects the notion of the exteriority of the environment altogether. As he argues, "the open world that creatures inhabit is not prepared for them in advance. It is continually coming into being around them. It is a world, that is, of formative and transformative *processes*. If such processes are of the essence of perception, then they are also of the essence of what is perceived" (2011, p. 117). The weather, then, is 'not so much an *object* of perception as what we perceive *in*, underwriting our very capacities to see, hear and touch' (p. 130). It is for that reason that he criticises the scopic predisposition of ideas about material agency, which, he argues (after Peter Pels), is rooted in a kind of *animism*, "a way of saying that things are alive because they are animated by something foreign to them, a "soul" or … spirit made to reside in matter. … Whatever its source may be, this animating principle is understood … as *additional* to the material object on which it has been bestowed" (p. 28).

From this point of view, the great problem of invoking the weather as part of theatrical performance is that (except perhaps in the case of durational performances outdoors, such as the Ghost Dance itself) the weather is brought inside, and made subject to the kind of objectification of which Ingold is suspicious, as it implies a 'furnished' landscape: "… perhaps because we are so used to thinking and writing indoors, we find it so difficult to imagine the inhabited environment as anything other than enclosed, interior space" (p. 119). While it attempted to acknowledge the atmospheric and environmental context of the events of the night of 9 February 1963, *Dawns Ysbrydion* could not be said to have acknowledged the weather as a medium to the extent that neither May nor Ingold, perhaps, would applaud. In ecodramaturgical terms, of course, its central conceit of substituting flour for snow was flawed, flour having a quite different material constitution and derivation from snow and requiring very different conditions for its production, refining and distribution. Even in relation to Ingold's contentions about the weather, this conceit was a significant concession to the visual and 'scopic'; however, in the show's defence, it did allow for the ground covering to retain a tactile and malleable (haptic) aspect, leaving traces of the performer's movements through it, and defying its definition as a simple 'object' in that sense (it also had connotations of combustibility, which was in keeping with the narrative thematic of the piece!). But the fact that the flour was alluding to snow at all may be significant when considering Ingold's argument above. Snow is perhaps one of the most interesting substances in relation to his proposition about weather as a transformative process rather than an object, since lying snow defies easy classification in this regard. It is the trace or detritus of what once fell from the sky as 'weather', and, in its continued need for particular atmospheric conditions for its sustenance (as the 1963 bombers found when making their final approach to the transformer, no doubt), it retains the properties which determined its original state.

In passing, it is worth contrasting *Dawns Ysbrydion*'s treatment of weather in a theatrical context with another notable recent production in Wales, Fevered Sleep's performance installation (produced with National Theatre Wales), *The Weather Station*. This show also brought the weather indoors, inviting audience members, who were admitted one at a time, to visit a nominally unpopulated house "which was full of weather and instruments and archives relating to the measurement of weather," such as "a bathroom covered in growing moss, a basement in which it was raining, one full of wind, one full of mist, light and an installation of televisions playing shifting cloudscapes" (Woynarski 2015, p. 30). Lisa Woynarski argues that this theatrical installation implicated humans in a relationship with the weather by bringing it indoors, so that "'human drama' in this instance is also more-than-human as human agency has extended to the atmosphere and biosphere" (p. 31). Her admiration for this production, though it may well have tempted Ingoldian scorn for showing weather inside a house, was based on its recognition of "the material agency of the more-than-human" and the creation of what she dubbed an "ecological anthropomorphism" (p. 24). This aspired to a kind of dwelling perspective on the show's basic action by complicating the relationship between audience and the 'furnished' environmental landscape of the house. Some of the items found in it, such as the moss, were not mere representations of particular kinds of weather but were indexically linked to the atmospheric conditions which had been created for their sustenance inside the house. However, such an appeal to materiality may have been compromised here by the reliance on metaphor—albeit an 'affective' one—which she herself ascribed to the work: "The image of a house full of different weather conditions," she notes, "is a metaphor for the way in which humans are affecting and shaping the climate and geological age" (p. 21). For all its haptic qualities of presence (it was certainly available as a 'somatic space' in Pearson's sense), it was only as an allusory image that it could ultimately claim to link the human and the bio-spherical.

I will attempt a summation and deal (albeit briefly) with the question of writing as well as theatre/archaeology's positing of precipitated traces, as promised in the introduction. It is difficult to argue in a piece such as this (or indeed, in the kinds of poetic texts which I tried to produce for *Dawns Ysbrydion*) for a form of writing which can successfully mediate between the human and nonhuman. As Ingold notes, there is a fundamental problem to negotiate here, namely the fact that writing tends to produce the very conditions of contemplative 'interiority' which forces a division between the continuities which we discern between our human existence and the nonhuman environment of which we are a part. Mieke Bal has articulated this problem with reference to narrative, noting how related narratives—even academic ones—deposit an implicit agent of interiority, a linguistic subject, "a function and not a person, which expresses itself in the language that constitutes the text" (c1997, p. 16). This problem of narrative has also been raised as a challenge to accounts in nature-culture and environmental history, most notably by William Cronon (1992), who, in discussing contrasting accounts of the environmental history of the Great Plains of the United States, notes that, though they may be concerned to show "that human acts occur within a network of relationships, processes, and systems that are as ecological as they are cultural ... scholars of environmental history also maintain a powerful commitment to narrative form" (p. 1349). He adds that, whatever its purpose, narrative "cannot avoid a covert exercise of power" by making "the contingent seem determined and the artificial seem natural..." (ibid. p. 1350).

A means—although by no means free of its own problems—of negotiating the various complexities of the relationship between human and nonhuman, dwelling and interiority might be offered by dealing with writing, the fact of perception and materiality of the environment as *traces* of performance, in the manner proposed by Pearson and the archaeologist Mike Shanks in their work *Theatre/Archaeology* (2001). There, they describe performance as a matrix of relationships (between different kinds of objects, actions and embodied practices of presentation and reception) which are available within—but also, crucially, *beyond*, a particular time and space. Performance is not to be understood simply as a phenomenon defined by its original 'presence' in action but by its traces, by what it deposits or precipitates in the body of the performer, the memory of the spectator, the location of the presentation and in the objects or materials assembled in the facilitation of exposition. Such is the transitory nature of that moment of exposition that the present action is available for review only through its traces: indeed, it might be said that the immediate experience of performance as action in the awareness of the spectator *is* of the instantaneous precipitation of these traces.

In that sense, theatre/archaeology eases the human/non-human binary by allowing that *both* are available to human scrutiny only in retrospect, and only through material traces whose provenance and implicit narratives are suspended, contingent and evasive. Though Pearson and Shanks' vision of theatre and archaeology is rooted in materials, therefore, those materials are always the subject of an *encounter*, a continuous *oscillation* between the human and nonhuman, with the emphasis on the shift between the two rather than the implicit primacy of either.

To return, at last, to *Dawns Ysbrydion* and the night of 9 February 1963. It is a mark of the poignant *indifference* of the human/nonhuman binate within the central action of the bombing that the snow which augmented the drama of the occasion also retained traces of the bombers' approach and retreat from the scene. Those traces in their turn formed a key part of the evidence presented against them at their prosecution. And so an event, which was always intended not to be seen was testified to by its remains in the cognate nonhuman material, remoulded by human action, which then, as it thawed, seeped down into the not-yet-dammed river Tryweryn.

Notes

1 For readers fluent in the Welsh language, Williams (1979) is particularly recommended.

2 It is worth noting here that Liverpool's population as a city was falling during the 1950s, and continued to fall for the rest of the twentieth century. From it historical peak of 856,000 in 1931, it fell to 789,000 in 1951 and to 747,000 by 1961. The Merseyside region as a whole rose from 1,348,000 in 1931 to 1,386,000 in 1951, but then fell to 1,366,000 in 1961. Figures from Demographia website: www.demographia.com/db-ukcities.htm (accessed 31 August 2015).

3 *Bro* is a Welsh term which broadly equates to the idea of 'homeland', and is traditionally marked by a detailed knowledge of topography, genealogical ties and language patterns.

4 A phrase adopted by Pearson in a journal article with the photographer William Yang. See Pearson, M. and Yang, W. (2001).

References

Arons, W., and May, T. J. (eds). (2012). *Readings in Ecology and Performance*. New York: Palgrave Macmillan.

Bal, M. (1997). *Narratology: Introduction to the Theory of Narrative*. Toronto, London: University of Toronto Press.

Coleman, W. S. E. (2000). *Voices of Wounded Knee*. Lincoln, NE: University of Nebraska Press.

Cronon, W. (1992). A place for stories: Nature, history, and narrative. *The Journal of American History*, *78*(4), 1347–1376.

Hittman, M. (1997). *Wovoka and the Ghost Dance*. D. Lynch, (ed.). Lincoln, NE: University of Nebraska Press.
Ingold, T. (2000). *The Perception of the Environment*. London: Routledge.
Ingold, T. (2011). *Being Alive*. London: Routledge.
Lehmann, H-T. (2006). *Postdramatic Theatre*. London: Routledge.
Miller, D. (ed.) (2005). *Materiality*. Durham, NC; London: Duke University Press.
Pearson, M. (2010). *Site-Specific Performance*. Basingstoke: Palgrave Macmillan.
Pearson, M. (2011). Why performance? [PDF document]. Retrieved from AHRC Landscape and Environment Network website: www.landscape.ac.uk/landscape/documents/eventpapers/toolkit/1whyperformance.pdf.
Pearson, M., and Shanks, M. (2001). *Theatre/Archaeology*. London: Routledge.
Pearson, M. and Yang, W. (2001). You can't tell by looking... *Performance Research: A Journal of the Performing Arts*, 6(2), 31–8.
Smith, R. A. (1981). *The Moon of Popping Trees*. Lincoln, NE: University of Nebraska Press.
Solnit, R. (2004). *River of Shadows: Eadweard Muybridge and the Technological Wild West*. Harmondsworth: Penguin, [reprinted edition].
Stafford, B. M. (2008). *Echo Objects: The Cognitive Work of Images*. Chicago, IL: University of Chicago Press.
Thomas, R.S. (1993). *Collected Poems 1945–1990*. London: J. M. Dent.
Thrift, N. (2009). Pass it on: Towards a political economy of propensity. *Emotion, Space and Society*, *1*(2), 83–96.
Williams, O. (1979). *Cysgod Tryweryn* ['The Shadow of Tryweryn']. Caernarfon: Siop y Pentan.
Woynarski, L. (2015). A house of weather and a polar bear costume: Ecological anthropomorphism in the work of fevered sleep. *Performance Research: A Journal of the Performing Arts*, 20(2), 24–32.
Thomas, W. (2013). *Hands Off Wales*. Llandysul: Gomer.

Part II

Decolonizing research

7 Co-becoming time/s

Time/s-as-telling-as-time/s

Bawaka Country, including Laklak Burarrwanga, Ritjilili Ganambarr, Merrkiyawuy Ganambarr-Stubbs, Banbapuy Ganambarr, Djawundil Maymuru, Sarah Wright, Sandie Suchet-Pearson, Kate Lloyd, and Jill Sweeney

In our research together as Bawaka, an Indigenous homeland in the north of Australia, we have woven ourselves together, as family co-becoming with each other, co-becoming place and co-becoming time/s. We have done this in terms of what we know, what we do, how we do things, as well as who we are. We are five Indigenous Yolŋu women from north east Arnhem Land and we are three non-Yolŋu women from 'down south' in New South Wales. We are Laklak—inspiring elder—her three sisters Ritjilili, Merrki, Banbapuy, and their daughter Djawundil. We are teachers, community leaders, tourism business owners and hosts. We are Sandie, Sarah and Kate, human geographers based in Australian universities.[1] And we have co-become together as Bawaka with the three academics being adopted into family as sisters, mothers and grandmothers. Our doing together includes weaving baskets and gathering *gunga* (pandanus): dyeing, splitting, finding natural dyes, passing the *gunga* through our fingers, looping, threading, drawing together,[2] and, while doing so, talking, drinking tea, eating, caring for each other and for our children (Bawaka Country including Suchet-Pearson et al. 2013).

In these and other ways, we have made time/s, been time/s, lived time/s, attended to time/s as they communicate with us, and practiced being in the moment of co-becoming together, a moment that co-becomes with what has gone before and what will come. Time/s have become the eternal present/past/future, and we have woven ourselves into more-than-linear multiple patterns of co-emergence and responsibility, the patterns and rhythms of Yolŋu Law/lore, that underpin, allow and bring forth Country.[3]

In this chapter we discuss how our Indigenous-non-Indigenous, human-more-than-human research collaboration has grappled with 'the concept of time' in our research. We situate our work within Yolŋu ontologies of co-becoming. To do this means attending to more-than-human agencies and to our human-more-than-human co-constitution. This includes our co-becoming with and as time/s. At Bawaka, time/s are communicative, relational and agential. They show themselves in the material presences that come and go in place, and that indeed, co-become *as* place. This contingent and complex understanding of time/s presents unique challenges and possibilities in the conducting of nature-culture research.

Our chapter is structured around three aspects of our methodological contributions: *telling time/s*, *time/s telling* and *time/s as telling/telling as time/s*. These three contributions are co-emergent, are aspects of a methodology of attending and co-becoming. In this paper, by necessity, they are presented in a linear fashion but we have not done this to privilege or preclude any section, but rather, we use the subtitles as entry points to grapple with the notion of *doing, being and co-becoming time/s*. The first section, 'Telling time/s', discusses story-telling and argues that it is important to decolonize research by opening time up to multiplicity and non-linearity. 'Time/s telling' discusses the way time/s are not just told but can tell. Here we focus on attending to time/s as communicative, relational and agentic. And 'Time/s as telling/telling as time/s' situates our work within a conceptualization of co-becoming whereby we neither tell nor listen to time as a discrete and distinct entity, but understand ourselves as part of time/s themselves. Here, relationality is foregrounded as are human-more-than-human co-constitutions of time/s.

Methodologically, our chapter highlights opportunities to consider how to recontextualise and share histories, stories and experiences in ways that actively embrace the unique temporalities and agencies at work, and which infuse historical material with agencies and complexities that would not have been recorded or acknowledged in colonial/Western knowledge frameworks. We argue that recontextualising history through: 1) attending to absent presences, and 2) attending to what is present now and their relationalities (especially temporal relationships) can challenge colonial and two-dimensional versions of 'the past' with enriched and empowering approaches.

Telling time—stories-telling, narratives, linearity and authority

Much of our collective work is driven by the imperative to decolonize our research, but if we truly want to do this then we must also grapple with decolonizing 'time'. This has huge methodological implications. Mainstream forms of research rely on very strong, and usually unspoken, colonizing assumptions about 'time' (Perkins 1998; Natcher et al. 2007). Strang (2015) points out that Western societies have shifted away from close observance of the material environment as a means of telling and experiencing time/s, and have instead largely embraced time as a numerical measurement abstracted from place, spirituality and materiality. The notion of causality, so important within many academic disciplines, implies an inherently linear version of time in which one thing causes another in a strict sequence. Narrative and story-telling, too, imply an ordering, an account of events connected through time (Reismann 2005; Wiles et al. 2005). Indeed, the very singularity of the term 'time' implies universality. Time is awarded a coherent meaning, and grants academics the ability to uncover and author a particular authentic truth, and to predict 'the' future (Haraway 1991). As Natcher et al. (2007, p. 114) suggest, researchers assuming a singular, linear notion of time "may inadvertently be placing research participants in situations that are potentially offensive to a sentient world—a world that can be highly sensitive, vindictive, and dangerous."

Within our research together, time/s have asserted themselves in multiple non-linear ways: they have 'evoke[d] and unsettle[d]' (Hill 2015, p. 422), guided and challenged us. The non-Indigenous academics within the collective have tried to open themselves up to multiplicity, and all members of the collective have grappled at the interface of Yolŋu time/s and the linearity of time/s usually presumed within dominant academic story-telling. How is it possible to tell time/s in non-linear ways?

Working through this dilemma has been an important point of learning. During our time together at, with, and as Bawaka, for example, the academics have been told many stories about Bayini, a spirit woman of Bawaka.[4] However, it wasn't until they tried to write Bayini's story down for our co-authored book about weaving that the ontological challenges of dealing with multiple versions and ambiguous timescales became clear. Every 'time' they heard the story they would try subtly and respectfully to clarify and resolve some of the ambiguity by asking "So ... was Bayini a slave or a princess? Was she thrown overboard or did she jump? Which happened first?". They were confronted with having to decide which detail to include in the written version, what story to tell. In grappling with this they realised that it wasn't the slipperiness of the timescales associated with telling Bayini's story that was the problem, but rather the perceived need for a definitive story. Instead of adhering to colonial knowledge and linear narratives that seek to fix things in the past as history, in clear and measurable chronologies, as an objectively true record, they chose rather to respect multiple versions, to listen to what Bayini was trying to tell them, to listen, to respect, to feel, to be open to the ontology so that all those multiplicities could exist simultaneously. As Merrki explains as we write this chapter together, "even if Bayini would have been there a long time ago, we still believe she is there now. Not only her but our other ancestors. They still hunt with us, look after us, show us ways and that is why when we go there we always announce ourselves—to let them know we are coming".

So although one particular story sits in our book, it sits within a contextualized explanation of multiplicity. Further, our collaborative work co-becomes Bayini as we engage with, and represent her (for example right now), in different ways—such as hearing Bayini whispering in the bush, seeing her footprint in the rock, being aware of her eyes in the sand, and knowing she's there helping us to understand this. Bayini came to Bawaka in what can be called the past, but her footprint is here now: you can see her boat in the water, hear her in the bushes, she cares and looks after us, and has made Merrki and her mum tea. There is no real distinction between 'she is here now' and 'this happened then'. Through her refusal to conform to the academics' questions about her story, with their implied linearity and singularity, Bayini taught the academics that she is part of Bawaka, looking after us all, helping the children when they had a fever, helping Sarah to find her knife when out collecting *gunga*. She is a big part of us, enabling our relationships and our understanding of multiple and ambiguous timescales (Suchet-Pearson et al 2013). As Merrki reminds the academics, "We wish upon her, we give her presents, she gives us some. Give and take, she looks after us".

This was one of our meetings across ontologies (Bawaka Country including Lloyd et al. in press), as we began to understand the central importance of oral histories and practices of stories-telling in reproducing and sharing knowledge, cross-cultural learning and countering colonial knowledges (Bilous 2015; Strang 2015; Wright et al. 2012). The practice of retelling stories makes them active and present participants in the here and now, creating knowledge and law, infusing the material world with meaning that is immediately and actively connected to things that happened before and things yet to occur. Ontologically the Yolŋu world is alive with beings and actors who do not/no longer exist in Western ontologies, and in part this is because Yolŋu temporality is non-linear. Yolŋu temporality is something that is nourished and sustained, rather than something that exhausts itself continuously, such that the past is separate and distant.

While this has been an important aspect of decolonizing research, it has also brought us into some conflict with methodological approaches that gather power through their ability to control and organize knowledge[5] including into strict linear, chronological sequences. Some anthropological, archaeological, geographical and historical accounts base their authority on an ability to find the 'right' version of a story, and to fix events like Bayini's arrival at Bawaka to specific 'archaeological evidence' in ways that purport a 'right' and 'wrong' story. "To be history," it is claimed, "a narrative must establish a chronology and a location" (Fox 1980, p. 10). As such, incommensurate or multiple accounts may undermine the perceived authority/ies of our collective's work. This has been something to navigate, including on occasion through reviewer responses to our articles, or a chapter being withdrawn from a collection. Yet, it is something that we have increasingly been guided to tackle head-on, embedding our work in Yolŋu ontologies as we try actively and directly to destabilise colonializing narratives striving to document the 'true story' and chronology of events, and rather attend to our richly meaningful co-becomings with and as Bawaka.[6]

Time telling—time as communicative, relational and agentic

As we have grappled, through our collective work, to *tell time/s* differently, we have also tried to attend to the ways time/s tell, the ways time/s actively communicate. We have tried to attend deeply to the messages that Bawaka sends us, the way diverse co-becomings communicate with us as part of Country, as part of us.

This includes attending as Country tells time/s and makes time/s, orders life and connects us with each other and more-than-human kin as part of the complex, ever emerging, more-than-linear patterns, rhythms and responsibilities of Country.

As Laklak explains, talking of *walu*, which is the sun, and so much more:

> *Walu* tells us the time. Do you want to know the time? Look where *walu* is, look at the shadow, the shadow of the tree and the shadow of yourself. Or when it's about two or three o'clock you might hear a *garrukal* bird (kookaburra) singing, telling you the time, telling you it's nearly sunset. When the *garrukal* sings, the people remember that sunset is coming. … When you fish or get *ganguri* you always know there is *walu*, time. *Walu* tells you when to do these things. We use *walu* to measure things too. When we were little girls we'd use the *walu* at sunrise on *dharpa* (wood, tree) to measure ourselves. So *walu* is the sun, *walu* is the day, *walu* is time. *Walu* is very important for Yolŋu people; it is made by our ancestors.
>
> (Burarrwanga 2013, p. 106)

Rather than us (humans) 'telling time/s', here time/s—in multiple ways—are telling, as well as reminding, and reproducing, connecting us to ancestors, and remaking our connections to them anew as they stand with us now. Time/s in their co-becomings with Country, including as *walu* and the *garrukal* and *dharpa*, tell those that would attend what is happening, reminding them to undertake certain activities, and enabling important identity work that underscores the intimate connections between all living things that form (co-become) Bawaka—including the moon and the stars and the spirits. Laklak explains further:

> Yolŋu know in which part of the month the moon dies, when only stars light the sky. There is 'nothing moon', *bäyŋu ŋalindi*, until a new moon comes. Last night, in the early morning, we saw a little moon, *wirripikili*. When the new moon comes, it's very important, it's like the beginning of a new month. We spread the news, calling 'New moon!' Sometimes there's also sorrow and we cry because *ŋalindi* the new moon has come and it reminds us of our sadness for a loved one, maybe a sister who died in past months. *Ŋalindi* is a reminder. It has been a month since our loved one passed away. Or maybe we remember that next *ŋalindi* we have a tour group coming to visit.
>
> (Burarrwanga 2013, p. 111)

At Bawaka, time/s are communicative, relational and agential. Time/s are *walu* and *ŋalindi*; time/s show themselves in material presences that come and go in place, and in those who have passed away but are with us still. In our research, as a collective, we continue to try and deepen our ability to attend to *walu*, to *ŋalindi*, and to time/s in all their rich materiality. This we do differently, as Yolŋu and non-Yolŋu, and human and more-than-human. We have understood this as a *methodology of attending* (Bawaka Country including Wright et al. 2015b) that "stems from an awareness of our essential co-constitution as we care for, and are cared for by, the myriad human and more-than-human becomings that emerge together [to create Country]" (p. 269).

While conventional understandings of communication assume a pre-existing, already structured, objective 'out-there' world which we all inhabit, a *methodology of attending* sees multiple realities emerge from action in specific contexts and places and times. Here, we are called to become sensitive, communicative and alive within a more-than-human, co-constitutive world. We are called to do this in ways that extend existing methodologies that understand research as co-construction of knowledge by the human interviewer and interviewee (Nicholls 2009; Rose 1997). We privilege Yolŋu knowledge of communication and kinship reaching beyond human-centred research relationships to incorporate other kin, animals and place (Bawaka Country including Wright et al. 2015b; Wright et al. 2009). The methodology is underpinned by more-than-human networks of care and responsiveness and recognizes that research is always embodied within a multidimensional ethical context.

For the academics, *attending* means being guided by Yolŋu adopted grandmothers, mothers and sisters as they patiently point out active agencies of Country, and the ways those co-becomings communicate with us, including as expressions of time, in ways that would be otherwise largely missed. Sarah/Madirriny, for example, has been called to attend to Madirriny, the south wind/her Yolŋu name, in many ways: as a wind moving the *gunga* and inserting itself into our conversation, guiding us to move from one activity (splitting *gunga*) to another (weaving), from one discussion to another, shaping conversation and research direction; as an indicator of a new season, placing us in within the recursive patterns of co-emerging, vital and communicative more-than-human place-time. She has also been called to attend to Madirriny as a *yumalil/manikay* (songline), an aural expression of the wind itself played on *yidaki* (digeridoo). As Banbapuy points out, Madirriny, in these instances, is not separate from Sarah, her being, the co-becoming of Bawaka, our research, or from the *yumalil*: the song. These are not metaphors, but co-becomings within our collective, guiding us, refusing dichotomies, becoming with and as us. As *Madirriny* moves through and shapes our interactions, it also shapes our collaboration, encouraging us and showing us ways to communicate, to feel confident enough to decolonise understandings of what study and 'real' research looks, sounds, feels and smells like, so that Country can be acknowledged as an author on publications, and can be part of our collective, as one of us, in our co-becomings.

Time as telling/telling as time

Time is not abstract, it is not empty, it does not exist separately from relationships, from human and more-than-human worlds. It does not order worlds in a strict and linear, universalized sequence. 'It' does not exist as 'it,' a singular universal truth. What we have learnt from our work together is that we co-become with place-space as Bawaka (Bawaka Country including Wright et al. 2015), and we also co-become *as* time. But this is not a disordered co-becoming, since Yolŋu ontologies provide an underlying, complex pattern for these time-place relationships. But how have we come to think in this way? How are we grappling with these understandings of 'time'?

Reflecting on the methodological challenges that we encounter as our research collective works together, and specifically works towards three co-authored books (Burarrwanga et al. 2008; Burarrwanga et al. 2013; and our current book still in draft and tentatively entitled *Milkarri*), is illuminating. We've framed this learning process as co-becoming—within an ontology and methodology that is about relationships and relationality (Lloyd et al. 2012), in, through, with and *as* time. This has involved the academics letting go of their ontological presumptions, pushing them to the edge of their ontology and relinquishing their points of reference (for example past, present and future tenses, having a first word and last word in this chapter) and not expecting to be presented with an easy alternative, but rather embracing the opportunity to work together—human and more-than-human—to co-constitute meanings, to constantly co-become in intimate and responsive (and responsible) relationships with each other. And time is certainly meaningful—time is telling.

Time was *telling* for us as we wrote *Welcome to My Country* (Burarrwanga et al. 2013). For example, we wanted to talk about the importance of a particular period of time/history: the Bark Petition.[7] In reflecting on our writing about Laklak's life story (as chronologically presented through the book) we realise we were in fact bringing the past into the present—co-constituting time. But this didn't occur in a random, *atimely* fashion. Rather, there is a pattern—an ordering to the book's structure and messages—to our learning and writing together. The book and our learning is underlain by patterns—cycles, recursivity, relationality—Yolŋu kinship relationships (*gurrutu, yothi-yindi*) which weave direction and hold everything together.[8] And there is an ethics to these complex patterns, a human-more-than-human kinship responsibility of nurturing and giving birth to each other which also bind what we could learn together, and what we couldn't, what we couldn't share, how things could and couldn't be connected and structured in our book. These lessons were learnt as the academics were told again and again about *gurrutu* relationships; through word of mouth; through diagrams on butcher paper and in the sand; through embodied knowing as they were referred to by their kinship titles and tried ever so hard to respond correctly; through drafts of book chapters and corrections in margins; through clear boundaries of what can't be discussed or included; through what aspects of Country Bawaka chose to reveal; on which paths and patterns we were directed.

And in our latest work together on our new book *Milkarri*, we find that we are telling *as* time; we are grappling with (and gee is it *hard*) and co-becoming *as* time. By telling as time we are bringing time into being—being at Bawaka, walking through the sand, listening to the women crying, writing right now. ... we are bringing time into being, a more-than-human embodied on-going co-becoming.

The academics are (ever so slowly) learning that *Milkarri* is one aspect of an English concept 'Songlines'. These conceptualisations of complex time-place relationships can so easily be caught up in clumsy Western singularisations of time and space (such as the Dreaming), yet the Yolŋu women/Bawaka Country are enabling the academics a deeper understanding. And all are co-becoming time as Bawaka Country, and grappling with understandings in which the past is always in the present and in which singing (chanting-crying-dancing) actualises and realises this intense co-becoming (see Figure 7.1). Hence the methodological challenges of learning this also become methodological challenges of how to present what we've learnt. We are currently struggling with the linear constraints of the written medium (our book) and are thinking through multimedia, embodied ways of sharing our learning—opening ourselves up to non-linear and multiple means of 'telling as time/time as telling'—bringing 'us' humans into multiple, more-than-human co-constitutions of time, realising that as we share our learning we are concurrently and very real-ly bringing into being Country, bringing into being time/s.

Figure 7.1 *This is time*: some of the collective co-becoming time as they work on this chapter in Adelaide. As Banbapuy states: "in Western ways time means money, for Yolŋu, time means catching the tides or doing what needs doing" (in this case writing this chapter in 'time' for the deadline)

Towards the exit point/beginning again

> The spiral in a woven basket shows the life cycle—tells us how babies grow, their arteries and their selves, both in the womb and as growing children and on through the generations. As the basket grows, so a baby girl grows up until she's a young mother and she'll learn and have babies, and the knowledge and growth keeps spiralling. The woman becomes a grandmother and the cycle keeps going. There's no actual death, life keeps spiralling on through the generations.
>
> (Burarrwanga et al. 2008, p. 15)

Linear conceptions of time, the time of university research, the time of Western narrative and ideas of cause and effect do not mesh easily with Yolŋu time/s. In our research collective, we have tried to do, tell and become time differently. As we have woven baskets together, we have woven our lives together. The spirals of the basket are our life cycles, and the life cycles of our human and more-than-human kin. And these spirals hold knowledge to keep and to share. Time/s are marked by and made by these spirals and by the fundamental recursive, intergenerational patterns and relationships that underpin and create life.

Attending to time/s in this way presents a myriad of methodological challenges. It is not possible to stand outside time, even though there are illusions of doing so, just as it is not possible to stand outside place, or outside research, or our relationships with each other, our more-than-human kin. To impose a separate, linear time on a rich world of co-becomings, is to (re)colonize in profound ways the rich, vibrant and vital, more-than-human, communicative worlds of Indigenous ontologies. As we have tried to decolonize our work, we have changed our ways of doing, writing and *being* research in many ways. One way is that the 'we' of our research has expanded, so that 'we' is no longer only a human 'we' but one that includes Bawaka. 'We' is the rich timescape/more-than-human relational kinscape of Bawaka. We understand that *we* cannot stand apart from Bawaka or any of its co-constituting patterns and relationships, including those that make, tell and are time/s. We are still very much in the process of grappling with the meaning of our co-constituted selves, and of navigating our research journeys.

For us, centering Yolŋu ontologies through methodologies of attending and co-becoming, means exploring diverse temporalities and how these make present or absent a range of actors, meanings and relations—including, in the case of Bawaka, time/s as actors, as beings that have voice and that join things together in complex and more-than-human, rather than linear, ways. Attending to Yolŋu conceptualisations of time/s means attending to the role of relationality in shaping place, as well as challenging nature-culture separations and homogenous understandings of the world in which we live. The many layers of Bayini make the 'past' present, actively destabilising colonialising narratives and approaches that often consign a meaningful Indigenous landscape to a mysterious and unreachable past.

In *Milkarri*, in Yolŋu songlines, when we cry and sing *ŋalindi*, we say that *ŋalindi*, the moon, drags the tide in, the water. *Ŋalindi* co-becomes with us (human and more-than-human), *ŋalindi* communicates with us, *ŋalindi* tells stories and is told through story and through ceremony. Time/s are now, here, these are time/s. Time/s are our relationships with each other and with and as Bawaka, our weaving together, our gathering of *gunga*, our sitting on the beach, our writing and your reading. In this paper, we have marked each section with various aspects of *ŋalindi* to centre time's cyclical and communicative nature and its material presence. We are working together on attending to time/s, letting times tell, listening and learning, taking ourselves to the edges of our knowing/being/becoming. In this paper, we have invited you on this journey, we have invited you to feel and become part of *ŋalindi's* pull.

Notes

1 Non-Indigenous academic Jill is a human geographer who has worked with and supported the collaboration's work at various points, contributing valuable insights into the relevant literature and applying it to the collaboration's work.
2 See Burarrwanga et al. (2008).
3 Country is a word in Aboriginal English which includes not just the territorial, land-based notion of a home *land*, but encompasses more-than-humans and all that is tangible and non-tangible and which become together in an active, sentient, mutually caring and multidirectional manner in, with and as place/space and time (Bawaka Country including Suchet-Pearson et al. 2013; Bawaka Country including Wright et al. 2015).
4 Bayini is much more than 'a spirit woman of Bawaka' but she is also this and so this is how we will begin to introduce her.
5 For a critique of 'strong' or 'paranoid' theory that controls, reifies and organizes in this way, see Sedwick (1997) and Stewart (2008).
6 Archaeologists, historians, geographers and anthropologists such as Porr and Bell (2012), Hill (2015) and Rose (2000) are similarly engaging with these issues.
7 The Yirrkala Bark Petitions, prepared by Yolngu elders, are the first Aboriginal documents recognised by the Australian Parliament, and are thus the first documentary recognition of Indigenous people in Australian law (Burarrwanga et al. 2013).
8 For a more detailed explanation of the patterns of *gurrutu* see Bawaka Country including Wright et al. (2015), and of *yothu-yindi*, Burarrwanga et al. (2013).

References

Bawaka Country including Lloyd, K., Wright, S., Suchet-Pearson, S., Burarrwanga, L., Ganambarr, R., Ganambarr-Stubbs, M., Ganambarr, B., Maymuru, D. and Hodge,

P. (in press). Meeting across ontologies: Grappling with an ethics of care in our human-more-than-human collaborative work. In Scott and Hicks (eds). *Educating for Sustainability in Unsustainable Environments*. East Lansing, MI: Michigan State University Press.

Bawaka Country including Wright, S., Suchet-Pearson, S. Lloyd, K. Burarrwanga, L. Ganambarr, R., Ganambarr-Stubbs, M., Ganambarr, B., Maymuru, D. and Sweeney, J. (2015a) Co-becoming Bawaka: Towards a relational understanding of place/space *Progress in Human Geography*, 40, 455–475. DOI: 10.1177/0309132515589437

Bawaka Country including Wright S, Suchet-Pearson S, Lloyd K, Burarrwanga L, Ganambarr R, et al. (2015b) Working with and learning from country: Decentring human author-ity. *Cultural Geographies*, 22(2), 269–283.

Bawaka Country including Suchet-Pearson S., Wright S., Lloyd K., and Burarrwanga L. (2013) Caring as country: Towards an ontology of co-becoming in natural resource management. *Asia Pacific Viewpoint*, 54(2), 185–197.

Bilous, R. H. (2015) 'All mucked up': Sharing stories of Yolŋu–Macassan cultural heritage at Bawaka, north-east Arnhem land. *International Journal of Heritage Studies*, 21(9), 905–918.

Burarrwanga, L., Ganambarr, R., Ganambarr-Stubbs, M., Ganambarr, B., Maymuru, D., Wright, S., Suchet-Pearson, S. and Lloyd, K. (2013) *Welcome to My Country*. Allen and Unwin, Melbourne.

Burarrwanga L., Maymuru D., Ganambarr R., Ganambarr B., Wright S., Suchet-Pearson S., and Lloyd K. (2008) *Weaving Lives Together at Bawaka, North East Arnhem Land*. Centre for Urban and Regional Studies, Newcastle, Callaghan: University of Newcastle.

Fox, J (1980) Retelling the past: The communicative structure of a Rotinese Historical narrative'. *Anthropology*, 3(1), 56–66.

Haraway, D. 1991. *Simians, Cyborgs and Women: The Reinvention of Nature*. London: Free Association Books.

Hill, L. J. (2015) Human geography and archaeology: Strange bedfellows? *Progress in Human Geography*, 39(4), 412–431.

Lloyd K., Wright, S., Suchet-Pearson S., Burarrwanga, L. and Bawaka Country (2012) Reframing development through collaboration: Towards a relational ontology of connection in Bawaka, North East Arnhem Land. *Third World Quarterly*, 33(6), 1075–1094.

Natcher, D. Huntington, O., Huntington, H., Chapin, F., Trainor, S., and DeWilde, L. (2007) Notions of time and sentience: Methodological considerations for Arctic climate change research. *Arctic Anthropology*, 44(2), 113–126.

Nicholls, R. (2009) Research and Indigenous participation: Critical reflexive methods. *International Journal of Social Research Methodology*, 12(2), 117–126.

Perkins, M. (1998) Timeless cultures: The dreamtime as colonial discourse. *Time and Society*, 7(2), 335–351.

Porr, M. and Bell, H.R. (2012) 'Rock-art', 'animism' and two-way thinking: Towards a complementary epistemology in the understanding of material culture and 'rock-art' of hunting and gathering people. *Journal of Archaeological Method and Theory*, 19, 161–205.

Riessman, C. (2005) Narrative analysis. In N. Kelly, C. Horrocks, K. Milnes, B. Roberts, D. Robertson. *Narrative, Memory & Everyday Life*. University of Huddersfield, Huddersfield: 1–7, http://eprints.hud.ac.uk/4920

Rose, D. B. (2000) To dance with time: A Victoria River Aboriginal study. *The Australian Journal of Anthropology*, 11(3), 287–296.

Rose, G. (1997) Situating knowledges: Positionality, reflexivities and other tactics. *Progress in Human Geography*, 21(3), 305–320.

Sedgwick E (1997) Paranoid reading and reparative reading; or, you're so paranoid, you probably think this introduction is about you. In E. Sedgewick, (ed.) *Novel Gazing: Queer Readings in Fiction* (1–40). Durham, NC: Duke University Press.

Stewart K (2008) Weak theory in an unfinished world. *Journal of Folklore Research*, 45, 71–82.

Suchet-Pearson S., Wright S., Lloyd K., Burarrwanga L., and Hodge P. (2013) A footprint in a rock: entwining lives and co-constructing 'the field'. In Australia, in Johnson J. T. and Larsen S. C. (eds) *A Deeper Sense of Place: Stories and Journeys of Collaboration in Indigenous Research* (21–40). Corvallis, OR: Oregon State University Press.

Strang, V. (2015) On the matter of time. *Interdisciplinary Science Reviews*, 40: 2, 101–123

Wiles, J., Rosenburg, M. and Kearns, R. (2005) Narrative analysis as a strategy for understanding interview talk in geographic research. *Area*, 37(1), 89–99.

Wright, S., Lloyd, K., Suchet-Pearson, S., Burarrwanga, L. and Tofa, M. (2012) Telling stories in, through and with country: Engaging with Indigenous and more-than-human methodologies at Bawaka, NE Australia. *Journal of Cultural Geography*, 29, 39–60.

Wright, S., Suchet-Pearson, S., Lloyd, K., Burarrwanga, D. and Burarrwanga, L. (2009) 'That means the fish are fat': Sharing experiences of animals through Indigenous tourism. *Current Issues in Tourism*, 12(5), 505–527. doi: 10.1080/13683500903042907

8 *Dibaajimowinan* as method

Environmental history, Indigenous scholarship, and balancing sources

Lianne C. Leddy

I can clearly remember picking blueberries with my grandmother on a hot summer day up on the rock behind my grandparents' house. This activity had several purposes that went beyond the obvious goal of pie making: *Nokomis* was trying to teach me the importance of patience, to learn my place in the world as an *Anishinaabe kwe* (woman), and that being part of a community means that everyone needs to contribute. She also told me stories (*dibaajimowinan*), both up on the rock and at her kitchen table. These were funny stories about her children growing up, stories about family history and cousin connections, and a distressing story that framed her life's work as an *Anishinaabe kwe* active in community politics as a long time councillor and school board trustee. These are lessons that have not just informed me as a person, but also shaped the reason I wanted to pursue Indigenous history and events that had happened at home on *Anishinaabe* territory.

The resulting project, which originally took shape during my graduate studies, was a case study that interrogated Cold War uranium extraction in a global market and military context and its effects on the Serpent River First Nation, an *Anishinaabe* community on the north shore of Lake Huron in Northern Ontario. It is here that I need to position myself as an *Anishinaabe kwe* and both a citizen of the Serpent River First Nation and someone who grew up in Elliot Lake, Ontario, a nearby settler uranium mining town. In the study, I made connections between Indigenous and treaty rights, dispossession, and environmental devastation using both archival sources and oral history (Leddy 2011). While the overall project was an environmental and Indigenous history of uranium mining in Northern Ontario framed by Indigenous-settler relations, the process itself turned out to be a story on its own which will be discussed here: it is one of decolonizing research practices that reflects on oral history and *Anishinaabe* stories, the importance of humility and the learning process as a whole, and asserting legitimacy as an Indigenous historian. The balance between archival sources, which were records kept by colonial bodies in Canada such as the Department of Indian Affairs, and oral history interviews with Indigenous knowledge holders can be difficult. They represent two very different views of history, and have not traditionally been seen as equal by the scholarly historical community. Reconciling these different views of stories is a process that takes compromise, particularly when one is trying to practice decolonizing methods and at the same time meet the expectations of the historical profession.

As the editors of this collection have noted (p. 4), "decolonizing nature-culture-history research … requires conducting research differently, both at the archives and beyond." One of the keys to accomplishing this is to understand that Indigenous ways of knowing are reliant on oral history and storytelling, and are not considered new or experimental in our communities. Western sources such as archives and newspapers are necessary research tools, but they must be used with caution and balanced with Indigenous sources. When Indigenous histories are explored from Indigenous perspectives, and, in particular, when Indigenous knowledge holders' voices are heard, we gain a better understanding of the intersections of environmental and Indigenous history, and of the lived experiences of Indigenous peoples. Yet, as Jo-ann Archibald (Q'um Q'um Xiiem, Stó:lō) has written, "going out to the field to 'talk story' takes time, patience, openness, and the will to keep talking with one another in order to learn how to engage in story listening and to make story meaning; none of this is unproblematic. It is hard work" (Archibald 2008: 126–127). This work is not always accepted by mainstream disciplines, particularly when Indigenous researchers are conducting work 'from home' and also seeking acceptance in academic departments.

While oral history may be a relatively new research method for Western-trained scholars, it is an essential part of Indigenous research. Part of the newness of these stories for settlers, even when they are in fact very old, has a lot to do with the fact that, to use a phrase by Paulette Regan (2011), they unsettle the settler within. As just one current example, the Truth and Reconciliation Commission heard testimony from more than 6000 witnesses, and its summary report, released in June 2015, has resulted in not only an extensive record of residential schools, their legacies, and recommendations to address this history, but also a national research centre and archive.

Truth-telling exercises have controversies, and the TRC spawned a debate about whether or not to save the records gathered through witness testimony. The director of the National Centre for Truth and Reconciliation, Ry Moran, has called the preservation of records pertaining to the IRS system as "a sacred obligation" (http://umanitoba.ca/centres/nctr/director.html). But this has been a controversial discussion, as the CBC (Canadian Broadcasting Corporation) has reported that some claimants made their very personal and painful statements regarding their experiences at residential schools with the promise that they would be destroyed. Some survivors have worried that their stories will be made public, and that they did not know that would be the case (www.cbc.ca/news/politics/residential-school-survivors-fear-testimony-could-be-made-public-1.2644349?autoplay=true). On the other hand, Terri Brown, a residential school survivor has said, "it's the true record of what happened to us, once it's destroyed, it's gone forever" (www.cbc.ca/news/aboriginal/fate-of-documents-detailing-abuse-at-residential-schools-undecided-1.2681138). The courts will decide the outcome of this controversy, but the purpose of discussing the TRC here is to talk about the importance of stories, truth-telling, and evidence.

Dibaajimowinan, or "stories," are those that document everyday events and news. On the other hand, *aadisokaanag* are, according to Cory Silverstein and Zeek Cywink (2000), "traditional tales and legends." These are stories that are sacred: they frame our traditions and our ways of seeing ourselves in Creation. The lines between these two different types of stories are not always clear, and both are relatively new to settlers and Western-trained scholars, as are the ways they are transmitted. Silverstein and Cywink argue that among some *Anishinaabek*, "*aadisokaanag* are more strictly limited to sacred stories that tell about spirits in the time before the present world, while *debaajimowin* encompasses all stories that refer to events within living memory" (Silverstein and Cywink 2000: 41). But these types of stories are not always clearly delineated in the interview process, especially when they are told in English. As *Anishinaabe* scholars Jill Doerfler, *Niigaanwewidam* James Sinclair and Heidi *Kiiwetinepinesiik* Stark (2013) have written, "*Anishinaabeg* stories, we believe, can create and transcend, affirm and deny – they have the abilities to do all sorts of things, sometimes at the same time" (Doerfler et al. 2013: xxvii). An interview can include contemporary events, personal events from an Elder's youth or early marriage, or stories about our responsibilities to the world around us. They can be funny, moral, or angering, but they can also tell us how to live our lives.

In a 2008 interview I conducted with Serpent River First Nation Elder Arnelda Jacobs, she shared her expertise that stemmed from working with youth to reclaim *Anishinaabe* language, culture, and artistic forms. But she also told me about the importance of the Serpent to our community and its role as a protector. There was no break in the interview when she told me something sacred to delineate it from everyday events in the 1950s. I am not a fluent *Anishinaabemowin* (*Anishinaabe* language) speaker, and I was not raised with the language of my ancestors because of colonial historical patterns and pressures. The fact that the knowledge holder was a fluent speaker added an additional layer of complexity to the interview process, especially when it came to my understanding important concepts. Arnelda left it up to me to see and understand the importance of what she was telling me in English, even though she likely knew that the stories featured beings and ideas that are best described in *Anishinaabemowin*. Several times during the interview I asked questions because I struggled to understand something, and we worked together towards clarity. Ultimately, though, it was (and continues to be) my responsibility to learn respectfully and to make sense of what Arnelda shared with me.

The most meaningful part of the research process was talking to Elders in the community. It was through their words and experiences that I was best able to explore the sense of loss and grief that came as a result of uranium extraction: from loss of traditional pursuits to loss of life, they shared their experiences with me, and it has been humbling to learn from them. Humility is one of the Seven Grandfather Teachings, a set of expectations for how *Anishinaabek* live our lives and relate to one another. While it is a good thing to demonstrate humility from an *Anishinaabe* perspective, it can be difficult to work through when one is a junior scholar. As scholars, we are meant to be experts, and sharing authority with knowledge holders is a relatively new way of seeing traditional 'sources.' But when research is also a part of one's identity reclamation, as it has been for me, it is difficult to balance humility with scholarly expectations. I grew up in a settler town that has always been proud of its mining history, and I am from both communities—my mother is from Serpent River First Nation and my father is from an Irish-Canadian mining family. This research process has enabled me to learn more about both community histories, but also about the particular ways in which Serpent River First Nation was affected by waves of extractive industries ranging from timber to uranium. I learned more about the Elders I knew from community events and visiting with my grandmother. The gradual understanding that this would be a journey that went beyond this one university graduate degree, this one book, was difficult to reconcile with the realities of graduate school where the project has an end date. Research relationships should not have such strict end dates.

Changing that perception took time, and I had to learn to embrace the journey as a learning experience. As scholars, we are expected to be the experts in our chosen areas of study. We conduct research, defend our findings to other experts, and compete for job interviews, all the while proclaiming to know all there is to know about our subject areas. This is incompatible both with the teaching of humility and the realities of sharing authority with Indigenous experts. It sometimes felt like I should have been more of an expert already and that I should have known more than I did. *Anishinaabe* scholar Sheila Cote-Meek (2014) has documented the ways in which Indigenous students and professors experience and resist trauma in the classroom, and while her work focuses on the post-secondary environment, it may also have relevance for Indigenous research, too. One student's story of feeling inadequate in the classroom resonated for me: "I felt really … for myself, I *felt really dumb*, that I *didn't know my own culture*, my own history. I didn't know the facts, I really didn't know the facts. And I was really embarrassed about that" (Cote-Meek 2014, p. 120). It was humbling to recognize how little I knew about my own community history until it was time to start writing it in a way that is respectful, inclusive, and decolonizing.

One of the challenges I continue to face as a Western-trained scholar is to decolonize my own thinking, even when I endeavour to write from an Indigenous perspective. Like many of the Indigenous scholars who have come before me (Absolon 2011; Anderson 2011; Kovach 2009; Smith 1999; Wheeler 2005; Wilson 2008), I have spent time reflecting on my own position as a researcher and trying to ensure that my work is done respectfully, all while still trying to meet the rigorous expectations of peer-review in the historical community. When I brought my completed dissertation back to my First Nation upon defending it in 2011, the Elder present at the community gathering spent a lot of time differentiating between traditional *Anishinaabe* knowledge based in land, language, and community, and Western book-based knowledge, like that found in my dissertation. This has always served to remind me how knowledge is transmitted between generations in culturally-appropriate ways, and that true knowledge is based in land and language because it reflects Indigenous ways of knowing. Kathy Absolon *Minogiizhigokwe* eloquently positioned herself as an *Anishinaabe kwe* and "re-searcher" in her book *Kaandossiwin: How We Come to Know* (2011): "[d]oing a research project on Indigenous re-search methodology is an act of decolonization as I claim my own Aboriginality and Indigenous knowledge. For example, I now see that Indigenous ways of searching were taught to me a long time ago and those teachings occurred in the bush" (Absolon 2011, p. 19). Employing Indigenous research methods is a crucial component of reclaiming our ways of knowing and transmitting knowledge, and can also be a powerful act of identity reclamation. These methods are also important if we are truly interested in respectful and engaged research with Indigenous communities.

This decolonization and reclamation can take many forms. Melissa K. Nelson (*Anishinaabe*/Métis) expanded on *Anishinaabe* scholar Gerald Vizenor's idea of trickster consciousness to overcome what she calls the "split-head society": the Cartesian way of understanding one's body and mind, and our relationship to Creation. Vizenor (as cited in Nelson 2008) defines trickster consciousness as "a comic liberator that craves chance, surprise, difference. The trickster is a healer in a fragmented world. The trickster denies singularity, monocultures, and completion. The trickster is communal, sensuous, erotic. The trickster is going to help us get to our next place" (Nelson 2008, p. 291). In other words, learning one's Indigenous language, reclaiming art forms, and using new technologies to convey old ways can all be ways to embrace trickster consciousness. Writing a community environmental history requires the examination of written evidence in addition to other sources, because, as Donald L. Fixico has argued about Indigenous positionality in historical research, "writing from home also means that things happen in a certain way, so there is not always written proof of such occurrences. This situation is also a part of Native logic and Indigenous ethos" (Fixico 2009, p. 558). More Indigenous scholars are articulating the importance of reclaiming language and art forms such as song, beadwork, and ceremony, but there is still resistance to these methods by those who claim they are less rigorous or less professional.

To what extent is writing environmental and Indigenous history from Indigenous perspectives seen as a legitimate pursuit by the academy? Mary Jane Logan McCallum has pointed to a significant exclusion, arguing "Native people's history is now considered a legitimate topic of historical labor by the historical profession. Yet Indigenous professional historians remain anomalies within it" (McCallum 2009, p. 524). In 2009, McCallum, remarking on the exclusion of Indigenous scholars from Canadian history, stated there was "not a single tenured Aboriginal professor working in any history department in Canada" (ibid., p. 528). Indeed, while this statistic has slightly improved over the last five years, many of us trained in history are still isolated in Indigenous Studies programs throughout the country, while the vast majority of historians teaching and researching Indigenous history are of settler backgrounds. Whether the reason for this is because Indigenous Studies as an interdisciplinary field is more welcoming to Indigenous academics, or because Indigenous historians are seen as too controversial or lacking scholarly legitimacy, is debatable. As Cote-Meek has pointed out, "[i]n the academy, Native academics are often confined to places designated as being Native, such as Native studies departments or Native student services; it is only more recently that we have seen more Native academics being hired in other disciplines" (Cote-Meek 2014, p. 72). The incongruous nature of being an Indigenous scholar is one that many of us have to reconcile in our practice (Mihesua, 1998). As our institutions and disciplines are often responsible for the construction of this incongruity, it is both timely and necessary to emphasize the positive effect decolonizing research can have on these divisions.

At the same time, decolonizing research needs to overcome a history (and in some cases, present-day reality) of academics extracting information from Indigenous communities. First Nations scholars have long critiqued the ways in which Indigenous communities have been seen as laboratories for research tourism. In 1969, Vine Deloria, Jr. (Sioux) took aim at "anthropologists and other friends" in his seminal work, *Custer Died for Your Sins*. Nearly fifty years later, his critique that "the massive volume of useless knowledge produced by anthropologists attempting to capture real Indians in a network of theories has contributed substantially to the invisibility of Indian people today" (p. 81) is still one shared by many Indigenous communities. Deloria also criticized the culture of research grants and cites one anthropologist who spent "close to ten million dollars studying a tribe of less than a thousand people!" He wondered what could have been accomplished with that money given to the communities themselves, reaching the ultimate conclusion that "there would have been no problems to study!" (p. 93). Not much has changed in the past half-century, with the exception of the fact that Indigenous peoples ourselves sometimes occupy the position of researcher. I was reminded when I read Deloria's critique of a comment from a colleague at a SSHRC (Social Sciences and Humanities Research Council of Canada) funding workshop several years ago, who, upon learning I was an Indigenous researcher, remarked upon my higher chances of success in the funding competition. Whether it is research grant success or perceived professional protection or tokenism, the position of Indigenous scholar is complex. As one anonymous faculty member told Sheila Cote-Meek, "I find that when I'm involved in projects, it's often as an add-on, you're an afterthought still [...] You're singled out. So I find that it's institutional, and sometimes it's very subtle. I've had non-Native faculty say things to me like, 'you got away with that because you're Native' (ibid., p. 84). A recent opinion piece that appeared in University Affairs suggested there was bias in the awarding of SSHRC doctoral awards, ending with "... I'm predicting a good year for students studying children or Indigenous peoples" (Crymble 2012, www.universityaffairs.ca/opinion/in-my-opinion/is-sympathy-for-certain-topics-influencing-sshrc-doctoral-awards).

The perception, and perhaps the reality, that Indigenous topics are lucrative when it comes to research funding raises questions about how that research is conducted and for what purposes. Namely, what level of engagement do Indigenous communities have in these research projects? Are these research relationships reciprocal? And, crucially, is this research of benefit to Indigenous communities? Certainly, the proliferation of research ethics boards (REBs) at Canadian universities, as well as Tri-Council policies, were meant to mitigate the concern that Indigenous peoples may be exploited by settler scholars. But to what extent is chapter nine of the *Tri-Council Policy Statement: Ethical Conduct for Research Involving Humans* (2014) an effective instrument of decolonization? It forces scholars to rethink their assumptions about academic freedom, a long-held belief that our research questions and methods should be protected from political or economic influence. In fact, when one colleague informed his REB that he had a data sharing agreement with a federal government department which required that he share his results with that department in advance of publication, concerns were raised about academic freedom and the integrity of the research. In that case, the emphasis was on ensuring that the partner not have any undue influence on the analysis or findings or be able to restrict publication. I was struck by how very different the expectations are in Indigenous research. In contrast, those of us working with Indigenous communities have a duty to co-design and execute research projects, and to ensure that the results are not only shared, but that there are built-in ways that the research will benefit those with whom we work.

At the same time, archival research is not without its methodological and ethical challenges. Because of the international origins and implications of my own project involving uranium extraction, the topic is one that has been widely covered in the written record, be it in archives or in the news. Archives provide a wealth of information about First Nations communities in Canada, but the very existence of these records is a result of colonial policies, surveillance, and an uneven power structure with Indian agents and administrators controlling and commenting on every aspect of Indigenous lived experiences. Whether it was counting Catholic and Protestant converts with an aim to reduce 'heathenism,' tallying the number of children attending residential or day schools, applauding the number of men working menial jobs, or judging women's household industry and sexuality, the correspondence was extensive. Record Group 10 at Library and Archives Canada (LAC), as just one example, is distressing evidence of a colonial past and present whereby (mostly) male colonial agents sought to convert, control, and assimilate First Nations people in Canada. Reading 'against the grain' of these records is necessary, and historians have been doing that since the 1970s. But the words of Indigenous peoples in these records, especially until the mid-twentieth century, are generally relegated to x-signatures, the odd letter from a chief often written by someone else, or an agent's recollection of a conversation.

Decolonizing critiques of archives are not new: Laura Ann Stoler has argued in her work about the Netherlands Indies that she sees "colonial archives as sites of the expectant and conjured – about dreams of comforting futures and forebodings of future failures" (Stoler 2009, p. 1). The documents contained within often say more about the people writing than they do the people under surveillance. Indian Affairs officials wrote about their hopes and expectations for the communities they attempted to control, which were often at odds with those held by First

Nations themselves. They wanted to encourage Christianity, industry, and Euro-Canadian notions of morality and personal conduct. When First Nations people failed to live up to these prescriptions, or, worse, when they actively resisted them, Indian agents complained to their supervisors. Indian agents controlled whether or not a band council resolution was approved, and they were gatekeepers for First Nations access to funds held in trust for them (Brownlie 2003).

From an historical point of view, and Indigenous critiques and caveats aside, archives contain information that was necessary for the colonial project as a whole, particularly when read carefully. As it turned out in my research, archival sources from LAC, the Archives of Ontario, and Laurentian University Archives proved to be particularly useful, as did newspaper articles from the period. These records contained everything from water quality reports reflecting environmental devastation, to extensive correspondence within different government branches about First Nations peoples, as well as letters written by the First Nation's leadership, especially from the 1960s onward. It was, however, jarring to *read* the contents of these fonds, not just because the material was often distressing, but because it came in a completely different format from the original story that I had been *hearing* since I was a girl. Here I was, a PhD student, sitting at one of the long tables at LAC in Ottawa on unceded Algonquin territory in a bright room overlooking Victoria Island and the Ottawa River, reading Noranda Mines' annual reports and government correspondence about mineral rights, timber, and community development. Indian agents provided reports to those in middle-management department positions, detailing the everyday lives of community members and the leadership's attempts to resist ongoing colonial policies.

In addition to RG10 records held at LAC, I was also interested in the papers of Dr. Rosalie Bertell, who was the founder of the International Institute of Concern for Public Health. She conducted studies at Serpent River First Nation in the early 1980s and her manuscript group is a record of her extensive research in the area of industrial and environmental impacts on health in many parts of the world (Rosalie Bertell Fonds, LAC, MG 31, K39, Vol. 36). Her papers contain data and reports on her health research in Serpent River First Nation and surrounding communities, commenting on occupational diseases, fetal death, and skin and eye infections. Her work as a social activist researcher provided the community with the scientific data the leadership needed to request environmental studies and land reclamation processes. It also represents the translation of intimate and devastating lived experiences into raw statistical data that, in the absence of a concerted federal and industrial response to what they saw as merely anecdotal evidence, could be used as a catalyst for change. She was interested in the effects mining and industrial pollution had on men, women, and children in the community, as mining effluent had threatened the Serpent River watershed and a sulphuric acid plant had been established in the community to facilitate uranium processing. Dr. Bertell wanted to find out the extent of the impact these activities and resulting pollution had on community members' health and wellbeing. She was interested in the occupational health and safety aspects of men's work, but she was also concerned about women's health.

Anishinaabe beliefs tie water to women's wellbeing and life-giving roles, and see pregnant women as a door between our world and the spirit world. Kim Anderson's work on Indigenous women's life stages and her interviews with Indigenous Elders has told us more about traditional views of pregnancy. It is "a sacred time; a time to honour the spirit that was coming as well as the mother who carried that spirit" (Anderson 2011, p. 43). While the new Cold War-era activities of the uranium industry coincided with the transition many women made from relying on midwives to giving birth in hospitals, women clearly remembered and retold their sense of loss and grief in the face of miscarriage, still births, and children born with disabilities when Dr. Bertell completed her study in the early 1980s. As Saulteaux Elder Mosom Danny Musqua told Anderson, in Saulteaux societies, the child "had to have a sense of belonging through the mother, and the woman had to have a sense of pride because she was contributing to the life of the community. She was bringing in new life, and she was treated special" (ibid., p. 43). Indeed, pregnancy and loss were not hidden, shameful secrets borne by individual women alone. Rather, they were losses grieved by families and the whole community because children were so highly anticipated and welcomed. *Anishinaabe kwewak* (women) remembering and retelling their personal experiences could be seen as a decolonizing act, especially as other aspects of Indigenous parenting have historically been and continue to be under attack (Lavell-Harvard and Corbiere Lavell 2006). Bertell's work, then, assisted in a cultural translation to work for reproductive—and therefore holistic environmental—justice.

As an historian, I read these reports with interest because it was one of the first comprehensive health studies of the community. It was also the kind of evidence that peer-reviewers like to see to support claims of environmental devastation and reproductive compromise. As a community member and granddaughter of one of the participants, though, I immediately drew the sad connections between the reports and the lived experiences that the numbers so coldly represented on the page. This was one of the first times I began to think critically about my position in this work. While it was exciting to conduct my own research independent of any particular assignment, the responsibility I felt to tell the story well was weighty. And, I thought to myself, how do these two types of learning, these different ways of communicating, fit together in one person, particularly since that person is part of a large family and community that shared (and continues to live with) the same environmental legacies? There was no immediate answer to this at either the research or writing stages of the process, although I did try to work out some of these issues in an article very early in my career (Leddy 2010). Trickster consciousness, then, takes time to formulate.

While I struggled with how to balance archival work and oral history and the stories they told, newspapers, particularly the local *Elliot Lake Standard*, helped to establish a timeline of events. Furthermore, in some cases, they provided insight into public and mining companies' opinions about environmental damage and charted patterns of change over time. Put simply, when uranium was booming, economic progress was considered paramount. When the market busts came, an emphasis was placed on the importance of fishing and hunting tourism, and by extension, clean lakes and rivers. Daniel Francis (2011) interrogated the ways in which dominant society historically constructed the "imaginary Indian" in Canada, and Anderson and Robertson (2011) have argued that newspapers in Canada have been responsible for the colonial construction of Indigenous peoples as inferior, even to this day. But in news coverage related to this project, the late-1960s marked a pivotal turning point as the number of news stories related to First Nations issues grew, and many of them were empathetic to the struggles of dynamic community leadership. As Scott Sheffield found in *Red Man's on the Warpath: The Image of the "Indian" and the Second World War* (2004), constructions of "Indianness" were indeed crafted by dominant society, including public servants and newspapers, but at certain points during and immediately after the war, Indigenous peoples themselves were able to influence these images (Sheffield 2004, p. 177). In the Cold War context at Serpent River First Nation, Chief Bill Meawasige, for example, harnessed the power of media attention to the community's advantage. This media trend continued into the 1980s, when community leadership openly resisted government inaction on environmental devastation through active protests and blockades. Just as Sheffield found that both internal and external events shaped these constructions in the mid-twentieth century, so too did the rise of environmentalism and 'red power' movements in the 1970s and onwards.

It is finding the balance amongst these various sources that has posed a challenge to writing about environmental history. There are parts of the monograph that rely exclusively on archival sources while the words of community knowledge holders frame other sections. There is sometimes a clear division of sources that challenge my own attempts to decolonize research. It can be difficult to effectively balance these two different types of evidence, and while they can be combined to tell a more comprehensive story, they constitute two very different *ways* of conducting research, and, indeed, of understanding the world. This separation speaks to the question of how to accomplish decolonizing research: while I had attempted as an historian to write a comprehensive story that integrated my source bases, at times this simply was not possible. The story as a whole was comprehensive, but its chapters, its pieces, felt incomplete. Archival and newspaper sources could provide a timeline, but as Donald L. Fixico writes, "Indian history is perceived differently by Indians close to their traditions as opposed to academic historians [...] Indian history of the former is conveyed in the oral tradition via stories where 'experiences' are more important than 'events'" (Fixico 2009, p. 553). Trying to reconcile two different perceptions of history was an uneasy, but necessary, compromise needed to fit the established criteria of scholarship in the historical discipline.

This chapter has examined environmental history as told in stories: who gets to tell them, how they are told, and why. Reflecting on Indigenous methodologies necessitates an understanding of how all of these elements work together: the complexities of conducting research as an Indigenous researcher, and doing so with Indigenous communities in a way that is helpful. In some cases this has meant at the very least ensuring the results and resources are available to community members. In my case, it also meant the dissertation was a resource for Elders to use for context in writing their own stories, which they did in *Connected to the Land: Stories from the Serpent's Band* in 2013. I was also happy to maintain the community relationship by offering assistance in proofreading the collection (Janice Gamble and Elder Tea 2013).

The goal is to engage in research that is decolonizing, in a way that honours the ways our communities have always transmitted knowledge. In telling the story about uranium and community resilience, I sought to convey our story from our own perspective, and in honouring the words of Elders as much as possible, I sought a balance between archives and community. Sometimes this posed challenges—some parts of the story necessitated a strong reliance on archival and news sources, and others allowed me to highlight the words of Indigenous knowledge holders. Most of all, since I first embarked on this research journey nearly a decade ago, I have come to recognize and embrace that the research journey is just as important as the findings themselves.

Acknowledgements

The author wishes to acknowledge that this research was supported by a doctoral SSHRC Canada Graduate Scholarship, and she thanks the Elders of Serpent River First Nation that she interviewed for their assistance and the editors of this collection for their helpful comments.

References

Absolon, K. Minogiizhigokwe (2011). *Kaandossiwin: How We Come to Know*. Blackpoint, NS and Winnipeg, MB: Fernwood Press.

Anderson, K. (2011). *Life Stages and Native Women: Memory, Teachings, and Story Medicine*. Winnipeg, MB: University of Manitoba Press.

Anderson, M.C. and Robertson, C. L. (2011). *Seeing Red: A History of Natives in Canadian newspapers*. Winnipeg, MB: University of Manitoba Press.

Archibald, J. (2008). *Indigenous Storywork: Educating the Heart, Mind, Body, and Spirit*. Vancouver, BC: University of British Columbia Press.

Brownlie, R. (2003). *A Fatherly Eye: Indian Agents, Government Power, and Aboriginal Resistance in Ontario, 1918–1939*. Don Mills: Oxford University Press.

Canadian Press (2015, June 19). Fate of documents detailing abuse at residential schools undecided. CBC News. Retrieved from www.cbc.ca/news/aboriginal/fate-of-documents-detailing-abuse-at-residential-schools-undecided-1.2681138.

Canadian Press (2015, May 15). Residential school survivors fear testimony could be made public. CBC News. Retrieved from www.cbc.ca/news/politics/residential-school-survivors-fear-testimony-could-be-made-public-1.2644349?autoplay=true.

CIHR, NSERC, and SSHRC (2014). *Tri-Council Policy Statement: Ethical Conduct for Research Involving Humans*. Ottawa: Secretariat on Responsible Conduct of Research.

Cote-Meek, S. (2014). *Colonized Classrooms: Racism, Trauma and Resistance in Post-secondary Education*. Blackpoint, NS and Winnipeg, MB: Fernwood Press.

Crymble, A. (2012, July 25). Is sympathy for certain topics influencing SSHRC doctoral awards? How else to explain different rates of success for different topic categories. *University Affairs*. Retrieved from www.universityaffairs.ca/opinion/ in-my-opinion/is-sympathy-for-certain-topics-influencing-sshrc-doctoral-awards.

Deloria, V. (1969). *Custer Died for Your Sins*. New York: Macmillan.

Doerfler, J., Sinclair, N. and Stark, H. (2013). Bagijige: Making an offering. In Jill Doerfler, Niigaanwewidam Sinclair, and Heidi Kiiwetinepinesiik Stark (eds), *Centering Anishinaabeg Studies: Understanding the World Through Stories*. East Lansing, MI and Winnipeg, MB: Michigan State University Press and University of Manitoba Press.

Fixico, D. L. (2009). American Indian history and writing from home: Constructing an Indian perspective. *American Indian Quarterly*, 33, 4. Pp. 553–60.

Francis, D. (2011). *The Imaginary Indian: The Image of the Indian in Canadian Culture*, Second Edition. Vancouver, BC: Arsenal Pulp Press.

Gamble, J. and Elder Tea (2013). *Connected to the Land: Stories from the Serpent's Band*. Education Department: Serpent River First Nation.

Jacobs, A. (2008). Personal communication. Serpent River First Nation, ON.

Kovach, M. (2009). *Indigenous Methodologies: Characteristics, Conversations, and Contexts*. Toronto: University of Toronto Press.

Lavell-Harvard, D. and Corbiere Lavell, J. (eds.) (2006). *Until Our Hearts are on the Ground: Aboriginal Mothering, Oppression, Resistance and Rebirth*. Bradford, ON: Demeter Press.

Leddy, L. (2011). *Cold War Colonialism: Serpent River First Nation and Uranium Mining, 1953–1988* (unpublished dissertation). Wilfrid Laurier University, Waterloo, Ontario.

Leddy, L. (2010). Interviewing Nookomis and other reflections of an Indigenous historian. *Oral History Forum d'histoire orale*. 30, special issue. Pp. 1–18.

McCallum, M. (2009). Indigenous labour and Indigenous history. *American Indian Quarterly*, 33, 4. Pp. 523–44.

Mihesuah, D. (1998). *Natives and Academics: Discussions on Researching and Writing about American Indians*. Lincoln, NB: Nebraska University Press.

Moran, R. (n.d.). Welcome from Director. *University of Manitoba National Centre for Truth and Reconciliation*. Retrieved from http://umanitoba.ca/centres/nctr/director.html.

Nelson, M.K. (2008). Mending the split-head society with trickster consciousness. In Melissa K. Nelson (ed.), *Original Instructions: Indigenous Teachings for a Sustainable Future* (pp. 288–297). Rochester, VT: Bear & Company.

Rosalie Bertell Fonds, MG 31, K39, Vol. 36. Library and Archives Canada, Ottawa, ON.

Silverstein, C. and Cywink, Z. (2000). From fireside to TV screen: Self-determination and Anishnaabe storytelling traditions. *The Canadian Journal of Native Studies*, 20, 1. Pp. 35–66.

Smith, L.T. (1999). *Decolonizing Methodologies: Research and Indigenous Peoples*. London and New York: Zed Books.

Stoler, L.A. (2009). *Along the Archival Grain: Epistemic Anxieties and Colonial Common Sense*. Princeton, NJ: Princeton University Press.

Wheeler, W. (2005). Reflections on the social relations of Indigenous oral history. In Ute Lischke and David T. McNab (eds), *Walking a Tightrope: Aboriginal People and Their Representations* (pp. 189–214). Waterloo, ON: Wilfrid Laurier University Press.

Wilson, S. (2008). *Research is Ceremony: Indigenous Research Methods*. Blackpoint, NS and Winnipeg, MB: Fernwood Press.

9 Giving and receiving life from *Anishinaabe nibi inaakonigewin* (our water law) research

Aimée Craft

Being a mother is a defining life role for *Anishinaabe* women. We are many other things but our role as mother is given to us directly by, and connects us to, Mother Earth. In ceremonies, we often go back to the relationship between mother and child to help us understand who we are as people and our purpose in life. One example is the sweat lodge that symbolizes the womb of a mother and the purification and new beginning that the ceremony can bring. From the *Anishinaabemowin* language we know that to be a mother is not only the role of a birth mother, but that of a mother's sisters (a child's aunts) as well. As mothers, aunties and close friends, we can be mothers to all of our children.

For me, the research relating to water that I discuss in this chapter is like having and beginning to raise a child. It is what I consider to be a step in a lifelong learning, teaching, sharing and caring role in relation to water and the teachings that flow from it. Children are born of water. This water is carried by their mothers for the express purpose of creating and bringing life. We are all made up in large part of water and we need water in our daily lives to sustain our bodies and spirits. Our life comes from and depends on water. I analogize this research and the phases of its development to the coming to life of a child through conception, birth and what we learn as young children. Vine Deloria reminds us that our "cultures are rich with ways of gathering, discovering, and uncovering knowledge. They are as near as our dreams and as close as our relationships" (Deloria 1996, p. 182). The research methodology described in the context of this water law research tells us how to grow and learn together. This is not a methodology or research design that belongs to me, but rather is an evolving participant-designed project. This methodology was gifted to all of the participants (Elders, students, other participants and me) through the process of working together, sitting together in ceremony, and reflecting on our purpose in a way that reflects *Anishinaabe* ways of being that are both historic and contemporary.

Conception: Purpose of the research and first thoughts

The *Anishinaabe nibi inaakonigewin* (ANI) research project was conceived as part of a larger research partnership, based at the Centre for Human Rights Research at the University of Manitoba, relating to clean water and sanitation in First

Nations communities. The larger project goal, funded through a SSHRC (Social Sciences and Humanities Research Council of Canada) partnership development grant, was to combine public perceptions, economic costs, and knowledge of law relating to clean water for First Nations. Advocacy, legal action, scholarly and practical tools were to be created for First Nations partners. The goals of the ANI project were twofold within the context of the larger research grant. First, the ANI research was to foster understanding and support policy changes related to clean water in First Nations communities. Second, it aimed to be an opportunity for the documentation, translation, and preservation of *Anishinaabe* cultural and knowledge systems, in this case related to water law.

Jurisdiction over and about water is unclear. It is shared between governments, divided between navigable waters and river/lake beds, and contested. Disputes about ownership and control are frequent. Rights and obligations related to clean water for First Nations are a political hot potato that gets circulated around and around and, of course, sometimes dropped. While law can provide many answers to complex socio-political issues, water and jurisdiction over and about water is a murky issue. Professor Karen Busby, principal investigator on the larger grant, researched constitutional and international law relating to clean drinking water and sanitation rights for First Nations. Busby found that there is a reasonable argument to make that the government has breached its constitutional obligation to provide clean water and sanitation as "essential public services of reasonable quality to all Canadians." Further, she determined that a United Nations General Assembly resolution (dated 2013) recognized access to water and sanitation as a human right that is essential for the full enjoyment of life and all human rights. She argues that, based on the resolution, other international law declarations and political commitments made by governments around the world, the human right to water and sanitation is now crystallized in international customary law (Busby, forthcoming 2016).

Within the legal research initiative, it was important to canvass not only Canadian and international law on clean water and sanitation rights, but to consider treaties and Indigenous legal traditions as well. The research I undertook was aimed at better understanding Indigenous legal theory through the realm of ANI while trying to extract legal principles that could assist with decision-making relating to clean water. In the next phase of my research, I will be working on how treaties address the question of water. In conceiving the research, many questions came to mind: would ANI have anything to offer the multiple decision-making processes related to water, among various forms of decision-makers (First Nations governments; municipal, provincial and federal governments; administrative bodies; environmental regulators) and contexts (industrial development, human rights complaints, legal actions)? Would it help reframe Canadian law or allow for shifts in legal and policy perspectives? Or could it even help fill the contested and disjointed knowledge about water in order for proper decision-making to take place? Could it shift priorities from ownership and the economic value of water to acknowledgement and protection of the spirit of water itself? What might Indigenous law have to offer in this murky well of competing jurisdictions?

This was an opportunity to redirect the energy I had been spending fighting about who gets to decide, and who takes the priority seat at the decision-making table, toward work on valuable knowledge translation that would inform those kinds of decision-making processes. So often, I had been in a room with First Nations, governments and industry each talking to, at, and across each other. Even in the cases where there was a genuine desire to listen, the ability to understand across profoundly different worldviews with radically different normative values (or laws) often resulted in an unsatisfactory and imbalanced final result, rather than an "ethical space" for innovative, cross-cultural and respectful process of inquiry (Ermine 2000). Scholar Shawn Wilson (2008) links the colonist agenda of control over land and resources to the continuous place of Indigenous people as the researched rather than the researchers working within their own Indigenous paradigms.

All legal traditions are centered on rights, obligations and responsibilities. Most legal systems exist in the realm of reciprocal and bilateral rights and obligations, often between the individual and the state. In approaching this research, my thesis was that *Anishinaabe* water law is focused primarily on responsibility, rather than rights, and that it is multi-layered, multi-dimensional and sourced from relationships among beings (human and non-human). This was meant to be a legal theory project with a goal of extracting *Anishinaabe* legal theory from *Anishinaabe* water law discussions. It turned out to be so much more.

While important inroads have been made to have Indigenous legal traditions recognized within the Canadian legal framework, questions and work remain about how to ensure that that space is filled with legitimate and thoughtful Indigenous legal content. Each Indigenous legal tradition has its method of making law, deliberating on the application of law and instructing on its own legal tradition. Researchers such as Val Napoleon and Hadley Friedland (2014) have developed a methodology of case briefing Indigenous stories in order to understand legal principles for a variety of Indigenous legal traditions. Employing Western legal techniques and frameworks for relating or understanding Indigenous laws can serve an important educational purpose. However, decolonizing methodologies based on traditional means of knowledge sharing in a non-static and forward-thinking way are, in my view, at the heart of the development of *Anishinaabe* legal theory. The ANI methodology evolved to ensure that traditions as well as modern practices and realities are integral to the project. Throughout my work, the Elders have reminded me that this is not simply legal theory, but rather our way of life.

The ANI research explores Indigenous law through a methodology of participant- and event-based research that is fluid in its boundaries, clear in its objectives and engaged with knowledge holders not as subjects but as researchers themselves. The original project design incorporated four annual four-day gatherings, which grew from the participation of a small faculty of Elders to later include community members and youth. From the start, the Elders were identified and treated as a "faculty of Elders." Their knowledge and experience relating to Indigenous laws and water made them the teachers. Each Elder was approached with an offering of tobacco and a request to be part of the project. They were advised of the format, objectives and research ethics. Each Elder consented in both the ways required by the Tri-Council of Canadian government research funders and in accordance with *Anishinaabe* ethics and protocols of research. The Elders were compensated for their time with honoraria and their expenses for travel were reimbursed. The payment of an honorarium is meant to be an acknowledgement of the time and effort that was made to prepare for the gathering and participation in the gathering itself. It is not payment. To send an honorarium after the fact would be antithetical to the purpose of the honorarium, and providing a cheque is often just as problematic. This is a source of tension between this type of Indigenous-led research and the financial administration of research funds at universities.

Generally, female Elders carry the knowledge and ceremonies relating to water and the men are responsible for supporting the women and speaking for the group on certain matters. Each of the men acknowledged where he received his water stories, songs and teachings, usually from his mother or grandmother. The Elders were both male and female, drawn from the four directions of what anthropologists refer to as the "Northern Ojibway" territory. From Western Manitoba (Keeseekoowenin, Rolling River and Sandy Bay), up the mouth of the Winnipeg River (Sagkeeng First Nation), east to Lake of the Woods (Ojibways of Onigamiing, Whitefish Bay First Nation) and Lake Huron (Sagamok Anishnawbek First Nation), and south to Roseau River (Roseau River Anishinabe First Nation). The faculty of Elders were selected for their base of knowledge, but what each member also brought to the research was a framing of the research itself and the ability to guide the process of learning through knowledge and experience. The Elders came together to develop a curriculum, share their strengths and gifts, and collectively make their knowledge accessible. *Centering Anishinaabeg Studies: Understanding the World through Stories* highlights the importance of relating through stories and the idea that various gifts extend beyond the boundaries of disciplines. In that collection, Niigaanwewidam Sinclair reminds us that *Anishinaabe*

> are a body of diverse people embodying countless subjectivities, experiences, and perspectives that together enact an autonomous language and literature which forms relationships among themselves and with others. What makes *Anishinaabeg* who they are, according to [Basil] Johnston, are their negotiable, multiple, and fallible truths – constantly emerging and based in principles of complexity – and their inherent and ongoing ties to one another in a trajectory of continued, collective existence.
>
> (Sinclair 2013, p. 89)

In the context of the ANI project, the Elders' work was not to compare and contrast but rather to identify, as a starting point, how non-Indigenous systems of law functioned and were allowed to be superimposed on Indigenous legal relationships with water. This was an important decolonizing process of acknowledging colonization, loss, oppression and resilience. The discussion could then turn to the task of explaining, elucidating, contextualizing and sharing *Anishinaabe nibi inaakonigewin*. Through means of knowing and transmitting knowledge such as song, language and storytelling, all in the process of *Anishinaabe* ceremonies (including pipe and water ceremonies), the faculty of Elders was able to weave a blanket of water stories that can ripple through the imagination, the lived world and the spirit world.

Birth: Law through our own epistemologies and ontologies

The existence of every person and every being is framed in the context of its environment and story (or stories) of Creation. In *The Truth About Stories: A Native Narrative* (2003, p. 10), Thomas King tells us that creation stories remind us "how the world was formed, how things came to be, for contained within creation stories are relationships that help to define the nature of the universe and how cultures understand the world in which they exist." *Anishinaabe* people have been shaped by their environment and have learned from their surroundings since placed on this earth by the Creator. As part of understanding and acting in relationship to and with water, we must seek methodologies that are considerate of our ways of knowing and being. Our law requires an understanding of history, nature, environment, people, relationships and an acceptance of a particular worldview. Indigenous research methodologies must be guided by the environment and "worldviews, beliefs, values, principles, processes and contexts. Indigenous methodologies are wholistic, relational, interrelational and interdependent with Indigenous philosophies, beliefs and ways of life" (Absolon 2011, p.22).

As we were thinking about *inaakonigewin*, we were reminded that as human beings, we were the last to be placed on the earth and the most dependent of all the beings in creation. As humans, we depend on a complex web of relationships in order to live well, to live our *mino-biimaadiiziiwin*. Through the relationships that govern our interactions among beings, we develop a sense of our normative values and legal principles. Shawn Wilson (2008, p. 87) notes that relationships with land and relationships with each other are equally sacred. In this ANI work, we were exploring our relationship to water, and corresponding responsibilities to other beings. Whereas, in other systems of law, water is treated as a subject or an object, often to be *owned* and *used*, in ANI *nibi* (water) is treated as an actor in a relationship. The research and gatherings were premised on the teaching that water has a spirit, that it is living and that we must conduct ourselves in relationship with it. However, the conversation could not be limited to water itself, given that water is in constant interaction with other beings (including each of us, our bodies and our spirits). In fact, water sustains all life. We also took time through words and ceremony to acknowledge the spirit of water and the spiritual beings that look after water and to recognize those things that are beyond the control or perception of human beings.

At the outset I wondered: "How does one research within the realm of the spirit? Should I engage with research in ceremony, song, language and storytelling? Would this help us understand our legal relationship with water and to each other?" One Elder, Florence Paynter, addressed these questions during our research by stating,

> Everything here has been done in ceremony, which is how we've been doing things for a long time. We should go down to the water and offer tobacco and our thoughts, so we can ask the water to bless us in helping her in what we're talking about. I had a dream about doing this. It was showed to me and told to me in my dreams: this is what you do. This is what we do as a people and what we've done for a long time. We also talk about looking to the youth to help us.

In thinking about the purpose of this research, I had to consider how we could be responsive to changing physical and natural environments, new forms of communication and a modern, fast-paced world. How could we anchor tradition and ancient laws while developing mechanisms for the transfer of knowledge among Elders, knowledge keepers and next generations? And how could this knowledge inform our individual and collective decisions and actions? The research had to have a decolonizing approach and purpose that permits respect for living tradition while ensuring that we as people continue to exercise agency in our relationships and legal norms. As Marie Battiste and Sakej Henderson remind us (2000, p. 72), an Indigenous nation "may share the right to enjoy and use certain elements of its knowledge, under its own laws and procedures, but it always reserves a perpetual right to determine how shared knowledge is used ... [which is] critical to the identity, survival, and development of each indigenous society."

Originally, one of the research goals was to engage in a process of deliberative law-making through the traditional system of governance in clans. The clan-based system is one in which groups of relatives gather internally within the nation, deliberate under the guidance of clan mothers and speak with other clans to inform broader group decision-making on particular issues. In addition, each clan contributes through its own particular gifts and responsibilities to the daily living and general well-being of the nation (Mcguire 2008). The research did not move in the direction of clan-based gathering or decision-making. This objective was presumptive in that without significant efforts across communities, there would not be a desire to assemble in clans for decisions. For many generations, colonial influences have attempted to distance *Anishinaabe* from neighbouring (as well as more distant) First Nations communities. Decision-making could not take place without first establishing a base of knowledge and comfort with each other in that knowledge. Engaging in an effort to rebuild without formalizing or forcing decision-making was an important act of decolonization in itself. That effort to bring communities and people together was reflected in the efforts to build a lodge in a shared territory, in order to ensure that all felt welcome and included. Instead of decision-making, the research resulted in active engagement within the gatherings and within families and communities. This engagement was purposive, transmissive and collectivized.

There was a generalized good feeling of being together, which generated an energy and commitment in the gatherings and thereafter.

The first sounds we make: The first gathering

We held five research gatherings in total, three of them over a June solstice. The first and the last were four-day gatherings in which we were in traditional teaching lodges. We worked from sunrise to the end of the day. We shared meals, discussed and often participated together in evening activities, including sweat lodges and, at the last gathering, creative work with clay, led by visual artist KC Adams.

The first research gathering was held in Roseau River, Manitoba, at a Midewiwin teaching lodge. Most of us stayed on site, camping in tents, while others stayed in a motel nearby. Food was brought to us by the local *Anishinaabekweg* (women) who made food with love and care for each of us in accordance with the *Anishinaabe* protocols of food preparation. Many of the Elders and knowledge keepers knew each other, and I had worked with each of them in the past. It was important to the gatherings to build trust with each other and with the students in attendance. At the first gathering, we did not enter the teaching lodge in accordance with all the protocols of the Midewiwin ceremonies. The spirit of the lodge provided guidance through a dream that I had on the first night and which guided us on how properly to honour the lodge and the sacred location of our gathering. Tobacco was presented to an Elder on the morning following the dream so that he could interpret it. Following the interpretation, a course of action was engaged and a ceremony was conducted to honour what had been shared with us through the dream. This was an important part of the methodology that guided the entire ANI research project and which grounded us in *Anishinaabe* epistemologies and relevant knowledge sources (Kovach 2009).

During the gathering, rain waters flooded Calgary and Winnipeg and storms approached and bypassed the sacred site of the lodge. On the last day of the gathering, rain poured down: a symbol of the cleansing power of water. It also reminded us of our vulnerability to the weather and our dependence on nature to sustain us. During each of the gatherings there was precipitation, welcomed as a reminder of what we were assembled to do, which was to work for the water. Often we will look to the environment and the elements (in this case the rain) as a guide for our actions—what many of us call natural law.

The format for the first gathering was generally a discussion circle. During the first day and a half, we focused on understanding what was meant by "law." This was an opportunity for mutual exchange among the students, Elders, knowledge keepers and me. We shared our stories—our understandings and perceptions of how provincial and federal laws relate to water, as well as some of the tensions we had experienced with those laws. "Stories remind us of who we are and of our belonging. Stories hold within them knowledges while simultaneously signifying relationships … In considering story as both a method and a meaning, it is

Figure 9.1 The rapids at Roseau

Source: Photo Katelin Neufeld.

presented as a culturally nuanced way to knowing" (Kovach 2009, p. 94). We tried to understand how *Anishinaabe* law might be different from Canadian law. In the afternoon of the second day, the discussion on law (in its technical sense) was completed and the subject turned to water with songs, stories, words and teachings relating to it. On the fourth and last day of the gathering, we discussed what *Anishinaabe nibi inaakonigewin* (water law) meant to each of the Elders. The gathering was completed with a water ceremony and offerings to the water, guided by the grandmothers and with the assistance of *queezence* (little boy water drum). All agreed that rather than wait until the following June to meet, a report of the gathering would be prepared and distributed and the group would meet in December to revise the report and decide on next steps for the research.

First steps

Winter is the time to tell stories. The second research gathering was held the following December in the University of Manitoba's ceremonial space at *Migizii Agamik* (Bald Eagle Lodge) to review the material and add to the stories, songs and teachings. Some stories are only shared at certain times of year, and generally winter is the time for telling traditional or spiritual stories. There are other forms of stories that we tell: place-specific stories, lived-experience stories, teaching stories, etc. That December, we sat in ceremony together, and shared a meal. Prior to this one-day gathering, I provided a summary report to each of the Elders, along with a full transcript for them to review and comment on. The report is rich in direct quotes from the Elders and provides preliminary findings from the research. The report was revised based on suggestions provided at this second gathering. In this way, we agreed on the beginning structures of the knowledge base and curriculum that would carry into the future gatherings. The preliminary report was presented at a water research conference at the university in June 2014 and at other water conferences outside of Manitoba, including internationally.

No formal translation was provided for the gatherings. Where it was appropriate, the speakers translated for themselves, or asked another person to do so. In some cases, no translation was offered, which at times excluded *non-Anishinaabemowin* speakers and less fluent speakers. In some cases, the use of a particular word or concept was discussed and, in others, words were deconstructed for their meaning. Without using *Anishinaabemowin* words, the full intent of the speaker might not be conveyed or heard. There are things that are only said in the *Anishinaabemowin* language, especially in speaking to the spirit and in conducting ceremony. Language and its structure give rise to a way of thinking and being. As Kovach puts it (2009, p. 61), "Language is a central component of Indigenous epistemologies and must be considered within Indigenous research frameworks." Basil Johnston notes that terms in *Anishinaabemowin* have "three levels of meaning" – "the surface meaning that everyone instantly understands. Beneath this meaning is a more fundamental meaning derived from the prefixes and their combinations with other terms. Underlying both is the philosophical meaning" (Sinclair 2013, p. 87). At the December 2013 meeting, Elder Niizhoosake Copenace expressed it this way: "Our language is the way we sound; kaanwe'ing. When we speak it, it is an expression of our spirits and our spiritual protectors that are speaking through our words." Often in ceremony, we will ask forgiveness for speaking other languages that are not those of our ancestors. By holding to our procedural requirements to conduct ceremony in the language, we are acknowledging that there are normative values and laws that are held within the language itself. Language and worldview are so intimately intertwined that in some cases, introducing a Western language would change the intent of the prayer, and in this case, the research intention.

At my suggestion, the Elders decided at the December gathering that some of their stories be made into an illustrated book on *Anishinaabe* water law, in both English and *Anishinaabemowin* (on the model of *Anishinaabemowin* language-learning books). The Elders then shared more stories to include in the book. The book is for youth, families, academics, lawyers, policymakers and everyone else as well. While the story book does not have the same purpose as accessing stories for the purpose of research, the book aims to be considerate of the *Anishinaabe* knowledge systems that underlie the telling and interpretation of the stories. While the book is still in the process of being put together, we are hoping to include recorded songs with the book in an audio CD format. Songs themselves are ceremony and an important part of ceremonies. Songs are also language and story.

Starting to speak

A year after the first gathering, the report was presented to a conference of science and humanities students, academics, researchers and Indigenous community partners, all of whom were working on water projects. The Elders met the day before the presentation, again at *Migizii Agamik*, and continued to share knowledge in relation to water. The document was taken into ceremony and blessed by spirit. One of the Elders shared a vision that he had been given to build a wiigwaam (teaching lodge) at a sacred place called Manito Api in Whiteshell Provincial Park. The site is an important place of gathering for *Anishinaabe* and other Indigenous people and has been for thousands of years. It houses ancient petroforms that tell stories of *Anishinaabe* creation and law. The Elders agreed that the wiigwaam was to be the learning institution of ANI.

The wiigwaam was built over a weekend in September 2014, just as summer was turning to fall. More than 30 people attended to help. Food was provided, and despite the rain that fell, all were in great spirits and happy to be doing the work. People came from many miles away, some traveling more than four hours each way to be part of the lodge building. The wiigwaam is a 120 × 30 foot free-standing sapling structure. When it is in use, it is covered with mosquito nets and tarps to keep the bugs, sun and rain out. Four days after the lodge was built, we entered into the lodge ceremonially and continued the water gatherings and discussions.

Finding our gifts and place in the world (vision)

In June 2015, around the time of the solstice for the third year in a row, we held a four-day ANI gathering. It was focused on knowledge dissemination and more than 150 people attended. Guest speakers were invited to complement the knowledge that the Elders were sharing, particularly younger people who had taken on responsibilities by walking with the water, learning the language, participating in ceremonies and encouraging youth to learn about *Anishinaabe* culture. They came to share their stories, help with the gathering and learn from the Elders.

> Stories shape how we see and interact with the world. They lend insight into the ways in which we see our communities, as well as how we see ourselves within these communities. The power of stories is found in their ability to outline and clarify the connections people have to their place, their people, and their history. Indigenous stories outline relationships – the relationships we have to one another, and the relationship we have to self.
>
> (Stark 2013, p. 260)

Many of us camped on site near the lodge (and in local accommodations), food was made for us by *Anishinaabekweg*, and children were an important presence at the gathering. Two community-based organizations (the First Nations Health and Social Secretariat of Manitoba and the Kenora Chiefs Assembly) partnered to host the gathering, supporting the opportunity to share ANI knowledge with youth. The FNHSSM (First Nations Health and Social Secretariat of Manitoba) supported Elders from its health knowledge program (Partner for Engagement and Knowledge Exchange) to attend the gathering and hosted one of the gathering feasts. Other important contributions were made by the Treaty Relations Commission of Manitoba and the Government of Manitoba. Cree artist KC Adams was on site to assist people in making small ceramic smudge bowls from clay based on traditional pottery practices local to the area. The bowls were then fired in the earth. On the second evening of the gathering, each of the participants were invited to place ceramic bowls filled with water on the ground. The water was taken from the river that flows by the lodge site. The ceramic bowls took on the life and the spirit of the water, along with the spirit of those who were participating in the ceremonial placement of the water-filled bowls. The ultimate form of the installation resembled the turtle petroforms that are located in and around the site where the lodge was built. This illustrated the connection among the land, lodge, people, spirit and water.

Figure 9.2 Ceramic bowls at the ANI gathering

Source: Photo by KC Adams.

The journey of life

A natural consequence of assembling this group of knowledge keepers was a celebration of the natural gifts each of us brought. All of us have varied and equally beautiful ways of knowing and transmitting knowledge. Each brought their love for the water and commitment to working for the water. Each of the Elders brought to the gatherings their knowledge and understanding, shared through many generations. These songs, language, ceremonies and stories were not methodologies, but gifts. The relationship that was forged among each of the participants and the Elders, and myself as a researcher with the Elders, emphasized the importance of relationship in research. While I was acting as a researcher, I was also an important part of the evolving research, bringing my knowledge of the ceremonies, songs, stories and worldview into the research process itself. In order to better identify and understand subtleties of Indigenous knowledge I had to rely on my existing base of knowledge and ceremonial practice. Without being rooted in the tradition and the ability to identify the subtle messages within the messages, Indigenous knowledge can be misunderstood and de-contextualized:

> It is an ethic that can be seen in the decisions and actions of a community and that is contained in the songs that Native people sing and the stories that they tell about the nature of the world and their place in it, about the webs of responsibilities that bind all things.
>
> (King 2003, pp. 113–14)

The methodology itself evolved based on the spirit of the gathering and the place in which we gathered each time. The method we used might be seen by some as a hybrid between key informant interviews and participant observation. I would suggest that it is both a new and very old methodology, an Indigenous methodology that consists of being guided by Elders and ceremony and being responsive to the land. It is rooted in listening to those who have sacred, ancient and cultural knowledge and allowing them to define the process for learning. A discussion/talking circle was the starting methodology but it evolved to include the use of various forms of knowledge and knowledge transmission: ceremony, song, language, and story. I do not consider these methodologies: the methodology is in how space was made for all of these forms of knowledge and knowledge transmission to come together for a variety of purposes (teaching, accessing legal thought, disseminating knowledge). What some may view as cultural indicators (language, song, ceremony and stories) are sources for understanding legal principles. They help us understand how we govern our interactions among beings.

In many Indigenous communities around the world, research has come to be a bad word. It has been an exploitative and colonizing practice (Smith 1999). Doing more good than harm through research is a constant concern of mine. Indigenist research can help shed some of the layers of colonialism that have been perpetuated through research. It can be a process of coming to know. Kathleen Absolon suggests searching from our own location, using our own ways:

> The focus, topic and questions surrounding the re-search are relative to Indigenous peoples' realities. The research is by nature related to Indigenous peoples' contexts: historical, political, legal, economical, geographical, cultural, spiritual, environmental and experiential ... As we re-search, we re-write and we re-story ourselves.
>
> (Absolon 2011, p. 21)

The ANI research was participant-led and developed, a decolonizing method which in part addressed colonization through the starting point of discussing colonial law in contrast with *Anishinaabe* law. Assembling this group (the "faculty of Elders") and ensuring that their ways of knowing were able to take space and prominence allowed for the information and knowledge to flow. It is an Indigenous cultural and land-based methodology that was participatory and open. The methods of transmitting knowledge were arrived at organically. Ceremonial protocols took prominence in our water work; we made all accommodations necessary to ensure that we had the time to do the work properly and to include ceremony throughout. There was participation from multiple *Anishinaabe* lodge traditions, families, ways of understanding and individual gifts. The research design was responsive to direction in ceremony and participation in ceremony was encouraged throughout all gatherings. Ceremony can be an important and sometimes defining part of Indigenous research (Hart 2009). Knowledge was not removed from the minds and hearts of the participants—they continue to control their knowledge and their ability to withdraw their consent to participate is ongoing. All of the gatherings were situated beside bodies of moving water. Even coding, analysis and drafting of the interim report were done in close proximity to water.

Part of this decolonizing method aims to allow people to learn experientially rather than in a static, inaccessible and sometimes inaccurate form. It allows for individual and collective interpretation. The knowledge that was shared was not transferred in the sense of ownership. It remains *Anishinaabe* knowledge. However, these gatherings allowed others to be exposed to this knowledge and to experience ANI. Each of the gathering participants (including the Elders) found their own understanding and sense of responsibility. This understanding and responsibility will continue to evolve as knowledge and experience is acquired, and as the moments in these gatherings are remembered (including through the story book and other publications). What I had originally hoped would result in a clan-based law-making process was not the ultimate direction for the research. The format of gathering, providing mentorship, transmitting knowledge and encouraging people to engage in ceremony and action was the outcome, rather than overtly deliberative engagements and management initiatives. We have been sharing our knowledge and learning process at community engagements and conferences, including recently in New Zealand and Australia.

Figure 9.3 Importance of water at the ANI gathering

Source: Photo by Monica Mai.

Ceremony, song, language and story are means for assessing and identifying legal principles and helping us understand how we govern interactions among beings. The ability to extract legal theory and legal principles continues to be the role of academic research in collaboration with the faculty of Elders prepared to share their knowledge base. The research reinforced an *Anishinaabe* legal framework of decision-making and action based on responsibility (rather than rights and obligations). The next phase of research will consider the place of these legal theoretical underpinnings in a variety of decision-making processes. My hypothesis is that this framework may better serve collective interests than those aimed at individualism and protection of private property. That would require a societal shift and an acceptance (at some level) that Indigenous law and values may better serve common and collective interests. This work has in no way reached its maturity. I anticipate that it will be an ongoing, long-term process that will engage many generations into the future. We have no choice but to continue to be engaged and to act. Water is our collective responsibility. Water is life.

Acknowledgements

With many thanks and gratitude to the Faculty of Elders: Florence Paynter, Sherry Copenace, Darlene Courchene, Violet Caibaiosai, Allan White, Peter Atkinson, Harry Bone, Charlie Nelson, Dennis White Bird, Neil Courchene, and to the aunties from Roseau River who fed us so that we could do our work.

References

Absolon, K. 2011. *Kaandossiwin: How We Come to Know*, Halifax and Winnipeg: Fernwood Publishing.

Batiste M. and Henderson J. 2000. *Protecting Indigenous Knowledge and Heritage: A Global Challenge*, Saskatoon: Purich Publishing.

Busby, K. 2016. From resistance to recognition: Recent developments in Canada touching on international law and the human right to water and sanitation. *Canadian Journal for Human Rights*, forthcoming.

Deloria, V.J. 1996. *Red Earth, White Lies: Native Americana and the Myth of Scientific Fact*, New York: Scribner.

Ermine, W. 2000. A critical examination of the ethics in research involving Indigenous peoples. Unpublished Master's thesis. Saskatoon, SK: University of Saskatchewan.

Hart, M. A. 2009. For Indigenous peoples, by Indigenous peoples, with Indigenous peoples: Towards an Indigenist research paradigm. In R. Sinclair, M. A. Hart, and G. Bruyere (eds.), *Wicihitowin: Aboriginal Social Work in Canada*. Winnipeg: Fernwood Publishing.

King, T. 2003. *The Truth About Stories: A Native Narrative*. Toronto, ON: House of Anansi Press Inc.

Kovach, M. 2009. *Indigenous Methodologies: Chracteristics, Conversations and Contexts*, Toronto, ON: University of Toronto Press.

Mcguire, P. 2008. Restorative dispute resolution in Anishinaabe communities: Restoring conceptions of relationships based on dodem. Research Paper for the National Centre for First Nations Governance.

Napoleon V. and Friedland, H. 2014. The inside job: Engaging with Indigenous legal traditions through stories. In Lucero T. and Turner D. (eds.), *Oxford Handbook on Indigenous Peoples' Politics*. Oxford, UK: Oxford University Press.

Sinclair, N. 2013. K'zaugin: Storying ourselves into life. In Doerfler J., Sinclair N., and Stark S. (eds.), *Centering Anishinaabeg Studies: Understanding the World through Stories*. East Lansing and Winnipeg: Michigan State University Press and University of Manitoba Press.

Smith, L.T. 1999. *Decolonizing Methodologies: Research and Indigenous Peoples*. New York: Zed Books.

Stark, H. 2013. Transforming the trickster: Federal Indian law encounters Anishinaabe diplomacy. In Doerfler J. Sinclair N., and Stark S. (eds.), *Centering Anishinaabeg Studies: Understanding the World through Stories*. East Lansing and Winnipeg: Michigan State University Press and University of Manitoba Press.

Wilson, S. 2008. *Research is Ceremony: Indigenous Research Methods*, Halifax and Winnipeg: Fernwood Publishing.

10 Decolonizing intellectual traditions

Conducting research and telling our stories in a 'Mi'gmaq Way'

Interview with Fred Metallic

Fred Metallic is a member of the Listuguj Mi'gmaq First Nation, which is located in the south western part of the Gaspé Peninsula (Québec, Canada), surrounded by the Appalachian Mountains and bordered in the south by the Restigouche River (www.listuguj.ca). He wrote his PhD dissertation in Mi'gmaw, and defended it in his own community. Here he speaks with Jocelyn Thorpe, one of the editors of this volume, about his process of research and writing.

Jocelyn Thorpe: What were your objectives in researching and writing your dissertation? What did you want to know and what did you want to change? What did you learn in the process?

Fred Metallic: To answer those questions, I first need to talk about how I conducted my research. In 2005 when I returned to Listuguj, I began working in the community in the research sector of a comprehensive claims organization. Then, in 2008, I took a one-year leave from my work to conduct research with Elders and knowledge holders. It was necessary to take a leave for different reasons. First, on a practical level, I needed time to conduct the research and do the writing. But more than that, as a researcher it was important that this work not be enmeshed with the politics of land claims. As you can imagine, land claims can be very contested issues both within the community and outside. I felt it was important for participants to know that they weren't answering questions for a land claim that could be contested in a court of law. Often times, Indigenous peoples—Elders and knowledge holders—tell their stories in the context of a land claim or resource development project. Too often these stories are used to prove Aboriginal and title rights. However, I have heard Elders who have testified end up trying to educate judges on Aboriginal ways of knowing and understanding the world around them. Indigenous people not only have political rights to land, but they also have intellectual rights: the right to speak their language and to tell their own history.

In terms of what has been written about Mi'gmaq people, generally speaking, we have been the objects of study and our ability to speak for ourselves, about ourselves, and in our own language and in accordance with our own intellectual traditions, has been limited. Also, in terms of language, most of the writing in the academic community is in either French or in English. When I was conducting research for my MA thesis, I came across Eleanor Johnson's Master's thesis from 1992, which was written in Mi'gmaw. Johnson's work inspired me, and then in 2002, I wrote my MA thesis entirely in the language.

For the research that I conducted for my PhD dissertation, entitled "Ta'n Teli'gji'tegen 'Nnu'igtug aq Ta'n Goqwei Wejgu'aqamulti'gw," my objective was to address the gap in knowledge about Mi'gmaq people—our political history within our territory from our own critical perspectives and in the language of our nation, Mi'gmaw. As Mi'gmaq, we have our own history, political traditions, and intellectual traditions. As a researcher, my first objective was to listen to the Elders and knowledge holders and facilitate a process whereby they could express their own understanding of Mi'gmaq history, rights and responsibilities, and our changing relations with our territory and with each other. Then, I had to represent what I heard in the writing of the dissertation.

When I started my research, I already had a well-established relationship with Elders and knowledge holders in the territory. I established a committee in the community comprised of well-respected knowledge holders and linguists. This committee provided guidance and advice on how best to present myself to Elders, how to listen, and the ethical protocols of conducting research in the communities. This committee was in addition to the academic supervisory committee that I also had in place at York [University in Toronto, Canada], which was comprised of academics and one community Elder.

Before starting my research, I submitted my proposal to the Mi'gmaq Grand Council Ethics Watch and I presented my proposal to the Gespe'gewa'gi Tribal Council, and I received approval from the former, and a letter of support from the latter. Even though I had already successfully received permission from the university to conduct research as per the Tri-Council guidelines, I also believed it was important for the communities and their representatives to know about the research that I was proposing to conduct, and that they should have some say as to whether or not that research should be conducted in the territory. In my presentation to the Tribal Council leadership I explained my objective for the research and also how I was planning to proceed, in particular around the use of Mi'gmaw language. The leadership emphasized in their letter of support that the Mi'gmaw knowledge shared in the Mi'gmaw language would not be translated without the expressed written consent of the Mi'gmaq.

Then I began to speak with, interview, and participate in many different ceremonies and gatherings in the Gespe'gewa'gi district. I interviewed individuals in their homes, and many of these interviews were videotaped. It was important to videotape interviews because many of these Elders were getting up in age, many of them only spoke Mi'gmaw, and I felt that they should have something from the interview, some testimony that the family could retain in the process. Also, language is disappearing quickly and so it is important to have oral records of our language and Elders speaking. I wanted to get a deeper understanding of people's life histories, their views and perceptions about Mi'gmaq rights and responsibilities to territory, environment, and relations with each other and also with the Canadian government. I conducted a total of twenty-five interviews, the majority of them in Mi'gmaw.

When I think about "what did I want to know," for me it had more to do with the process of research. It was less about wanting to know something particular, and more about different ways of knowing and establishing a relationship with Elders and learning to respect their knowledge and distinct ways of knowing and sharing knowledge. I was very fortunate to hear their stories in our language, about the challenges they faced. Many spoke about times when government policy was overtly violent and coercive in comparison to what it is like today. I had grandmothers tell me about how frightened they were to go out at night as kids fearing that the Indian Agent would "grab them" and send them to residential school, knowing what was happening to kids who were sent there. I heard grandfathers talk about having to fish at night, without lights, knowing that someone was out there was looking for them, trying to catch them; that they were being criminalized for trying to take care of their families. These are powerful stories in our language. Too often these stories are sanitized when translated to English. It's common now to talk about the "residential school era," but it's different to hear in the language about the fear and the shame people felt.

I travelled throughout the territory, speaking to Elders, listening to their stories about the past, relations today, and what they hope for in the future. In some ways, the story of Gespeg'ewa'gi, the dissertation, unfolded through the Elders' telling of their own experiences and understandings of the colonial imposition and of our continued assertions of our rights and responsibilities to our territory and also to each other. So what did I want to change? I would say that the story that I wanted to present would not be a story of "emptiness and inferiority," whereby we are just victims of history, assimilated and conquered. I did not want to reproduce the story or textbook about the impact of colonialism. I wanted to show that we actually have an understanding and vision for our territory and that it involves many of our language speakers, knowledge holders, Elders, and that there is a story in Gespeg'ewa'gi that is distinct from the story that we hear in schools or from the Canadian government.

I learned that our stories not only capture the violence of colonization, but that they also embody resilience. What I found in speaking to Elders was that even when the Elders were sharing their stories about what had happened, their stories were coming from a place of strength. I never once felt that people were making excuses or apologizing. Even though they talked about what life was like and how government policy had impacted their lives, they still spoke from a place of strength. And, I guess for me I wanted to know: how do you keep that sense of strength within you to continue to resist and to create some resurgence so that we can have a better relationship than the one we're having today with the people from outside our communities? That is what I wanted to know and what I wanted to change and it was really about telling a different story than the stories that we have been taught.

Jocelyn Thorpe: Were there other reasons you wanted to write your dissertation in Mi'gmaw, and what were the challenges and/or rewards in that process?

Fred Metallic: In the community, there is resurgence to reclaim land, our language, our history, our culture and identity. Around the time I was conducting my research, the Tribal Council passed a Declaration stating that Mi'gmaw is the official language of the Gespe'gewa'gi territory. The majority of the interviews that I conducted with Elders and knowledge holders were conducted in Mi'gmaw and I had already written my Master's thesis in Mi'gmaw. So I really had some strong convictions about presenting the work in the language. I consider myself fortunate to be a fluent speaker. For me, because I have the language and the academic training, and also the support from community and leadership, I felt it was my responsibility to do the work in the language because I have access to the language. And I often hear that we all have our bundles and we all have our gifts and no two individuals share the same gifts and no two bundles are the same. And I think for me, doing it in the language with the community, with the leadership, with the university, was really an attempt to bundle: bundle this and to use this as a tool, as medicine, to heal the past and to forge a new relationship where the principle of respect is paramount. After listening to the people share with me their experiences, their hopes, and their aspirations I couldn't then translate their words and stories into English.

In terms of the challenges, although I am a speaker, I don't have formal training in writing and reading Mi'gmaw. All of my formal education has been conducted in English and most of my work has been in English. To now write in Mi'gmaw required me also to relearn how to read and write in my own language. I worked with Elders and linguists who helped me in the writing process. Our language is very complex. There are numerous prefixes and suffixes connected to root words that convey complex and particular meanings. We have animate and inanimate words. Most words are verbs and communicate actions rather than objects or the privileging of "I" in conversation. It was difficult to write in Mi'gmaq knowing that sometimes my thinking was still governed by the cognitive rules of the English language. What helped me in the writing was the fact that I started to produce films about people's experiences based on their testimonies. The images and the songs that I used in the films really helped me to communicate on the page and make that transition into a verb-oriented world. The actual dissertation is a written text, but in development is a sort of multimedia work because what you see on the page is not the whole product. I'll talk about that later on.

Jocelyn Thorpe: Can you tell us about your dissertation defence?

Fred Metallic: When my supervisor, Anders [Sandberg, co-editor of this volume], and I discussed the dissertation defence, I came to the realization, or I guess we both did, that the defence would have to take place in the territory. But the decision wasn't just made by Anders and me or the university; it was also decided by an Elders' Council convened in the community of Listuguj to discuss the location of the defence. One of the things that I found out early on in the process was that even though I as the researcher was conducting the research, the community had a lot to say about who I spoke to, how I would speak to people, and community members actually were more involved in governing the project than just me. So, when the issue of the location of the defence came up, there was an Elders' Council convened and I went there and I gave them my presentation again and they concluded that these are stories that come from the territory and that these stories needed to be shared in the territory. It was their way of saying that the defence had to take place in the territory. Once the decision was made about the location, it then required the academic members of the committee to travel from Toronto, and also from Winnipeg, to take part in the defence. Yet, by having the defence in the community, community members also had the opportunity to participate. In addition to the seven-member academic committee, there were also: Community Elders' Committee members; members of the Gespe'gewa'gi Ethics Committee; members of the Grand Council, which is the traditional political body of Mi'gma'gi; Chief and Council; and linguists. At the actual defence there were twenty-one people formally sitting at the table. In addition, surrounding this circle there were about a hundred to a hundred twenty-five people. Many were research participants and they participated as listeners in this defence. The defence was open to the community so there were many different people in attendance, including children and youth. We felt strongly that it was important for community members to participate in the process. In my community, similar to other First Nations, knowledge is still collectively (rather than individually) held, so participation in all aspects of the research process is important to maintaining our collective intellection oral traditions as Mi'gmaq.

The defence started around noon with a song and prayer. I went through each chapter. The external examiner translated questions and answers for English-speaking members of the committee and for all English-speaking participants. That portion of the defence lasted about five hours. The committee then convened and during this time a feast was being prepared. The announcement was made by Dr. Anders Sandberg that I had successfully defended my dissertation, and after that there was a feast. What I remember the most about the event is the amount of support from the people: from those who organized the venue, the Elders who prepared the food, the women who sang the songs and offered prayers, the young men who drummed, the young women who danced. There were also those who collected gifts that I later gave away at a giveaway. Even though the ceremony started around twelve o'clock in the afternoon, we didn't conclude until about ten or eleven o'clock at night. My overall feeling was that I had the opportunity to honour the stories that people had shared by ensuring that the defence was conducted in our language, in accordance to our traditions and cultural ways, respectively—both the academic community and also our own processes as Mi'gmaq. I really felt that it was about balancing our respective traditions and it felt that we had accomplished something together. There is a lot of commentary in First Nations communities about the relationship between educational institutions and First Nations' educational needs, and that relationships between institutions and communities needs to be strengthened. We need to have a better relationship with input from the First Nation community itself. I really felt that we had accomplished something together because it took many people from different places to put in the effort to change policies, to change regulations. For the community itself to have the university come here and participate was really an opportunity for us to get to know each other, and at the same time not have these differences become problematic in regards to conducting the defence based on these joint traditions.

Jocelyn Thorpe: Wow! What an exciting and inspiring event! What are your thoughts on academic research more generally? Is it possible to decolonize research related to nature/culture/history? If so, how?

Fred Metallic: I think that as a researcher it is important to respect the various traditions and protocols that govern relations within each Indigenous community. It is important to work with and recognize each community: their language, their unique history, their experiences with colonization, and their own aspirations. It's interesting because when I was reviewing the literature as to what's been written about the Mi'gmaq here in Mi'gma'gi it seems like there's a lot written by and about the Mi'gmaq in the southern portion of the territory, but there's not much literature about the Mi'gmaq in the northern portion of the territory. There's a lot written towards New Brunswick, Nova Scotia, P.E.I. [Prince Edward Island], and Nova Scotia is producing a lot of material, but we didn't see anything coming out of this part of the territory. So even though people are writing about the Mi'gmaq and people are reading about the Mi'gmaq, we felt that this part of the territory wasn't represented in the literature. Also, you've got to keep in mind that colonization unfolded differently in communities and the story that emerges from here is a little bit different than the stories that are being told in other parts of the territory. It was important for us to tell that story from here.

I have learned that it is possible to "bring in" the academic community to the process and find ways to creatively push boundaries and forge new anti-colonial relations. The academic community had to trust our process, and leave the "ivory tower" in Toronto and travel to our territory. Decolonization is not always easy; it takes time and patience. You have to be willing to listen and invest yourself in the process.

Jocelyn Thorpe: What are you up to these days? How does this connect to your PhD and/or decolonization?

Fred Metallic: I am currently the director of Natural Resources for the Listuguj Mi'gmaq Government. Our mandate is to develop a sustainable resource management regime. We are striving to reconnect people with the land, to take a more balanced approach in the way that resources are managed in the territory. We are also striving to develop resources, to work towards self-governance, and to rebuild our nation's economy and our wellbeing. I am also a part-time instructor for Cape Breton University in their Indigenous studies department. In 2013, the community of Listuguj started to offer an undergraduate program in Community Studies and I have been part of this program since the beginning. I teach undergraduate Mi'gmaq studies courses at the year one, year two, and year three levels. This work is very much connected with Indigenous aspirations for resurgence and rebuilding. I teach contemporary Indigenous thinkers such as Kiera Ladner, Leanne Simpson, Taiaike Alfred, Jeff Corntassel, Glen Coulthard, Sakej Henderson, Marie Battiste, Bonita Lawrence, Pam Palmater, and Stephen Augustine, to name a few! In terms of the approach, the students have also participated in talking circles, ceremonies, and feasts. We strive to offer accessible and relevant education. All the students are Mi'gmaw, and most are from the community of Listuguj. To me it's very rewarding to be able to teach in the community. Too often, our students need to travel to urban centers to attain a formal university education. Today, we are starting to offer university programs in the community with local instructors. The hope is that students will stay in the community, that the education that they receive will meet the needs of the community, and that we will "bridge the gap" between Indigenous and Western ways of knowing: this is decolonization. We can re-centre Indigenous learning and students have an opportunity to learn our history, our responsibilities and protocols, and they remain connected with their families, yet they also learn skills and traditions from Western traditions. They learn how to write essays and so on.

In Mi'gmaq, we say 'tet tle"eiultieg,' which means this is where we belong, in our territory. And it's one thing to say that; it's another thing to be connected through knowledge of a territory, with knowledge holders who have a very distinct understanding of that relationship and to have them share so openly and so freely their experiences and their own challenges, yet at the same time be happy to share that with you. It really helped to strengthen relations in the territory and it really, for me, helped me to understand what people are fighting for, what people are trying to defend. It's unfortunate that the way we conduct research doesn't necessarily allow for those very distinct stories to be told. Today we talk in English as though it's a neutral language and it's still very difficult to accurately convey stories from the territory that reaffirm or reinforce that sense of belonging and that sense of connectedness. And so we work with the English language and try to manipulate words so we can try more accurately to convey our vision for land, our vision for territory, our vision for our people, but I think it's been very difficult to put in words what that dissertation defence meant. I think all of us who took part in it and spent that whole day going through the process, we all walked away with a greater appreciation of that bundle that we were carrying. We all walked away with a sense of conviction about changing the relationship and learning to understand differences and appreciate those differences and not to be so hegemonic about things. And really try to create spaces and find spaces where we can have more creative expressions about things that we research and how we present that work. I mean at one point we contemplated whether or not this should even be written, and that would have been interesting but that would have been pushing the boundary too far for some people, but not for us. So we did settle on writing something.

As I mentioned, my project now is to try to transform the dissertation into a multimedia piece because there are about half a dozen short films produced about different topics. There was a twenty-minute film on residential schooling based on the testimonies that people provided and it's a powerful piece, but I want to put everything into one product, if I can just use that term. People want to hear, people want to see, people want to connect, and I think I'm going to try to do something with that piece that brings all those aspects together. It's interesting when we were getting ready for the defence, a few months before, the committee needed to see the text and I had to provide audio CDs so people could read and listen to the text and there were a lot of little things that needed to be done to ensure that people were engaged, that people understood what the conversation was going to be about and it was quite an experience. A lot of policies had to be changed. There was a lot of advocacy, and there was a national discussion going on, centred at York. There was a territorial discussion going on at the same time and the territory had to make declarations, big policies on research to ensure that there were provisions in place to secure access and success in research. So sometimes even though it seems the university is advancing in terms of having a better relationship with reconciliation, it's not just about that. It's also about First Nations making changes and being able to work with non-Indigenous researchers who are coming in. We all have to make adjustments in the way we conduct and participate in research. It is really interesting to look back at it now, all the things that needed to be done to ensure that when the defence came that we were governed by the principle of respect, and that differences were good. There's nothing wrong with that: being different. It's just that we need to create more spaces where we celebrate those differences.

Jocelyn Thorpe: It seems as though so much of the work that you're doing is focused on Indigenizing universities, for example by having university classes in communities and things like that. I wonder about your defence being an example of Indigenizing universities too, showing for example that speaking Mi'gmaw and respecting multiple ways of knowing are important. How were you able to have that conversation when not all of the academic members of the committee speak Mi'gmaw?

Fred Metallic: Every dissertation has to have an abstract and I conceded to provide an abstract in English. But I also provided a summary of each chapter in English before the defence. The external examiner was Stephen Augustine, who was the curator at the Museum of Civilization at the time. He is now the Dean of Unama'ki College at Cape Breton University. And so he was tasked with the duty of translating every question from English to Mi'gmaq, Mi'gmaq to English. I mean, I did answer some questions in English, but for the most part I wasn't obliged to answer in English. I was free to express myself in Mi'gmaq and it was nice to know that somebody else would be charged with the duty of having to translate my thoughts and words so that others could get a sense of what I was trying to communicate not only on the page but also in answer to those questions.

We made every effort to try to be inclusive. We made every effort to try to respect the traditions and the protocols and the procedures, whether it was producing an abstract in English, or producing summaries in English, or providing a translator. Interesting comment though, I'll share with you, at the defence. There was a member of the Grand Council who did ask, "Why is it that we have to have these defences and have English representation?" I mean I don't know how else to put that but he was saying, why can't we have these kinds of defences in our communities and so that we can evaluate ourselves how individuals will be positioned as knowledge holders. Because with knowledge comes power, they know that. And if you're going to be a knowledge holder who kind of has the authority to speak, they want to know who it is, they want to know what they're going to be speaking on, and from what position. And they were suggesting that maybe in the future we should have non-Indigenous researchers come to our community and explain in the Mi'gmaw language the work that they're going to be conducting and how that work is going to benefit Mi'gmaq people. Too often work is produced for the benefit of the general public, but Mi'gmaq understand who the 'general public' is: it's not usually them.

For those who did not have the advantage of speaking Mi'gmaw, and this includes both Indigenous and non-Indigenous peoples, we knew we needed to bridge the relationship. We needed to make sure that people who thought they were being excluded were being included and that they were participating in a Mi'gmaq defence and that we weren't trying to be dismissive. We were simply trying to acknowledge in the way we did that this is not about "us and them." It's really about adherence to the principle of respect and being inclusive, which is the nature of our ways: to be inclusive, and I think we've demonstrated that historically. We've never been about being exclusionary because otherwise nobody would have survived getting off those ships; they would have disappeared.

And I think other people have said one of the reasons we're losing our language is because we're too polite. You can have ten people speaking Mi'gmaq in a room and one English person walks in, and all of a sudden we start speaking English. That's the dominance of the English language and the dominance of certain ways of knowing. And so people—even though people say, "Well, I don't understand what you've written,"—they were able to participate, they were able to ask questions. I mean there's a lot of stuff written about Mi'gmaq people in the territory and nobody ever once asked Mi'gmaq, "Do you mind if I write this in English?" Nobody's asked that question before about being exclusionary and now that we're starting to write in Mi'gmaq, people are making allegations that we're trying to exclude people. Yet you know we have these public defences that allow people to ask questions and individuals could have raised their hand and spoken and said, "You know what, I don't agree with Mr. Metallic's opinion. I don't agree with Mr. Metallic's research. I don't think he conducted that research ethically. I don't think he did this, this, this, this, this. And therefore I don't think this defence should continue any further." You know that's a risk, but because of the responsibility, and I don't mean power in a hegemonic kind of way. That's the responsibility that comes with being a knowledge holder of the territory. Your words carry a lot of weight and your words can affect a lot of people at the most intimate level and you have to be careful in how you speak, how you present yourself, how you present the words on the page and the stories that you tell. It really has to reaffirm that sense of belonging. And it really has to do something about connecting people together again because what has happened in our territory because of colonization is that we've become very fragmented and we've become very protective in what we include in our bundles. We're trying to open the space back up so that we can talk about these bundles and our bundles so that we can continue to have better relationships and stop making allegations that are not healthy for us.

Jocelyn Thorpe: Well you've been clearing this path for a long time. Did you do your master's degree at Trent University?

Fred Metallic: Yes.

Jocelyn Thorpe: And what was it about?

Fred Metallic: It is entitled "Mi'gmawei Mawio'mi: Ta'n Teliassumsin aq Assumsultigw Listugujeweigatig." It is a story about political freedom, our collective freedom, and it's about having a conversation around leadership in the community and how we deal with distinction between personal freedom and our collective freedom. As people say, "We have a right" or "I have a right," and I agree. I've found myself standing on the front lines at the blockades. I've been a fishermen all my life and so I'm very well entrenched in this discourse on rights, and I'll be the first one to stand there and say, "We have a right to fish." But I also asked the question, "What's our collective responsibility to the fishery?" Yes, we have a right. But what's our responsibility? And by asking the question about responsibility … Well, let me ask you: how do you exercise your freedom as an individual as a Mi'gmaw? How do you live that, you know? And I'm not just talking about taking whatever you want, whenever you want it. This is, where's the line between what you need and what we require? And it's an attempt to try to get a conversation going in the community about individual rights and our collective freedoms and how we protect those freedoms. And then from that I think I learned a bit about how to write in Mi'gmaw, how to conduct research, how to present the research. And that thesis was presented at Trent and, incidentally, soon after that defence the policies at Trent changed. The new policy around language and graduate work, in my view, does not encourage students to explore and to conduct their graduate work in the language. The policy at Trent states that there needs to be faculty fluent in the language (reading and writing) who is able to supervise, that there be fluent internal and external examiners, and you have to give the university one year advanced notice if you are submitting your thesis/dissertation in a language other than English or French. There are conditions stipulated that I wouldn't want to go through. It's easier just to go to my committee, do it in English, jump through the hoops and take the well-established path.

But I've learnt from that: for me, Indigenous research is about relationships. It is important to connect with others in the academic community, in First Nations communities, with language speakers, with knowledge holders. It is important to establish connections and establish good working relations between Aboriginal communities and educational institutes. In my PhD work, what is interesting is that my supervisor does not speak the Mi'gmaw language, yet we were able to mitigate those barriers and learn to share our understanding and knowledge with each other. In Mi'gmaw, our language really helps you to understand that it is the space between the two, not the endpoints, that is important. It's not you, it is not I, but it is the relationship that is more important: the language privileges relations. Conducting research through the language is really an attempt to try to epistemologically push our relations so that we all experience that connection while research is being conducted, whether you're a speaker or not. What is more important is the relationship. I'm very thankful to have had Anders [at York] as my advisor, my supervisor, my mentor, and really my strongest advocate because if it weren't for Anders I don't think I would have gotten here. And I probably don't give him enough credit but he deserves more credit because he was willing to support a process without having the language.

Jocelyn Thorpe: I was reading Shawn Wilson's *Research as Ceremony* and he talks about how in Cree there's no word just for grandmother, that it's either "my grandmother" or "your grandmother." I was thinking about that as you were talking about language and relationships. The thing, or in this case the person, doesn't exist on its/her own, and that's such a different way of understanding than in Western thought, where the individual appears to be the central unit of importance. Similarly, in Western forms of research, the focus seems to be on the outcome or goal, and it's almost like it doesn't matter how you get there, although of course that is changing, thanks especially to challenges to dominant paradigms made by Indigenous researchers and community members. You seem to be saying almost exactly the opposite, that it doesn't really matter what the outcome is, that it's all about the process, the relationships, and the working together.

Fred Metallic: I have a colleague who's doing his PhD at the University of Waterloo and looking at the fishing communities in this region. And I give him some advice and sometimes he shows up and says, "Oh I want to talk to this one," or "I want to talk to that one," or "I want to meet with this group," and I say, "Hang on, hang on, hang on. Let's go talk to some people." It's taken about a year and a half. He's gotten to know the people, people have gotten to know him. People know what he's researching, why he's researching that, what he's trying to accomplish and it's taken a year and a half for him to have a relationship and to understand that research is not just about knowing. It's about how you come to know and what your responsibility is now. Once you've established that relationship, people don't share stories with you so that you can become the knower. People share stories to strengthen that relationship and to have that mutual obligation to each other: that we are now both invested in this. They say that there is a responsibility, you know. If I share a story with you—now you know and, because you know, you can't act like you don't know anymore. You can't be ignorant anymore. Now you're responsible to act on that. You're not just responsible to profess and to know, you have a social responsibility with that knowledge now in that you have to act on that. It's really to try to help your spirit grow and mature and to be able to carry the bundle that you carry. And it's really about that. It's not about just being the knower and the status of knowing, it's really about the responsibility with those teachings that are embedded in those stories and for you to be more socially responsible about the bundle that you hold.

11 It matters where you begin

A (continuing) journey toward decolonizing research

Jocelyn Thorpe

In my last big research project (Thorpe 2012), I ended at a beginning. I concluded that the Canadian wilderness is a product of historical power relationships rather than of timeless nature. One culture of nature—a settler colonial one—disrupted another, Indigenous, system. For members of the Teme-Augama Anishnabai, my ending was simply an obvious starting point. N'Daki Menan, or "our land," existed before Canada: a homeland, not a wilderness, n'Daki Menan, not the Canadian nation.

It was a long journey, and a lot of work, to arrive at a beginning. The self-evidence of n'Daki Menan did not in fact seem self-evident to me, a non-Indigenous person who grew up in what I took for granted as part of Canada. But I should not be surprised by the journey I took to get to what members of the Teme-Augama Anishnabai knew all along, for I am well versed in feminist theory that highlights the radically contingent character of every part of what we understand as our "selves": our bodies, our ideas, everything. In Simone de Beauvoir's words, "the body is not a *thing*, it is a situation" (de Beauvoir 2009, p. 46), and Judith Butler argues that bodies become gendered through the "stylized repetition of acts through time" (Butler 2007, p. 188). What we understand as self-evidently *there* is self-evident only because of its—and our—situation in place and time. Canada is invented, but that does not make it a pretend place.

Acknowledging the politics of location, as Adrienne Rich (1994) calls it, does not mean forgetting about our individual and collective responsibility for knowledge production. Instead, it means accountability, rejecting the "god trick of seeing everything from nowhere" (Haraway 1988, p. 581) in favour of recognizing and acting from the understanding that all perspectives are partial, embodied physically in historical time and geographical space. Not all partial perspectives, however, possess the power of the status quo. The story of n'Daki Menan as part of the Canadian wilderness, with its mining, forestry and tourism activities controlled by the state, is a story with teeth, a story powerful enough to negate the Teme-Augama Anishnabai's claim to the same territory, understood differently. It remains important to tell stories about the history of the status quo, to reveal it as a partial perspective rather than an obvious truth, because its power rests in its appearance as the norm. The Canadian wilderness may have been my story, but it was not my story alone. Because it remains a powerful story, I need to keep un-telling it.

Yet there is a danger in endlessly repeating the same story, even from a critical perspective. It re-centres colonial encounters—the "beginning" of the story, as many of us learned—as the main event, and in so doing re-centres the privileged bodies for whom terms such as Canadian wilderness signify not dispossession, but home. Bodies such as mine.

It matters where you begin. Cree scholar Shawn Wilson puts forward the position that foundational to Indigenous research methodologies is relational accountability, premised on the underlying perspective that "reality *is* relationships or sets of relationships" (Wilson 2008, p. 73). In *The Truth About Stories: A Native Narrative*, writer and academic Thomas King convincingly argues that the "truth about stories is that that's all we are" (King 2003, p. 2). While it may appear that an understanding of reality as stories is quite different from a consideration of reality as relationships, they are two sides of the same coin. Stories make possible or prevent different kinds of relationships from forming; it is *relationships* that stories can make, or break, or hold. Literature professor Ted Chamberlin indeed contends that this is what all stories do: "hold some people together and keep others apart" (Chamberlin 2004, p. 227). Neither is it all about people. The story of the Canadian wilderness, for example, authorizes some relationships between people and land while outlawing others.

Beginning with relationships is different from beginning with wilderness. When I began with wilderness, I found hidden colonial relationships and the understanding that the subject I thought of as me, the person curious about the history of Canadian wilderness, was as much a product of those colonial relationships as was the region I set out to study. I ended at a beginning because I had the privilege of not losing my land, of not being taken away from my family and sent to residential school, and therefore of not understanding that wilderness is always already a relationship of theft and appropriation. I had to ask about wilderness because I could not see the ties that bound it to the history of colonialism, but the ties were there all along.

With relationships, there is less pretending. And as Indigenous people have said for generations, we have serious relationship problems in the territory now known as Canada. Recently, the Canadian Broadcasting Corporation (CBC), decided to close comments on stories about Indigenous people until it has had time to review and change how comments are moderated (Fenlon 2015). CBC's acting director of digital news explains that while a number of subjects elicit "worrisome" comments, stories with Indigenous content result in a disproportionate number of comments that violate the CBC's guidelines for respectful debate that does not cross the line into hate speech and personal attacks (Fenlon 2015). Attacks against Indigenous people do not take only the form of hate speech, but of physical violence as well. Indigenous women are six times and Indigenous men seven times more likely to be murdered than non-Indigenous women and men, and Indigenous women made up twenty-one per cent of murdered women last year, while comprising approximately four per cent of the population (D'Aliesio and Grant 2015).

The statistics are becoming more widely known—particularly about the nearly twelve hundred Indigenous women and girls who have gone missing or been murdered in the last thirty years—and the numbers are staggering, but there are also hopeful signs of change.[1] The newly elected federal government has followed up on its election promise to launch an inquiry into missing and murdered Indigenous women and girls, something families have identified as a necessary step toward a society where Indigenous women and girls can live without the threat of violence ("Families Hopeful, Advocates Unsure" 2015; Mas 2015). Family members of missing and murdered Indigenous women, as well as Indigenous women artists and activists, have created an enormous shift in mainstream consciousness from one of ambivalence at best to one of indignation, concern and desire for action (Sterritt 2015). In a packed conference hall in June of 2015, the Commissioners of the Truth and Reconciliation Commission on Indian Residential Schools released their findings and calls to action. Of the many standing ovations commissioners' recommendations received from an audience full of residential school Survivors, their families and supporters, and interested Canadians, the recommendation of an inquiry into missing and murdered Indigenous women received the largest, longest ovation (Fedio 2015). The recent launch of the first stage of this inquiry, which includes listening to families' views on the design and intended outcomes of the inquiry (Mas 2015), indicates the success of Indigenous women in ensuring that this issue receives the attention that it deserves. Walking With Our Sisters is just one example among many of an action led by Indigenous women that has helped move us to this time of hope and potential change. Walking With Our Sisters is a commemorative art installation made up of a pair of moccasin vamps (or tops) for each missing or murdered Indigenous woman. Viewers take off their shoes and walk beside the beaded vamps, which are arranged on the floor in a winding path. Like each woman who has died or disappeared, each set of vamps is unique, and the sheer number of vamps calls attention to the scope of the loss, with the unfinished character of the moccasins mirroring the unfinished lives of the women they represent ("A Commemorative Art Installation" 2015).

It seems only to make sense that an inquiry into missing and murdered Indigenous women would place at the centre the very people most affected by these women's disappearances and deaths. Yet when it comes to other forms of research related to Indigenous peoples, the track record of non-Indigenous researchers has been very poor. Research by and large has perpetuated rather than challenged colonial relationships, in part through research methods that claim objectivity but in fact are rooted in western epistemologies that recognize as valid only culturally specific forms of knowledge (Smith 2012). Cree/Saulteaux scholar Margaret Kovach contends that one political challenge of conducting Indigenous research "is dealing with the … fundamental epistemological difference between Western and Indigenous thought," a difference which makes research challenging for Indigenous scholars who wish to have their work guided by their own cultural epistemology, but find they cannot when it is excluded from university settings (Kovach 2009, p. 29). This is why, as Sami scholar Rauna Kuokkanen argues, rather than "mainstreaming" Indigenous students, universities need to mainstream Indigenous worldviews and philosophies (Kuokkanen 2007, p. 2). Yet by suggesting that alternative frameworks for knowledge exist, Indigenous methodologies set up a challenging task: encouraging non-Indigenous scholars "to engage in reflexive self-study, … to adjourn disbelief and, in the pause, consider alternative possibilities" (Kovach 2009, p. 29). Nishnaabeg writer and scholar Leanne Betasamosake Simpson thinks that the effort required to "Indigenize the academy" would be better spent "valuing and recognizing our individual and collective intelligence on its own merits and on our own terms" (Simpson 2014, p. 22).

Simpson asserts that the colonial refusal "to recognize and support the validity, legitimacy, rigor and ethical principles of Nishnaabeg intelligence … should be met with Indigenous refusal" to fight for inclusion into an inherently flawed system (ibid. p. 22). Cracks in the flawed system exist, however, for example in the forms of postcolonial studies, gender studies, Indigenous studies, and indeed environmental history, all areas of scholarship that exist today because of the efforts of scholars and students in the past to expand our collective understanding of the nature of knowledge, who can produce it, what it is that we need to know, and why. Underlying the challenges and changes to western academic knowledge is the question of who is "we"? In his piece called "A Place for Stories," historian William Cronon grapples with the challenges that postmodern theory poses to the storytelling work of environmental history, asking whether "nature and the past [are] infinitely malleable in the face of our ability to tell stories about them" (Cronon 1992, p. 1370). He concludes that they are not, and that our storytelling practices are bounded in at least three ways, one of which is that historians "write as members of communities" but also "for a community of other scholars—some very different from me in their backgrounds and biases" (ibid. p. 1373). According to Cronon, this community, the "we" he implies, is in the best position to remind him of the "excluded facts and wrong-headed interpretations that my own bias, self-delusion, and lack of diligence have kept me from acknowledging" (ibid. p. 1373). But for him the role of the community extends beyond a policing function: "We tell stories *with* each other and *against* each other in order to speak *to* each other" (ibid. pp. 1373–4).

Following Cronon's lead, I might recognize that the "we" of this chapter includes fellow researchers, whether students or faculty members, interested in questions of decolonizing historical research that considers relationships between human and nonhuman beings. And although "we" too may be quite different from one another in our backgrounds and biases (and indeed, "we" *need* to be diverse in order to notice one another's biases), we might agree that decolonizing research is vitally necessary, tied as it is to Indigenous sovereignty and futurity (Tuck and Wang 2012, p. 35). Surely "we" do not want to conduct research that maintains injustices rooted in colonialism, even as some of us have benefitted from such injustices. The reality of missing and murdered Indigenous women makes it impossible to forget that lives are at stake; as Minister of Indigenous and Northern Affairs Carolyn Bennett recently articulated, "racism and sexism in this country kills" (Smith 2015). Given our diverse backgrounds (and biases), however, we need to think seriously about who to listen to and how to listen when it comes to decolonizing nature-culture-history research. If colonialism has operated in part through the creation of a narrow "us" understood by that "us" as superior to the "them" or "other" (see, for example, Said 1978), then decolonization must challenge that pattern by exploding hierarchical notions of self and other, and by taking seriously the voices and knowledges of those "othered" by colonial knowledge systems.[2] As feminist poet and scholar Audre Lorde has memorably written, "we have, built into all of us ... old structures of oppression, and these must be altered at the same time as we alter the living conditions which are a result of those structures. For the master's tools will never dismantle the master's house" (Lorde 1984, p. 123).

Just as it only makes sense for Indigenous women, families and community members to be at the centre of the inquiry into missing and murdered Indigenous women, so too it only makes sense for Indigenous scholars and methodologies to be at the centre of decolonizing research. Decolonizing approaches to research, including in my own past work, that use critical theory to take apart colonial discourses may reveal how colonial relationships operate and become naturalized, but they leave nothing in their place and tend to centre the very histories and power relations they mean to challenge. Deconstruction is necessary, but so too is building an elsewhere, or pointing to elsewheres that already exist. Beginning with an Indigenous paradigm, or "relational worldview," is one way to listen to and respect the lead of Indigenous scholars who have much to teach all of us about Indigenous thought and approaches to research that are "not extractive and [are] accountable to Indigenous community standards" (Kovach 2009, pp. 34, 29). I am well aware of the danger of "settler moves to innocence" in which settlers like me attempt to assuage feelings of guilt about their/our complicity in colonialism without having to give up land, power, or privilege (Tuck and Wang 2012, p. 10). And yet I hope there is room for settler scholars to think *with* Indigenous scholars in ways that do not aim for innocence or reinforce "settler adoption fantasies" in which "the Native (understanding that he is becoming extinct) hands over his land, his claim to the land, his very Indianness to the settler for safe-keeping" (Tuck and Wang 2012, p. 14).

I begin again, then, not from a place of innocence, or wilderness, but from a place of embeddedness: embeddedness in the power relationships I seek also to change; embeddedness in the place and time that I inhabit and cannot live outside, even as I focus the energy of my work on building with others a decolonized world I cannot quite imagine, in which human and nonhuman forms of life can truly flourish; and embeddedness in relationships with humans and nonhumans that are a product of our location in time and space and that also produce us, often in deeply uncomfortable ways. (I do not know all the animal and human labour-related sources of my breakfast this morning, never mind of my outfit. My white daughter has a better chance of surviving to adulthood than does my neighbour's Indigenous daughter of the same age. These are truths and privileges I do not want, yet which are not easily or individually undone.) Embeddedness and guilt are not the same thing. The latter is immobilizing, not to mention self-centred. The former is merely a recognition of the complex ways that (colonial) power works through our bodies. In her book chapter on the body and the reproduction of femininity, feminist philosopher Susan Bordo adopts a Foucauldian analysis of power to argue that we embody gendered power relations through the organization and regulation of time and space and our movements in and through these domains (Bordo 2003, p. 165). She thus views "our bodies as a site of struggle, where we must *work* to keep our daily practices in the service of resistance to gender domination, not in the service of docility and gender normalization" (ibid. p. 184). Of course, Bordo writes about practices of femininity—the time and energy spent serving and thus reproducing a sexist status quo—but the idea translates well to decolonizing research. How do we organize our bodies in time and space? What are we doing (and who is "we"?), with whom, and in the service of what? Embeddedness does not mean searching for an escape route, for an innocence that does not exist, but instead means continuing to learn, move and change within the world of relationships we inhabit, toward a world of relational accountability described by Wilson (2008) as fundamental to Indigenous research methodologies.

Beginning again is a humbling experience, as is recognizing that a project I worked hard on and care about deeply may reinforce and not simply challenge the colonial power relations I examined. One story in particular remains with me. In the early-twentieth century, the government of the province of Ontario began to require that people who worked guiding tourists in what the government called the Temagami Forest Reserve (part of Teme-Augama Anishnabai territory) needed to carry a licence, which could be revoked if guides did not follow provincial laws (Thorpe 2012, p. 52). When I first learned about the licencing system while conducting archival research, its significance seemed clear: it was a colonial imposition, designed to force Teme-Augama Anishnabai guides to do the work of protecting the timber resource in the interest of the government that had seized control of it.

Yet Vicki Grant, the granddaughter of Teme-Augama Anishnabai guide Donald McKenzie, offered a different perspective on her grandfather's role both as a guide and as a keeper of the land.[3] He wore his guiding badge with pride, for it symbolized the Ontario government's respect for his knowledge and recognition of his authority to guide outsiders through n'Daki Menan. Of course Donald McKenzie had many roles beyond that of guide, and he did not consider that his guiding licence limited his ability to use the land according to Teme-Augama Anishnabai principles. While guiding, Donald and his fellow guides gave respect to the colonial government, but they also continued to use the land in the ways they always had, which included taking animals to feed their families and obeying their own system of government rather than the province's game laws. Vicki relates stories about her grandfather and his friends and relatives playing hide and seek with game wardens, stories that are funny and smart (Donald was never caught), but that also show that Teme-Augama Anishnabai men and women never forgot that they remained the true caretakers of the land.

Vicki's story remains with me not because I think one version of events is true and the other false, but because I now understand the mistake of my initial certainty about the truth of the version I deduced from my archival research. In that version, colonial power relations appeared the only salient ones, and Teme-Augama Anishnabai guides seemed mere victims. Vicki's story reveals a more complex picture of relationships, including those among guides, tourists and territory that challenge colonial dynamics. Guides such as her grandfather Donald McKenzie actively participated in their lives, a fact that matters now as much as it did then. The meaning of the licence becomes fixed not in its imposition but in the stories told about it over time, and it is unfair and incorrect to pin down its meaning in a way that ignores not only the agency but also the personhood of Donald McKenzie and his fellow guides. This is the power and the danger of stories that Thomas King describes (2003, pp. 9–10).

Beginning again is a humbling experience, but that is not a bad thing. Indeed, humility may just provide a perspective from which to begin research, a concept central to Indigenous thought. As Lianne C. Leddy puts it in her chapter in this collection, "Humility is one of the Seven Grandfather Teachings, a set of expectations for how *Anishinaabek* live our lives and relate to one another" (see p. 95). The Truth and Reconciliation Commission (TRC) has as its logo seven flames that represent the seven sacred teachings: love, respect, courage, honesty, wisdom, humility and truth. Each TRC national event was dedicated to one sacred teaching, and the Québec national event was dedicated to the teaching of humility: "knowing oneself as a sacred part of Creation" (Truth and Reconciliation Commission of Canada 2013). Leddy points out the challenge of working from a place of humility when as a scholar one is expected to be an expert (see also Hunt 2014.) The teaching of humility fundamentally challenges the notion of expertise by taking as a starting point that knowledge belongs not to individuals, but "to the cosmos of which we are a part" (Wilson 2008, p. 38). In his piece in this collection, Mi'gmaq scholar Fred Metallic agrees that research is not simply about knowing, but about "how you come to know and what your responsibility is now." He continues, "If I share a story with you, now you know and, because you know, you can't act like you don't know anymore ... It's not about just being the knower and the status of knowing, it's really about the responsibility with those teachings that are embedded in those stories" (see p. 130). If indeed we humans make up just one small part of the universe, we cannot possibly be experts, and must understand that we will always have more to learn than what we know. This way of thinking might be humbling, but it also makes possible learning. In Richard Wagamese's 2012 novel *Indian Horse*, the grandmother of protagonist Saul Indian Horse tells him, "We need mystery ... Mystery fills us with awe and wonder. They are the foundations of humility, and humility, grandson, is the foundation of all learning" (Wagamese 2012 p. 65).

With Newfoundland Innu artist and research collaborator Joanna Barker, I now carry on a different research project, one that examines the history of present-day relationships among Indigenous and non-Indigenous Newfoundlanders and the territory in which they live. We take to the centre of the work Mi'kmaq perspectives and stories, as well as the lessons learned thus far on a journey toward decolonizing research: begin with relationships, not wilderness, because those actually exist; centre Indigenous knowledge holders, knowledges and methodologies, because their marginalization has allowed for the perpetuation of injustices rooted in colonialism to continue to the present day; come from a place of humility rather than a desire for innocence, because I/we have much to learn, and because, like it or not, we are implicated in the power relations that we need to alter; and understand that learning means responsibility to act in accordance with that knowledge, because we are bound by the stories we know. I do not mean to suggest that this list is somehow complete or that with it I can go forth and decolonize research straightforwardly or on my own. Rather, in mapping this journey thus far, I hope to collect and reflect upon what I have learned in the past number of years in the hope that it will be useful as well for others travelling a similar path.

When I initially began research for this chapter, I searched in many academic areas for "tips" on how to decolonize research: in Native Studies, anthropology, history, archival studies, geography, postcolonial studies, and gender studies, for example. Each book or article I read led me to others, until the pile got so high that I understood why people might read only in their chosen discipline. Somewhere in the cycle of reading, adding to the pile and reading some more, I realized that I would need at some point to stop reading and write, even if that meant that the writing was incomplete, uninformed by all I had not yet read. But of course this is always the case with writing, reflective as it is of one's thinking at a particular moment in time, informed by the insights of others. And as Indigenous scholars have made clear, many things cannot be learned in books, and must be learned through oral tradition, through connection with territory, and through listening (Absolon 2011). My search for tips led back to humility in the form of the understanding that, cliché notwithstanding, learning really is a lifelong journey. I did, however, notice common elements in my reading across disciplines on decolonizing research, central among them a recognition of the situated or partial character of all knowledge and sources, as well, often, as a commitment to conducting research collaboratively.

Legal geographer and critical Indigenous scholar Sarah Hunt suggests, for instance, that one starting place for destabilizing dominant ways of knowing and making space for Indigenous knowledges "might be accepting the partiality of knowledge. Its relational, alive, emergent nature means that as we come to know something, as we attempt to fix its meaning, we are always at risk of just missing something" (Hunt 2014, p. 31). Similarly, the contributors to *Reading Beyond Words: Contexts for Native History* (Brown and Vibert 2003) recognize that neither historical documents nor oral history opens up a clear pathway to the truth about the past, showing instead the significance of cultural context in shaping the interpretation of events (see in particular Cruikshank 2003, p. 442). Postcolonial and feminist scholars as well as archival theorists have similarly challenged the commonsense notion of archives as containers of an objective past, arguing that the archival record is limited, implicated in the colonial history that produced it, and a necessary subject of study itself, even as archives can also be read unconventionally (see for example Bastian 2006; Brownlie and Korinek 2012; Burton 2005; Stoler 2002). As historian Adele Perry observes, a postcolonial practice of history "embraces rather than denies the interpretive challenges posed to mainstream historical methodology by the Indigenous archive, alternative ways of reading the written one, and the simple admission that the ways we know the colonial past are not only multiple but necessarily and unevenly partial" (Perry 2005, pp. 343–344).

Geographers Sarah de Leeuw, Emilie Cameron and Margo Greenwood warn that research collaboration in the form of participatory and community-based research between Indigenous community members and non-Indigenous academics will not necessarily work to decolonize research, instead insisting that critique of "best practices" must be incorporated into any such research, and suggesting that friendships outside of research relationships may better support the decolonization of research (Leeuw et al. 2012, p. 180; see also Hunt and Holmes 2015). Fraught with challenges though they might be, research relationships among differently located community members and researchers, including those between Indigenous scholars and the communities to which they may also belong, have the potential to result in stories more complete than research that results from engagement with the mainstream record alone (see for example Lloyd, Wright, Suchet-Pearson, Burarrwanga and Bawaka Country 2012; Watson and Huntington 2008). In one collaborative project, for example, the Haida Repatriation Committee, along with staff of two museums in England and two academic researchers, came together to study Haida "material heritage" that, through the history of colonialism, ended up in British museums. The book resulting from the collaborative research shows the quite different relationships with this "material heritage" between the Haida and museum staff members of the research team. As Jaalen Edenshaw, a member of the Haida Repatriation Committee, explains, some members of the British public "felt the museum was a house of science and that it was holding the remains and the Haida 'artifacts' in the name of science. They felt it belonged to them. Of course I told them that it was our collection and that they were merely housing it for us for the time being" (Krmpotich and Peers 2013, pp. 129–130). Although the book brilliantly shows from a number of perspectives what is at stake when it comes to the politics of museum collections and knowledge production, it also reveals the acuteness of the concern expressed by de Leeuw, Cameron and Greenwood (2012) that collaborative research has the potential to re-entrench colonial relations. The unfair burden placed upon members of the Haida Repatriation Committee becomes evident when one considers the mere fact of their having to travel across North America and the Atlantic Ocean to view, to handle, and to attempt to repatriate objects, and indeed the bones of ancestors, taken from their communities at the height of imperial exploitation and displayed for generations as "artifacts" of a "disappearing" culture.

In her work on interpreting historical treaties between First Nations and representatives of the Queen of England, Indigenous legal scholar Aimée Craft writes that throughout the research process, it is necessary to ask "have we got it right?" and "will we ever know?" and to check in with "the Elders and our communities." She further states, "we need to be following our practices that help us seek guidance, and we need to be humble and do what we can on behalf of the people" (Craft 2014, p. 18). Non-Indigenous scholars are differently implicated in the history of colonialism we all share, but it is incumbent on us as well to work with humility toward decolonization. Geographer Juanita Sundberg suggests that in order to decolonize posthumanist geographies, non-Indigenous scholars need to reflect critically on our own "epistemological and ontological assumptions," to learn to learn about "multiplicity," and to engage seriously with "Indigenous epistemologies, ontologies, and methodologies" (Sundberg 2014, pp. 39–40). Such a process cannot take place without taking the lead from Indigenous scholars and knowledge holders who have made decolonizing research into a pressing issue today. Given the historical context in which we engage in this research, Indigenous and non-Indigenous people alike, it is little surprise that it is a journey rather than a quick skip. Yet if, as Fred Metallic suggests in his chapter in this collection (see p. 122), the process is more important than the destination, then perhaps we are already on our way.

Acknowledgements

The author wishes to acknowledge that she was able to write this chapter as a guest on Treaty 1 territory, the original lands of Anishinaabeg, Cree, Oji-Cree, Dakota, and Dene peoples, and the homeland of the Métis Nation. She acknowledges as well the financial support for her research of the Social Sciences and Humanities Research Council (SSHRC). For their insightful and helpful comments on earlier drafts of this chapter, she thanks Joanna Barker, Aimée Craft, Janice Dodd, Vicki Grant, Rae Hutton, Lianne C. Leddy, Deborah McPhail, Stephanie Rutherford, Anders Sandberg, and members of the Critical Environments Research Group at the University of Manitoba.

Notes

1 In the introduction to their edited collection on Indigenous men and masculinities, however, Robert Alexander Innes and Kim Anderson (2015) contend that much less attention has been paid to the conditions facing Indigenous men than women, although Indigenous men face "distinct gender and racial biases that cause many to struggle" (p. 4). Innes and Anderson ask poignantly in their introduction's title, referencing the Walking With Our Sisters art installation described below, "Who's Walking with Our Brothers?" Their collection as a whole does however show that important work in this area is indeed taking place.

2 I don't pretend this is an easy or straightforward process, a topic famously addressed by Gaytri Spivak in "Can the Subaltern Speak?" (1988).

3 Vicki Grant is also my good friend and teacher, and an amazing bridge and connection maker. She originally took me on in 2006 as a student in n'Daki Menan, and has been stuck with me ever since. I am incredibly grateful for her guidance, wisdom and

friendship, and not at all surprised that a story she told me ten years ago gives me pause today. I know I am only beginning to learn what she has been teaching me all along. I thank Vicki for reading this chapter and for offering her insights, suggestions and words, and particularly for filling in the story of Donald McKenzie as a keeper of the land and as a provider for his family.

References

Absolon (Minogiizhigokwe), K. E. (2011). *Kaandossiwin: How We Come to Know*. Halifax, NS: Fernwood Publishing.

Bastian, J. A. (2006). Reading colonial records through an archival lens: The provenance of place, space and creation. *Archival Science*, 6(3), 267–284.

Bordo, S. (2003). The body and the reproduction of femininity. In *Unbearable Weight: Feminism, Western Culture, and the Body* (Tenth Anniversary Edition), 165–184. Berkeley, CA: University of California Press. (Original work published 1993.)

Brown, J. S. H., and Vibert, E. (eds). (2003). *Reading Beyond Words: Contexts for Native History* (2nd edn). Toronto, ON: University of Toronto Press.

Brownlie, R. J., and Korinek, V., (eds). (2012). *Finding a Way to the Heart: Feminist Writings on Aboriginal and Women's History in Canada*. Winnipeg, MB: University of Manitoba Press.

Burton, A. (ed.). (2005). *Archive Stories: Facts, Fictions, and the Writing of History*. Durham, NC: Duke University Press.

Butler, J. (2007). Performative acts and gender constitution: An essay in phenomenology and feminist theory. In H. Bial (ed.), *The Performance Studies Reader* (2nd edn), 187–199. New York: Routledge.

Chamberlin, J. E. (2004). *If This is Your Land, Where are Your Stories?: Finding Common Ground*. Toronto, ON: Vintage Canada. (Original work published 2003).

Craft, A. (2014). Living treaties, breathing research. *Canadian Journal of Women and the Law*, 26(1), 1–22.

Cronon, W. (1992). A place for stories: Nature, history, and narrative. *The Journal of American History*, 78(4), 1347–1376.

Cruikshank, J. (2003). Discovery of gold on the Klondike: Perspectives from oral tradition. In J. S. H. Brown and E. Vibert (eds), *Reading Beyond Words: Contexts for Native History* (2nd edn), 435–458. Toronto, ON: University of Toronto Press.

D'Aliesio, R., and Grant, T. (2015, December 16). OPP says 180 Indigenous homicide victims in recent decades. *The Globe and Mail*. Retrieved from www.theglobeandmail.com/news/national/opp-says-48-indigenous-people-missing-in-recent-decades/article27779667.

De Beauvoir, S. (2009). Biological data. In *The Second Sex* (C. Borde and S. Malovany-Chevallier, trans.), 21–49. New York, NY: Vintage Books, a division of Random House, Inc. (Original work published 1949).

De Leeuw, S., Cameron, E. S., and Greenwood, M. L. (2012). Participatory and community-based research, indigenous geographies, and the spaces of friendship: A critical engagement. *The Canadian Geographer*, 56(2), 180–194.

Families hopeful, advocates unsure as MMIW inquiry gets underway. (2015, December 8). *CBC News*. Retrieved from www.cbc.ca/news/canada/manitoba/manitoba-mmiw-inquiry-reaction-1.3355842.

Fedio, C. (2015, June 2). Truth and reconciliation report brings calls for action, not words. *CBC News*. Retrieved from www.cbc.ca/news/politics/truth-and-reconciliation-report-brings-calls-for-action-not-words-1.3096863.

Fenlon, B. (2015, November 30). Uncivil dialogue: Commenting and stories about Indigenous people. *CBC News*. Retrieved from www.cbc.ca/newsblogs/community/editorsblog/2015/11/uncivil-dialogue-commenting-and-stories-about-indigenous-people.html.

Haraway, D. (1988). Situated knowledges: The science question in feminism and the privilege of partial perspective. *Feminist Studies, 14*(3), 575–99.

Hunt, S. (2014). Ontologies of indigeneity: The politics of embodying a concept. *cultural geographies, 21*(1), 27–32.

Hunt, S., and Holmes, C. (2015). Everyday decolonization: Living a decolonizing queer politics. *Journal of Lesbian Studies, 19*(2), 154–172.

Innes, R. A. and Anderson, K. (2015). Introduction: Who is walking with our brothers?. In R. A. Innes and K. Anderson (eds), *Indigenous Men and Masculinities: Legacies, Identities, Regeneration* (3–17). Winnipeg, MB: University of Manitoba Press.

King, T. (2003). *The Truth About Stories: A Native Narrative*. Toronto, ON: House of Anansi Press Inc.

Kovach, M. (2009). *Indigenous Methodologies: Characteristics, Conversations, and Contexts*. Toronto, ON: University of Toronto Press.

Krmpotich, C., and Peers, L. (2013). *This is Our Life: Haida Material Heritage and Changing Museum Practice*. Vancouver, BC: UBC Press.

Kuokkanen, R. (2007). *Reshaping the University: Responsibility, Indigenous Epistemes, and the Logic of the Gift*. Vancouver, BC: UBC Press.

Lloyd, K., Wright, S., Suchet-Pearson, S., Burarrwanga, L., and Bawaka Country. (2012). Reframing development through collaboration: Towards a relational ontology of connection in Bawaka, North East Arnhem Land. *Third World Quarterly, 33*(6), 1075–1094.

Lorde, A. (1984). Age, race, class, and sex: Redefining difference. In *Sister/Outsider: Essays and Speeches by Audre Lorde* (114–123). Freedom, CA: The Crossing Press.

Mas, S. (2015, December 8). Missing and murdered Indigenous women: 1st phase of public inquiry outlined today. *CBC News*. Retrieved from www.cbc.ca/news/politics/missing-murdered-inquiry-1.3355492.

Perry, A. (2005). The colonial archive on trial: Possession, dispossession, and history in *Delgamuukw v. British Colombia*. In A. Burton (ed.), *Archive Stories: Facts, Fictions, and the Writing of History* (325–350). Durham, NC: Duke University Press.

Rich, A. (1994). Notes toward a politics of location. In *Blood, Bread, and Poetry: Selected Prose 1979–1985* (210–31). New York: W. W. Norton & Company. (Original work published 1986).

Said, E. W. (1978). *Orientalism*. New York: Vintage Books Edition.

Simpson, L. B. (2014). Land as pedagogy: Nishnaabeg intelligence and rebellious transformation. *Decolonization: Indigeneity, Education & Society*, 3(3), 1–25.

Smith, J. (2015, December 8). Liberals to consult victims' families on inquiry into missing, murdered Indigenous women. *The Star*. Retrieved from www.thestar.com/news/canada/2015/12/08/ottawa-to-begin-consulting-families-on-inquiry-into-missing-murdered-indigenous-women.html.

Smith, L. T. (2012). *Decolonizing methodologies: Research and Indigenous peoples* (2nd edn). London: Zed Books. (Original work published 1999).

Spivak, G. C. (1988). Can the subaltern speak? In C. Nelson and L. Grossberg (eds), *Marxism and the Interpretation of Culture* (271–313). Urbana, IL: University of Illinois Press.

Sterritt, A. (2015, November 20). A movement rises. *Opencanada.org*. Retrieved from www.opencanada.org/features/movement-rises.

Stoler, A. L. (2002). Colonial archives and the arts of governance. *Archival Science*, *2*(1), 87–109.

Sundberg, J. (2014). Decolonizing posthumanist geographies. *Cultural Geographies*, *21*(1), 33–47.

Thorpe, J. (2012). *Temagami's Tangled Wild: Race, Gender, and the Making of Canadian Nature*. Vancouver, BC: UBC Press.

Truth and Reconciliation Commission of Canada. (2013). *Québec National Event, April 24–27, 2013* [Brochure]. Truth and Reconciliation Commission of Canada.

Tuck, E., and Yang, K.W. (2012). Decolonization is not a metaphor. *Decolonization: Indigeneity, Education & Society*, *1*(1), 1–40.

Wagamese, R. (2012). *Indian Horse*. Vancouver, BC: Douglas & McIntyre.

Walking With Our Sisters. (2015). A commemorative art installation for the missing and murdered Indigenous women of Canada and the United States. Retrieved from http://walkingwithoursisters.ca/about.

Watson, A., and Huntington, O. H. (2008). They're *here*—I can *feel* them: The epistemic spaces of Indigenous and Western knowledges. *Social & Cultural Geography*, 9(3), 257–281.

Wilson, S. (2008). *Research is Ceremony: Indigenous Research Methods*. Halifax, NS: Fernwood Publishing.

Part III

Senses and affect

12 On narrative, affect and threatened ecologies of tidal landscapes

Owain Jones and Katherine Jones

> All stories are true. I know that now.
> Unwritten memories and bodies
> Subside like pit villages.
> But they leave small spaces
> Where poppies and dandelions grow
> Amongst resurgent grass
>
> (from 'Burials' by Adam Horovitz 2011)

This is a story about tides, about tides in Bristol, about two characters that were found in the mud of the Severn Estuary—Peri and Proxi. It is also a story about telling stories as a method for the creative accounting of eco-social histories, presents and futures, and material and non-material entanglements of such through space–time. And it is a story about loss: of ecologies, of ecocide, of getting lost, and of finding a way through stories.

As so much of life is, this is an experimental mixing. Tides mix things—fresh and salt water, land and sea. In unsettling boundaries and definitions, they invite a focus on inter-relationships, and on flux and change. Intertidal landscapes are in constant motion, change and cycles, denying fixity and stasis. As such, they unsettle both linear thinking and linear understanding that, in spite of movements to break away from them, continue to shape historical and geographical accounts in many areas of thinking.

This chapter is tidal in its approach—ecological. It tells stories of stories through the tides, an ebbing and waning, from solid to fluid and back again with all that comes in between. It embraces the constant motion of life and understanding, and attempts to bring this into a textual representation. In short, it is a story of an ecological approach to eco-social storytelling.

Narrative #1: The power of stories

There has been a veritable burgeoning of what we term "literary geography" in the last few years. This is geography where the process of research and its reporting becomes reworked and all wrapped up in the process of making and 'telling a story' and 'telling a story well'. Often the aims are avowedly literary, with the quality of writing (language-structure-delivery) brought very much to the fore in methodological and ontological terms. This, for example, is exemplified in Fraser MacDonald's paper, "The ruins of Erskine Beveridge" (2015), the opening line of which is: "This is a narrative essay, the animating purpose of which is stylistic as much as analytic. It is a story; and, unusually for academic geography, the story is primary" (MacDonald 2015, p. 477).

This quotation reflects a wider 'narrative turn' in the social sciences and art and humanities. And the point and potential of this turn is to create narratives which can capture—or, better, channel—the elusive subtleties of everyday becoming with all its affective richness. Here, more or less, we are interested in stories, in *ecologies of stories*, and how to seed new threads of narratives into that ecology which *carry demands to know and live differently*.

Harari (2015), in his recent history of humanity, makes storytelling the human characteristic which has shaped the last 2,000,000 years and developed science, culture and politics as we know them:

> Over the years, people have woven an incredibly complex network of stories. Within this network, fictions … not only exist but also accumulate immense power.
>
> (Harari 2015, p. 31)

> Since large-scale human cooperation is based on myth, the way people cooperate can be altered by changing the myths—by telling different stories. Under the right circumstances myths can change rapidly.
>
> (ibid., p. 32)

But …

> Telling effective stories is not easy. The difficulty lies not in telling the story, but in convincing everyone else to believe it. Much of history revolves around this question: how does one convince millions of people to believe particular stories about gods, or nations, or limited liability companies?
>
> (ibid., p. 31)

Tides #1: The tides: I, tide

I (OJ) have been researching and writing about tides for some years now because I love tides. I grew up with them (and saw the destruction of tidal lands). Tides—the grand movements of the Earth's oceans and seas due to the intersecting forces of interplanetary gravity and the spin of the earth—cannot be stopped, or even fractionally altered, by humans, but the ecologies of tides are being degraded in many ways, as is so much of our biosphere.

The Lamby was an area of intertidal grassland used for summer grazing of livestock on the east side of the city of Cardiff (Wales, UK), where the river Rhymney runs into the Severn Estuary, close by the once great docks of the city. As a farming family, we drove cattle and sheep there each spring for summer grazing, attended to them each month at higher tides, and then returned them to the farmstead in autumn. The Lamby was a beautiful and wild area of the city with birds (and bird watchers) and boats and walkers and horse riders and youth on scrambler bikes, and people harvesting driftwood. It was destroyed by building new seawalls along the seaward margin of the land and using it for a huge landfill and refuse reprocessing site.

The Taff Estuary, on the other side of the city, where the river Taff made a large tidal estuary as it opened out onto the yet larger Severn Estuary, was a larger, equally ecologically and culturally rich wild area in the city with birdlife and nature reserve status. This was drowned by building a barrage across the mouth of the estuary, thus creating an area of permanent high fresh water. This was not for power generation but for real estate value generation. This was, and is, claimed to be a great example of urban development with large areas of gleaming, high-value residential and commercial premises created along a new—non-muddy—waterfront.

The lower tidal reaches of the Taff river no longer flow, but form a ribbon lake running back up into the city. A river that does not flow anymore, with all the cultural and ecological implications of that! My memories are haunted by the loss of these places (see Jones 2005, 2015). Through the weaving of my personal history with that of the tide, my sense of loss becomes a mourning for ecologies of place that once were. The destruction of this ecology—this mixing of waters, of creatures, of the cyclical social rhythms of season and tide and human activity—is felt as a deeply personal loss, even if I am but one of the many things and people affected by it. I am now concerned for other particular tidal landscapes I know, and for tidal landscapes more widely. Like so much of nature, they are falling under the march of "ecocide"[1] through the process of modern 'development'. (OJ end)

Most recently, tides have flowed into our arts- and humanities-based project, "Towards Hydrocitizenship" (UK), and its Bristol case study, WaterCityBristol.[2]

Ecology #1: Ecology and narrative

Ecology is fundamentally about the *radical* interconnectedness of things—the interrelations which make up the ongoing practice of life. Life is not in objects or individuals; it is in the interactions between them.

This is ecology taken in a broad rather than narrow sense of 'pure biology'. The nature-culture divide does not exist within this wider sense of ecology, and nor do the other dualisms that have beset modernity for example the subject-object dualism. As Latour (1993) famously said, these divides are the corrosive tools of oppression within modernity, that are wreaking terrible harm on us as individuals, our communities, our beautiful biosphere home.

> In contrast with those who believe they may live detached from nature, environmentalists recognize that they are part of a community made of multiple relationships-both biotic and abiotic. This is an ecological concept: ecology explains that all organisms are connected in a network which makes what is called an ecosystem. Interconnection is a touchstone of ecology; everything that occurs in any given part of an ecosystem will affect the rest, many times in unexpected ways. According to the environmentalist and physicist Barry Commoner, *the principle that 'everything is connected to everything else' must be considered the first basic law of ecology*. While it might be more intuitive to trace those connections at the local scale, they are not less strong and real at the global level.
>
> (Armiero and Sedrez 2014, p. 1)

And this is a radical principle, or should be:

> The really subversive element of ecology rests not on any of its more sophisticated concepts, but on its most basic premise: inter-relatedness. But the genuinely radical nature of that proposition is not generally perceived, even, I think, by ecologists. To the western mind, *inter-related* implies a causal connection … but what is actually involved is a genuine *intermingling* of parts of the ecosystem. There are no discrete entities.
>
> (Evernden 1996, p. 93)

Complexity is more or less about the density of interaction and non-linear connections within systems. In one sense the more complex a system is, the 'easier' it is to move from one point to another in terms of connection. But finding one's way is the challenge—it is easy to get …

Lost … and (then) found …

The brilliant, groundbreaking writer W. G. Sebald said it was a revelation when he realized that "everything is connected across space and time" (Sebald 2013, p. vii). Thus to approach any subject one can pick a random starting point and then try to find one's way to your subject. Or one could start somewhere—anywhere—and soon be someplace else. No thing/no place is without interest, implications, possibilities, politics. The world is a hyper-complex and a fluid set of ongoing entangling non-linear processes. This is more or less what Ingold (2012) terms the "meshwork" of life. The meshwork is an expression of ecology. And as Tim Morton (2010) says, the ecological crisis lets us see that "everything is interconnected. This is the ecological thought" (Morton 2010, p. 1).

Reading Sebald can be a somewhat disorientating experience of momentarily wondering where you are, where you are going and why. The experience of that echoes the process of his method (see below). Yes, it's all ecological—yes, everything is connected—but this is not immediately, or even after a period of contemplation, graspable. Much is hidden—elusive—maybe even illusive. Thus to be active in the world in this way is often to feel—to be—lost. This is one challenge of affective theory/method, which seeks to wander into the realm of lived complexity. Seigworth and Gregg (2010) say, more or less, of affective-based approaches to research, if you are not experiencing some feeling of lostness and disorientation you might not be doing it right!

Lost is good, that is part of the process. In sympathy with Sebald's claim that creativity can only come through deliberately embracing chance and serendipity, the writer Duncan Fallowell (2015) asserts: "Unless you are prepared to get lost you can never discover anything new or grow, and that is true as an individual or indeed of whole societies."

Rebecca Solnit (2006) wrote a whole lovely book about the importance and affordances of being lost. Modern knowledge dislikes lostness, seeks to avoid and eradicate it. But embracing it and leaning to move differently means you are more open to the world and more likely to know it. Referring to Walter Benjamin, Solnit suggests "to be lost is to be *fully present*, and to be fully present is to be capable being in uncertainty and mystery" (Solnit 2006, p. 6, emphasis added).

So here is Sebald on all this; please excuse the length of the quote but we do feel this is worth sharing. This is Sebald in conversation with Joseph Cuomo, in front of a live audience in 2007 (in Schwartz 2007, pp. 94–95):

> [On the writing of *The Rings of Saturn* and the "research methods" used] I had this idea of writing a few short pieces for the feuilletons of the German papers in order to pay for this extravagance of a fortnight's rambling tour. That was the plan. But then as you walk along, you find things. I think that's the advantage of walking. It's just one of the reasons why I do that a lot. You find things by the wayside or you buy a brochure written by a local historian, which is in a tiny little museum somewhere, which you would never find in London. And in that you find odd details which lead you somewhere else, and so it's a form of unsystematic searching, which of course for an academic is far from orthodoxy, because we're meant to do things systematically.
>
> But I never liked doing things systematically. Not even my PhD research was done systematically. It was always done in a random, haphazard fashion. And the more I got on, the more I felt that, really, one can find something only in that way, i.e., in the same way in which, say, a dog runs through a field. If you look at a dog following the advice of his nose, he traverses a patch of land in a completely unplottable manner. And he invariably finds what he's looking for. I think that, as I've always had dogs, I've learned from them how to do this. [Audience laughter.] And so you then have a small amount of material, and you accumulate things, and it grows; one thing takes you to another, and you make something out of these haphazardly assembled materials. And, as they have been assembled in this random fashion, you have to strain your imagination in order to create a connection between the two things. If you look for things that are like the things that you have looked for before, then, obviously, they'll connect up. But they'll only connect up in an obvious sort of way, which actually isn't, in terms of writing something new, very productive. So you have to take heterogeneous

> materials in order to get your mind to do something that it hasn't done before. That's how I thought about it.
>
> Then, of course, curiosity gets the better of you. For instance, this whole business about this atrocious Chinese civil war in the nineteenth century, which we know so little about in the West—I knew nothing about it—I'd found that remark in a tiny little booklet written, I think, in 1948, which was still there for sale, that this little local train which ran around there [over the River Blyth in England] had been destined originally for the court of the emperor of China, which was a very bizarre, erratic fact. And then of course you wonder which emperor, and you go to the *Encyclopedia Britannica* from 1911 and you rummage around there, and it goes on like this. Which is the most pleasurable part of the work, as you uncover these things and move from one astonishing thing to the other. The actual writing, of course, is a different story. That's far from a pleasant occupation. [Audience laughter].

Narratives created in such a way are 'alive' to the ecology of the world, are flexible enough to get into all that complexity and follow it, *and make* paths through it. We are all on individual paths through the ecological matrix of unfolding life. As we proceed, we encounter, and react, and influence and move on. As every single thing does. Ecocide gets into the meshwork, kills bits of it, cuts it up, colonizes—parasitizes—the webs of connection, controls what it needs, degrades or removes what it does not.

The great thing about narrative as method is *complete freedom* in terms of space and time. The famous cowboy story cliché "meanwhile, back at the ranch" is a simple example of how a narrative can arbitrarily and instantly jump to any place and any time at the insertion of the simplest link, in the manner of a wormhole. Even if there was no (apparent) link between the two elements, the narrative link connects them ecologically. All stories are true. Or, as Picasso said, "everything you imagine is real."[3]

Of course the challenge is to have the imagination to make 'good wormholes', to tell stories compelling enough to get listeners to follow in imaginative, emotional and affective terms down the wormhole, or to use one of the most brilliant examples from literature, down the rabbit hole.

Importantly, as the writer Emily Woof (2015) tells of her own 'method' of similar haphazard wanderings through possible nodes, scenes and connections in the construction of narrative, we need to 'trust our subconscious' to make good and creative decisions about what we do and where we go in narrative construction. More or less that is what we are doing here.

Stories are the key fabric by which we construct the world individually and collectively. We will keep on with these three themes that make up this story about tides, about the ability of stories to creatively capture the complex ecology of the unfolded (history and unfolding world). If we were good enough story tellers we could say one story and these threads would emerge through the skill of telling. It is hard to throw off the academic habit.

Narrative #2: Ecology and ecocide

We live in a crowded, chaotic ecology of narratives. These are all the stories we gather and inflect—media, social media, family, work, communities—and ones we make ourselves (of ourselves) in all these settings, and share with ourselves and each other. As the eminent neurobiologist Antonio Damasio (2000) shows, our senses of self, senses of place, nation and so on, are essentially narrative constructions based in and upon memory within culture, but always on the move.

At present this ecology of narratives, which essentially is the living tissue of culture, has some particularly pernicious 'invasive species' in it, basically the narratives and (resultant) practices of industrial globalized capitalism and the Enlightenment/modernist ideologies, theologies and sciences it is in league with. As the Invisible Committee (2009, p. 16) put it, "two centuries of capitalism and market nihilism have bought us to the most extreme alienations—from ourselves, from others, from worlds ... It's with an entire anthropology that we are at war. The very idea of man."

A quick example: Nike, the footwear, sportswear/clothing/accessories multinational corporation, spends $100 per second, globally, on what they term "demand creation" (Forester 2014). And one needs only to imagine that multiplied by all the capitalist agents operating globally, and all the state and theocratic powers, to think of the *narrative force and noise* we all, to varying extents, live in—*the narrative oppressions we live under*. "Brands" are forms of lifestyle/identity narratives seeking to capture people's narratives of themselves. The corporations, the theologies, the states, the media are continuously pumping their narratives of self and lifestyle into the ecologies of narrative we live within. It is a common observation now that modern politics is all about narrative.

This is not the Enlightenment (as we are so often told); this is a new form of Dark Age, or as Heidegger put it, a time of the "World's night" (see Malpas 2008, p. 279), precisely because ecocide is well under way. The World Wildlife Fund's 2014 report which states that 52 per cent of the world's biodiversity has been lost between 1970 and 2010 is the starkest testament to that. Ecocide is also bound up with the 'death of affect' itself, as predicted by J. G. Ballard as the potency and force of human life is colonised by fabricated commodified emotions of, say, 'reality' TV. In 1989 Felix Guattari stressed that thus far mainstream politics has entirely failed to address the issue of ecocide, which is, quite literally, killing us and an ecological collective:

> Political groupings and executive authorities appear to be totally incapable of understanding the full implications of these issues. Despite having recently initiated a partial realization of the most obvious dangers that threaten the natural environment of our societies, they are generally content to simply tackle industrial pollution and then from a purely technocratic perspective, whereas only an ethico-political articulation—which I call ecosophy between the three ecological registers [the environment, social

> relations and human subjectivity]—would be likely to clarify these questions.
>
> (Guattari 1989, p. 17)

Twenty-five years and this situation has, if anything, worsened.

Tides #2: Telling the tides

Tides have been called 'the pulse of the earth'. We don't actually think the analogy bares too much scrutiny, but it is eloquent about the earth as a living system, which clearly it is as Lovelock has shown in the Gaia hypothesis. Tides are the fundamental physical process of the waters of the earth's oceans rhythmically, vastly moving in response to the cosmic forces of interplanetary gravity (earth, moon, sun) and the spin of the earth itself. Like in any complex body, the rhythms are also complex with crescendos and diminuendos of high and low waters and differing durations and patterns such as diurnal, semi-diurnal, and mixed tides, and all sorts of local (in space *and* time) variations.

So how to 'represent' these, and the richness associated with them and the risks they face? We need compelling narratives. One obvious example is the work of Rachel Carson, who as well as writing the epoch defining *Silent Spring* (1962), also wrote about the oceans and the tides. Carson was an eminent biologist and ecological conservation pioneer, whose first love was for the oceans, their shores and their tides. She filled her stories with scientific facts, personal observations and the observations of lay people. The titles of her trilogy of books about these subjects capture that love: *Under the Sea Wind* 1941; *The Sea Around Us* 1951; and *The Edge of the Sea* 1955 (Figure 12.1). It is the last of the trilogy that focuses on the tides most fully, not least as tides come into their own—and are most obvious in some senses at least—in coastal margins, even if they affect all the oceans.

Figure 12.1 The Rachel Carson "Sea Trilogy"

Rachel Carson was a public scientist whose work attracted sustained and even personal attacks from those seeking to protect established technologies and their related economies. In a new methodological article on storytelling/narrative as a (feminist) method of inquiry, Stone-Mediatore (2016) turns to Carson as an example of an academic and scientist who made such an impact because she was a great storyteller. Her impact as a public figure is gestured by her portrait being on a US postage stamp of "Great Americans" (1981). More recently, a major international environmental humanities research center located in Munich, Germany, has been named after her.[4] But despite Carson's (and others') compelling narratives of ecological destruction, the aggregate direction of travel of modern globalized society is still strongly on the path of ecocide. This is why Isabelle Stengers (2015, pp. 10–11) concludes:

> We do, however, know one thing: even if it is a matter of the death of what we have called a civilization, there are many manners of dying, some being more ugly than others. I belong to a generation that will perhaps be the most hated in human memory, the generation that "knew" but did nothing or did too little (changing our lightbulbs, sorting our rubbish, riding bicycles …). But it is also a generation that will avoid the worst—we will already be dead. I would add that this is the generation that, thirty years ago, participated in, or impotently witnessed, the failure of the encounter between two movements that could, together, perhaps have created the political intelligence necessary to the development of an efficacious culture of struggle—those who denounced the ravaging of nature and those who combated the exploitation of humans. In fact, the manner in which large environmental movements have adhered to the promises of "green" capitalism is enough to retroactively confirm the most somber of suspicions. But the retroactive justification should not erase the memory of a missed opportunity, of a blind division from which the capitalist sirens haven't failed to profit. Capitalism knows how to profit from every opportunity.

Figure 12.2 Rachel Carson US postage stamp, Great Americans series 1981

Of tides, Carson (1989, p. 149) famously said, "there is no drop of water in the ocean, not even in the deepest part of the abyss, that does not know and respond to the mysterious forces that create the tide." And through the effects tides have on all ocean waters, the ecologies of the oceans and intertidal areas are shaped by tides in terms of the habits of creatures and plants in terms of cycles of life, relationships, migrations and so forth.

Sea creatures are highly attuned to rhythm, current, pressure, sound, the movement of material through water, and so the tides are of general and often particular significance to them. Clearly these life forms heed the patterns and rhythms in the water but also, it seems, can somehow sense the varying gravitational pull of the moon itself. As Lyall Watson (1973, p. 22), in his somewhat strange book *Supernature* states, "every living marine animal and plant is made aware of the rhythm [of the tides]."

Watson then goes on to offer some examples of this. He reports on experiments conducted by Carson on *Convoluta* (a very small marine flatworm) in a laboratory far from the sea, where the worms still cycled through rising from mud and back into mud in tune with the absent tides. Watson also reports on the work of the US biologist Frank Brown, who, in experiments starting in 1954, took oysters to labs far from where they were collected to observe their behaviours and rhythms. For two weeks, after a journey of 1000 miles from Long Island Sound to a suburb of Chicago on Lake Michigan, the oysters opened to feed at what would have been the time of high tide at the home. But after that, their rhythms slowly changed to a pattern that would have suited tidal rise and fall in their new location.

Ecology #2: Towards ecologies of narrative and ecological citizenship

Here is a very important point about the ecological citizenship 'movement'. It is not the case that the various thinkers from Leopold and Naess onwards want to make modern humans ecological citizens. *Humans are always and inevitably ecological.* Of course we are—through the food, energy, water we consume, through the waste we produce, the atmospheric gases we inhale and exhale, in how we are always embedded bodily in the environment, in how our bodies are more an assemblage of organisms and heterogeneous DNA than in any way pure and unified. *So the challenge of the ecological self is not to become ecological*; it is to recognise this ecological basis of our being that has been hidden from us, denied to us, by a range of ideological trajectories. It is also to recognize the connections/processes we are embedded in utterly and always, and—as far as possible—the demands and consequences those connections bring, for us and others.

Unfortunately that means at present looking (or trying to look) into the knowledge abyss that separates culture (modern humans) from nature, and the consequences therein. This throws up a number of significant challenges, but these can be divided up as follows: Firstly, the unknowability of the (modern) ecological self. The nature of modern life—globalised, consumption based, highly complex technologically and economically—means that the networks we are in ecologically are extraordinarily opaque, complex, long and entangled. It is almost impossible genuinely to confront this. The mobile phone, for example, contains such a complicated 'footprint' of rare metals and other resources that the understanding of that and how these are derived and processed is daunting. The same applies to all complex products and everyday activities such as shopping in a modern supermarket. Movements such as 'Follow the Thing' and 'Fair Trade' are worthy interventions in this kind of challenge in which the consequences of at least some forms of consumption actions are traced. But as modern citizens living lives based upon very high levels of very complexly produced consumption, it is very hard to engage with the full implications of one's ecological footprint.

Secondly, the other equally challenging thing is that at present, the modern human's ecological 'footprint' is—collectively—simply devastating. Even if one's own ecological footprint is unknowable, it is not hard to get some sense of collective consequences. The ecological crisis is "so obvious now" (Morton 2010), yet almost totally ignored by mainstream politics and culture.

We feel a sense of this can be felt with just some effort of imaginational groundwork. For example, go to a large supermarket, walk down the tinned fish aisle, and then imagine all the other supermarkets and shops in the entire world, and the huge infrastructures delivering this tinned fish—and the consequences. Linking this to some knowledge of the overfishing of the seas and ecological consequences thereof obviously helps.

What is knowable is that, collectively, modernity is the bringer of 'ecocide' as defined by Guattari. Although it is difficult to trace out and act upon the particular connections we are bound into, the aggregate results of modern society are only too clear. Climate change and the crashing of terrestrial and marine biodiversity are the BIG STORIES of our age. To become self-aware in terms of ecological self is to put oneself in a very ethical and emotionally challenging position. Of course, one can live as 'greenly' as possible, but as citizens or subjects (taxpayers) we are bound into capitalist, militarized nation-state systems, which are very hard to step outside of. We will be, as the above quote from Stengers suggests, the hated generations who know – but failed to act.

Thirdly, to become ecologically self-aware, as far as is possible, and to confront the challenge of this, should sit with a commitment to try to adjust one's life to a more ecologically benign/creative footing. If one has to have political and spiritual 'foundations,' there are good reasons why ecological self-awareness should be those. This is not a new claim. It is akin to a point of the Deep Ecology platform item eight: "Those who subscribe to the foregoing points [the previous seven points of the platform] have an obligation directly or indirectly to participate in the attempt to implement the necessary changes [changes in the fundamental vision and practice of humans in nature particularly the abandoning of notions of human exceptionalism]."[5] It is also parallel to (eco)feminism's profound assertion that "the personal is the political." In other words, there is no private and public in politics, and flows of power and responsibility stream between the body and the body politic.

Becoming more ecologically aware (which is what ecological citizenship/self rests upon in the first instance) might put a strain on personal happiness and relationships. We are lucky enough to have a pretty good (separate) community and family lives. But taking part in our communities puts us into many positions where we are witnessing people who we respect and have affection for who are not, at present, concerned with their ecological embeddedness. They are interesting in their social, cultural and consumption-embeddedness in society (i.e. their lifestyle, their cars, their holidays, their homes). Of course, we are interested in all that too; it seems unavoidable in the way we live now. But at present many of those we witness are practicing liberal capitalist consumption-based self-becoming. Many are very kind and thoughtful, but like most Enlightenment souls, the ecological basis of becoming and its consequences is pretty closed off to them.

We need a new canon of stories, or a shift in the *reach* center of gravity of the ecologies of stories we live in and by, to change how modern people see themselves in the world. Stories are being seized upon for social justice purposes. For example:

> Social justice activists, artists and project leaders all over the world ... are insisting on the power that stories have to generate hope and engagement, personal dignity and active citizenship, the pride of identity, and the humility of human connectedness ... The beginning of the twenty-first century is an historical moment in which narrative is more broadly recognised than ever as a significant, simple, crucial vehicle for reawakening, disseminating and sustaining social justice impulses.
>
> (Solonger, Fox and Irani 2008, p. 1)

We need to tell ecological stories, of tides and everything else. From the above book—about a river and a community—comes this story. It is a sad one. Ecocide.

> Sadness lays hard on the land. Commodification and diversion of a most sacrosanct element, Water, the Zuni River, is not only an assault on the ecosystem and the people, but on the ability of our distinct culture to continue to grow and flourish, and of the Earth to regenerate and sustain us. Our lands are being plundered and our resources exploited for profit, with impunity. Precious watersheds that give birth to our lakes, springs, and streams and enable life in our communities are under attack
>
> (Wemytewa and Peters 2008, p. 22)

Tides #3: Affecting a story of tides

On 22 March 2015, the highest tide in twenty years was predicted at the Port of Avonmouth (Bristol), which stands at the mouth of the River Avon on the great Severn Estuary (UK). The old city port of Bristol stands some eight miles upstream, where the boats of conquest and empire once floated up and down on the high tide. Bristol is (was) a lunar city not a solar city following Lefebvre's elucidation of the rhythms of tidal cities (Lefebvre 2004, p. 91). In Lefebvre's eyes the 'feel' and 'character' of oceanic cities located on the Atlantic (macro-tides) and the Mediterranean (micro-tides) are different, as transmitted in their material arrangements, their habits and, particularly, their rhythms. All that together is more or less what affect is—the feel of a place—, or to be more precise all the interactions between bodies in place and bodies and place that make up it what it is.

In more detail, Thrift (2004) suggests:

> Cities may be seen as broiling maelstroms of affect. Particular affects such as anger, fear, happiness and joy are continually on the boil, rising here, subsiding there, and these affects continually manifest themselves in events which can take place either at a grand scale or simply as a part of continuing everyday life.
>
> (Thrift 2004, p. 57)

So how might this work in relation to tides in a city? In one of the most famous adventure novels ever written, R. L. Stevenson's *Treasure Island*, the ship of the story—the Admiral Benbo—is prepared for sail on the next high tide in the heart of Bristol with stores being purchased and rushed aboard. All the drama, excitement and nervousness of a hasty departure for an adventure are set out. But as the next high tide for sailing happens to be early morning, the port and its stores—and pubs—and dock hands are all-action right through the night. A lunar city.

Bristol and the wider Severn Estuary have some of the highest tides in the world. The sea-level rises as much as 14 meters in the space of six hours. Think of that: the sea itself rising up and lowering again. The implications are profound for the coast, the cities and all human and non-human life that live in the hinterlands of such tides. The wider research projects that we are a part of seek to explore creatively and develop ecological understandings of water in Bristol and how it connects people to each other, and people to nature, and the past, present and future of this, including the roles of the tide.

At the spring equinox of 2015, the tide was exceptionally high because, in all the regular Newtonian business of the heavenly bodies in motion that create the tides, there are subtle irregularities. The orbit of the moon around the earth, and that of the earth around the sun are not completely regular, and so, on occasion, distant forces 'conspire' to give an extra tweak to the more regularly rhythmic pulls that makes high tides.

At the autumn equinox of 2015, there was a "Blood Moon" to add to another very high tide, where the Moon glows a dull red in a full lunar eclipse. Eclipses: famous moments of mysterious affect. The Hydrocitizens project we are involved with, working with a number of partners, staged public performances to coincide with both these extra high tides, the latter called "Blood Moon" and held in Bristol's Planetarium. Telling the city stories of the tides, which for various reasons, including the port moving downstream, the shutting of the tides out of the center of the city, and the replacement of ferries with bridges, are less culturally/affectively and ecologically potent in the city than they were historically. This story, or set of stories, in essence, revolves around two characters, Peri and Proxi, who are the tides made flesh as old sea dogs, created out of Severn Estuary mud and sent to Bristol by an ancient pagan deity *Sabrina* to re-awaken the city to its tidal heritage.[6]

In the tides strands of our water research in Bristol, we have sought to create performative narratives to seed into the city's imagination through a series of events held as part of Bristol being 2015 Green Capital of Europe. Based upon the two characters Peri and Proxi, we co-generated a series of events in which the characters performed in locations around the city's waterways, in schools and other public events. Each event, and the wider gathering narrative of these characters, sought to generate affective narratives and atmospheres created by the performers *and* the settings. The content included oaths of allegiance to the river and tides; hydro-poetics; rituals (dipping hands into buckets of river mud); mud painting on walls; shared singing, walking and boat trips.

The skill of the performers, humour, wit and audience interaction all play a large part in these 'eco-entertainments'. The ongoing narrative of the anthropomorphized tides and their adventures are aimed at awakening stirrings of ecological senses of self and place in relation to the tides, the city and its hydrosphere. Their story invites interest and sympathy, and, combined with the embodied experiences, creates empathy. Films of other episodes of these characters' tidal adventures in Bristol can be found on-line.[7] The characters Proxi and Peri (after the proxigean and perigean tides) were co-created by an assemblage of community organisers, artists, and academics. The performances and interactions spurred new forms of expression, stories, of which one is included here.

Writing a short script for the characters Proxi and Peri is a way of finding new and imaginative ways to tell stories. It is made possible by the affective and emotional experiences contained in the performances, and in personal histories. The performances and characters become a new nodal point from which other stories can ripple outwards, reaching perhaps towards a place of 'ecological citizenship' through engaging the senses and sensibilities.

A key missing episode from Peri and Proxi's Bristol journey[8]

The Wedding-Guest sat on a stone:

> He cannot choose but hear;
> And thus spake on that ancient man,
> The bright-eyed Mariner.
>
> (Coleridge)

(Setting: Bristol Green Capital 2015 shindig on the Bristol docks by the Severn Shed Restaurant; moonlit night; high tide—but only down river—the docks are always high water since the great engineering of the 1830s)

You going back inside?
No – let's have another fag.
So then what happens?
Well, Sabrina whispers.
Yes, But then, she is always whispering.
Of course. The true gods only ever whisper – even though sometimes it is a roar – of wind; of water; of ripping ice. But this whisper is different – in an unknown tongue – a spell, a magic incantation – in the tongue of men. Men, mark you, not women (it is the men that need conjuring). A spell of creation. Audible to men-to-be-from-mud.
Men yet to be made?
Yes, men yet to be made – or maid?
No, I don't think so.
The worm-boiled mud stirs; a few wading birds take flight; the marsh grasses withdraw their roots … Just to be sure.
Shore?
Sure.
The tide is busy elsewhere – out –
And about?
Yes. It knows which side its bread is buttered on.
The mud bulges, madly but slowly. Bubbles and then lumps, then smooths again.
The cars hum-rumbling across the great bridge hum-rumble on, blind to the genesis below them?
Yes. And the people in them. The to-be-men start to form – first like crime scene outlines; then 3D; then flex – in the flat – sliding in – then ON the mud. Then fold – upwards, elbow by elbow, knee by knee. Rise from the hip. Grey out of grey; clutching at each other.
No eyes yet?
Eyes – but sealed by mud.
They lift head – and cock an ear to the near horizon of reeds and seawall. Sabrina whispers. Instructions.
Naked?
Sabrina? God no – the very thought – who could look on her, one would simply effervesce to nothing!
No, the men.
Oh! No. Their clothes drag out of the mud with them, like brown and grey melted cheese.
Cheese – the moon?
Not in this instance, just a coincidence.
They are new not-young men.
Well, what do new young men know?
Quite. They are new old men. Men of the sea, of the trade routes. Sea dogs. Men of the tides. With fish-tales – suitcases full of old letters, locks of hair – scrimshaw, Masefield men. Conrad men.

Masefield?
Masefield – you know!
Ah yes – of course "I must…
(together) We all must …
What about Marlowe?
That was another river. Another town. You can't blame the tides for what men float on them.
This town was bad enough. So then what?
They feel flesh and blood – their own – then each other's. And then drag their suitcases….
Sea chests?
No, suitcases. A bit odd, more like travelling salesmen, they drag them to the **seawall.**
They stop there. I'll be bound.
Well, yes, but just to look, on this occasion. It is an old friend.
Enemy, maybe.
Enemy, but something to be leaned on on a calm summer's night.
They stand and look downstream and upstream, across to the other side.
To Wales.
To Wales.
And then inland.
At the catchment, the reens?
The reens. Where the eels once went with them.
So many it made her silver in the moonlight.
She was so pleased. She preened!!
I know.
And the power stations and factories.
And roads, AND houses.
At risk of flooding?
At risk of flooding.
And the railway?
The railway – and The Station no less.
(together) Severnbeach
That is one problem solved.
One of many.
And Shirley's Café?
Breakfast first. Full English fry-up for two. As slippy as mud.
And nourishing.
If you say so.
Well, satisfying then.
Satisfying.
Then a slow branch-line trip to where the river is river – the Avon.
Shirehampton.
Yes – but they need tickets.
What's in the suitcases?

They need tickets.
We will come to that.
They need tickets.
Well, yes.
The train guard – she's a character – one of Sabrina's.
Oh right.
And, as if by magic…
By magic, you mean.
Well, yea, ok, but no one knows what magic even is these days. The train is full of characters.
Sabrina's characters
End of scene one at Severnbeach railway station

Scene 2 Train ride Severn Beach to Shirehampton
(To be performed and filmed)

Scene 3 (partial)
Sabrina sleeps. Part curled on her side. Breathing.
Curled like a feather – like a crescent moon?
Not really (sings) "just like a woman" – who said that?
Belly rising and falling with the moon's turns?
Yes, but not the Breath of the Moon as the Beade said.
No, but he was not that far off – a good stab I think.
Yates?
He knew Sabrina and the white goddess were – well – acquainted.
Sabrina dreams. Lying on her back, restless. Wild dreams of floods and cataracts, of when the world – the river – and the tides were young.
When she was elsewhere. Under ice?
Under ice – or under water!!

Acknowledgements

Thanks to the editors for their patience and support and to all those involved in the parts of the Hydrocitizenship project touched upon in this piece—and to the tides and our beautiful biosphere Gaia home.

Notes

1 Ecocide is a term used by Felix Guattari in his landmark book *The Three Ecologies* (first published in French in 1989). Guattari argues that the three interconnecting ecologies that make up the fabric of flourishing life—biodiversity, cultural diversity, individual physic diversity—are all falling to ecocide driven by a toxic combination of globalised capitalism, state power and oppressive ideologies.
2 www.watercitybristol.org
3 This quote is attributed to Picasso on many websites and blogs but the original source is not obvious.

4 A non-profit institution, The Rachel Carson Center for Environment and Society (RCC), is an international, interdisciplinary centre for research and education in the environmental humanities and social sciences. Founded in 2009, it is a joint initiative of Munich's Ludwig-Maximilians-Universität and the Deutsches Museum, with the German Federal Ministry of Education and Research.
5 To see the eight points of the Deep Ecology Platform, go to www.deepecology.org/platform.htm.
6 Personifications of the tides were co-developed with artists, a youth enterprise NGO, a filmmaker and a very skilled and experienced Bristol-based Street Theatre Company, The Desperate Men, who subsequently wrote the dialogue and played the parts.
7 www.youtube.com/watch?v=u79gnQD39x8, and www.youtube.com/watch?v=HunglU2KcXM
8 The various scripts for the characters Peri and Proxi have been written by the Desperate Men Theatre Company with input from others. This last episode is written by Owain.

References

Armiero, M. and Sedrez, L. F. (eds). (2014). *A History of Environmentalism: Local Struggles, Global Histories*. London: Bloomsbury.

Carson, R. (1962). *Silent Spring*. Boston: Houghton Mifflin.

Carson, R. (1989). *The Sea Around Us*. Oxford: Oxford University Press.

Damasio, A. (2000). *The Feeling of What Happens: Body, Emotion and the Making of Consciousness*. London: Random House.

Darwin, C. (1998 [1872]). *The Expression of the Emotions in Man and Animals*. London: Fontana.

Duncan, F. (2015). Speaking on *The Loss of Lostness*, BBC Radio 4, 10 Oct 2015, www.bbc.co.uk/programmes/b06gtfql.

Evernden, N. (1996). Beyond ecology: Self, place and the pathetic fallacy, in C. Glotfelty and H. Fromm (eds) *The Ecocriticism Reader: Landmarks in Literary Ecology* (92–104). Athens, GA: University of Georgia Press.

Forester, P. (2014). You won't believe how much Nike spends on 'demand creation'. http://uk.complex.com/sneakers/2014/07/nike-spends-3b-on-demand-creation.

Guattari, F. (1989). *The Three Ecologies*. London: Athlone Press.

Harari, Y. N. (2015). *Sapiens: A Brief History of Humanity*. London: Harvill Secker.

Horovitz, A. (2011). Burials (poem), in *Turnings*. Kirby: Headland Publications.

Ingold, T. (2012). *Being Alive: Essays on Movement, Knowledge and Description*. Abingdon: Routledge.

Jones, O. (2015). 'Not promising a landfall …'. An autotopographical account of loss of place, memory and landscape. *Journal of Environmental Humanities*, 6, 1–27.

Jones, O. (2005). An emotional ecology of memory, self and landscape. In J. Davidson, L. Bondi and M. Smith (eds), *Emotional Geographies* (205–218). Oxford: Ashgate.

Latour, B. (1993). *We have Never Been Modern*. Cambridge MA: Harvard University Press.

Lefebvre, H. (2004). *Rhythmanalysis: Space, Time and Everyday Life*. London: Continuum Books.

MacDonald, F. (2014). The ruins of Erskine Beveridge. *Transactions of the Institute of British Geographers*, 39, 477–489.

Malpas, J. (2008). *Heidegger's Topology: Being, Place, World*. Cambridge, MA: MIT Press.

Morton, T. (2010). *The Ecological Thought*. Cambridge MA: Harvard University Press.
Solnit, R. (2006). *A Field Guide to Getting Lost*. Edinburgh: Cannongate.
Solonger, R., Fox, M. and Irani, K. (2008). *Telling Stories to Change the World: Global Voices on the Power of Narrative to Build Community and Make Social Justice Claims*. New York: London; Routledge.
Sebald W.G. (2007). Joseph Cuomo, A conversation with W.G. Sebald (interview). In L. S. Schwartz (ed.) *The Emergence of Memory: Conversations with W. G. Sebald*. New York: Seven Stories Press.
Sebald W.G. (2013). *A Place in the Country*. London: Penguin Books.
Seigworth, G. J. and Gregg, M. (2010). An inventory of shimmers. In M. Gregg and G. J. Seigworth (eds.) *The Affect Theory Reader* (1–25). Durham: Duke University Press.
Stengers, I. (2015). *In Catastrophic Times: Resisting the Coming Barbarism*. Lüneburg: Open Humanities Press in collaboration with meson press.
Stone-Mediatore, S. (2016). Storytelling/narrative. In L. Disch and M. Hawkesworth (eds.) *Oxford Handbook of Feminist Theory* (934–954). London: Oxford.
The Invisible Committee (2009). *The Coming Insurrection (Semiotext(e) / Intervention Series)*. Cambridge, MA: MIT Press.
Thrift, N. (2004). Intensities of feeling: Towards a spatial politics of affect. *Geografiska Annaler* 86 B (1), 57–78.
Watson, L. (1973). *Supernature: The Natural History of the Supernatural*. London: Book Club Associates.
Wemytewa, E. and Peters, T. O. (2008). Zuni River—Shiwinin K'yawinaanne: cultural confluence. In R. Solonger; M. Fox; and K. Irani (eds), *Telling Stories to Change the World; Global Voices on the Power of Narrative to Build Community and Make Social Justice Claims* (15–22). New York: London; Routledge.
Woof, E. (2015). speaking on *Breaking News—Telling Stories in Fact and Fiction*, BBC Radio 3, Tue 10 Nov 2015. www.bbc.co.uk/programmes/b06ns0qc, accessed 18 January 2016.
WWF. (2014). *Living Planet Report 2014: Species and Spaces; People and Places*. Giland WWF.

13 Eat your primary sources!

Researching and teaching the taste of history

Ian Mosby

History has a distinct taste. Actually, it also has a distinct smell, feel, sound, and look to it, but, as a historian of food and nutrition, I always find myself coming back to the taste of history. No, I'm not talking about the musty, acrid taste of dust and mildew as you open up a long-neglected archival box or that weird metallic aftertaste you get after sitting in front of a microfilm reader for way, way too long. History can also taste like molasses, cloves, nutmeg, raisins.

History, in other words, can be delicious (or nausea-inducing, depending on which sources you use.)

At least this is what I tried to prove to my students when I recently had the good fortune to teach undergraduate courses on the history of food at the University of Guelph in Ontario, Canada. Not only did I want to teach my students about the versatility of food history as an entry point into the history of science, immigration, colonialism and gender, but I also wanted to prove to them that, as budding food historians, they should always make sure to *eat* their primary sources.

For instance, while lecturing on the history of the Canadian home front during the Second World War—my own particular area of expertise—I tried to communicate some of the ways in which social class determined Canadians' individual experiences of wartime rationing and shortages. I did this by presenting students with two cakes. One unmarked container held Canada War Cake—one of the most popular wartime recipes for a sugar-stretching, eggless, milkless, and butterless cake—based on a March 1942 recipe from the *Windsor Daily Star* (see Figure 13.1).

I also offered them a Depression-era cake called Poverty Cake—also often known as Economy Cake or Depression Cake—from a March 1930 recipe featured in the *Vancouver Sun* (see Figure 13.2).

The point of the exercise was a rather simple one. The cakes, for all intents and purposes, taste and look almost exactly the same. When asked whether they knew which one was Poverty Cake and which one was War Cake, the class was divided because, really, there weren't many tangible differences between the two. They both tasted good—if a bit chewy—yet neither was particularly sweet or decadent. Instead, they tasted like spiced, dense raisin bread. Both of them, I argued, told us something about the taste of war on the Canadian home front.

Canada War Cake

(Also from last war—contributed by Mrs. Graham)

Two cups sugar, 2 cups hot water, 3 tablespoons lard, 1 teaspoon salt, ½ teaspoon each of cloves and cinnamon, 1 package seedless raisins. Boil this together for 5 minutes and let cool. Then add 4 cups flour with 1 teaspoon soda dissolved (in 1 tablespoon water). I add also 1 teaspoon baking powder

Thank you for this unusual version of lemon pie—it was a great success with our family when I tried it.

Readers, I think you have been grand to respond to this appeal for wartime recipes as you have, but I feel that the resources of the majority of our cooks have not even been tapped. You know it is up to us to make this war effort count, and unless we try to the utmost (as our enemies are doing) there are dark days ahead. Send us your recipes that have been tried and that you know are economical and that will conserve Canadian food

Figure 13.1 Canada War Cake

Source: *Windsor Daily Star*, 14 March 1942.

Readers' Recipes

A prize of $1 will be awarded each day for the best recipe sent in by The Vancouver Sun readers. Address letters to the Recipe Editor, The Vancouver Sun. Write clearly on one side of the paper only.

Today's award for the best recipe goes to Mrs. S. H. Green, R.R. No. 2, New Westminster, B. C.

POVERTY CAKE

Boil together three minutes, 1 cup brown sugar, 1 cup hot water, 1-3 cup lard or cooking oil, 1 cup seeded raisins, ½ teaspoon nutmeg, 1 teaspoon cinnamon.

When cool add 2 cups flour sifted with ½ teaspoon each baking powder, soda and salt. Bake in square pan in moderate oven 35 minutes.

Dates or nuts may be used in place of raisins if desired.

As there are no eggs, milk or butter used this is a very economical cake.

Figure 13.2 Poverty Cake recipe

Source: *Vancouver Sun*, 18 March 1930.

For those who had suffered from long-term unemployment and underemployment during the 1930s, for instance, War Cake likely had a decidedly familiar taste. In fact, my own research has shown that many of the so-called "War" recipes printed during WWII were—like War Cake—often indistinguishable from the Economy, Economical, Poverty, and Low-Cost recipes that filled Depression-era cookbooks and newspapers. Such recipes therefore highlight the ways in which working-class and lower-income families didn't experience wartime rationing or shortages as a particularly significant shift in their daily eating habits. They were already skilled in stretching dear or expensive ingredients such as butter, sugar, tea, coffee, meat and preserves—the precise foods that were rationed in Canada during WWII—and were therefore well-prepared to make do with fewer of these items during the war in much the same way they had for years.

But for middle-class and wealthier families who had escaped the ravages of unemployment during the 1930s and therefore didn't personally experience it as a decade of hunger and want, the war truly did require drastic cutbacks in the kitchen. War, for these families, really did taste a lot more like sacrifice and austerity and, to that end, it was a unique culinary experience that harkened for them back to the previous war far more than the hard years of the 1930s.

Just as importantly, though, War Cake also tasted like Canada's precarious global supply chain. After all, sugar shortages were the result of Canada's dependence on overseas sources of sugar cane, a commodity whose supply was profoundly disrupted by unrestricted submarine warfare. While Canada did produce some of its own sugar in the form of sugar beet production, it was insufficient to maintain pre-war quantities of sugar. Canadians, in other words, were forced to do with less.

The same was true when it came to the cake's use of a small amount of lard rather than butter. Because fats were a key component in munitions production —and because Canada had long been a net importer of vegetable oils—butter became rationed and other sources of cooking fats became more and more scarce. It was for this reason that the government bombarded Canadians with ubiquitous propaganda images enlisting women, in particular, to "Save Fats and Bones to Help Smash the Axis!" (see Figure 13.3)—a message that helped bring home the need to make do with less, given Canada's commitment to total war but that also reflected Canada's profound dependence on overseas food production by the 1940s (for more context, see Mosby 2014).

Historical recipes, in other words, can be fun, tasty and helpful classroom props. But they can also teach students important lessons about the historian's craft, particularly when it is students themselves who do the cooking. That's why I designed an assignment that required each student to give a five-minute presentation on a recipe of their choice that in some way related to that particular week's readings and that they found in one of the more than 13,000 cookbooks housed in Guelph's fantastic Culinary Collections archive (Guelph 2015). In order to complete the assignment, students needed to use their chosen recipe to assess broader course themes and also to discuss some of the challenges they faced while attempting to recreate their historical recipe of choice.

Figure 13.3 'Save Fats and Bones' (government propaganda message)

Source: *The Cariboo Observer*, 27 March 1943.

The students quickly learned that, while many of these old recipes could be surprisingly delicious, they also posed a number of significant challenges. Many of the nineteenth and early-twentieth century recipes they found, for instance, often failed to include important information such as how long to cook something and at what temperature, how much a handful of sugar or a knob of butter might be, or what a slow oven refers to. And if students could actually read the handwritten recipes from old manuscript cookbooks, they often found that the books were designed for kitchen equipment that no longer exists outside a museum or, in other cases, that our highly processed, standardized modern ingredients act nothing like their historical counterparts.

The most important lesson I think that the students came away with was that primary sources—whether recipes, letters or census tracts—need to be read carefully and with a decidedly critical eye attuned to the source's particular historical context. Students learned quite a bit about what different forms of culinary literature said about women's lives, labour, and everyday life; the revolutionary impact of changing household technologies; and the industrialization of food and the transformation of print culture. But they also learned quite a bit about the limitations of primary sources more generally. When you misinterpret a recipe or fail to account for its historical idiosyncrasies, the result is immediate: the cake fails to rise, the meat is undercooked, or the soufflé collapses in on itself. When you fail to do the same for our reading of the census, on the other hand, the result is less immediately obvious but no less devastating: your data is simply wrong and your conclusions turn out to be only half-baked.

At the same time, this assignment firmly places taste at the center of the classroom experience. Because I encouraged students to bring in a sample of what they made for the rest of the class to try, class discussions often veered into the sensory. After all, people in the past—living in very different circumstances from ourselves—often had very different tastes and preferences than our own. Some students were surprised by the seemingly bizarre mixtures of sweet and savoury in mid-twentieth century Jell-O salads—a dish that has mostly disappeared from Canadian tables—while others discovered the hard way that foraged cat-tail roots taste like muddy rope or that a banana-sardine salad included in an early-twentieth century advertising cookbook was designed for a very different palate than our own (a flavour combination that I still think back upon with horror and revulsion).

But studying the taste of history is more than just a novel way of engaging with students. In fact, it is a key tool available to teachers for opening students' eyes to the profoundly important role that the senses have historically played in determining important changes to societies, empires, economies and environments. In fact, the fields of both food and environmental history are filled with important examples of this. Take sugar, for instance. As Sidney Mintz has shown, changing European preferences for "sweetness" helped drive not just the ecological transformation of large swaths of the Americas to make room for massive sugar cane plantations, but also spurred the devastating development of the Atlantic slave trade (Mintz 1986).

Taste plays a similarly important role in the environmental and social histories of other food commodities. John Solouri's research on bananas, for instance, shows how the taste preferences of American consumers, in particular, played a key role in the development of vast monocultures of the Gros Michel banana variety and—following the Gros Michel's almost total devastation by Panama disease in the first half of the twentieth century—the now ubiquitious Cavendish banana variety. This demand for a specific vision of what constitutes both a "good" tasting and "good" looking banana has had profound ecological, social and political consequences for the mostly South American countries where bananas were and are produced (Solouri 2002).

Stuart McCook's research on the global history of coffee cultivation tells a somewhat parallel story in which the spread of the plant disease coffee leaf rust helped to create a global division between regions producing what self-described coffee connoisseurs in North America would consider "good" and "bad" coffee—meaning arabica and robusta coffee varieties, respectively. McCook's research provides the important historical context for how North America's current preference for arabica coffee varieties is very much bound up in the global environmental, cultural and political histories of colonialism, capitalism and international development. "Good taste," after all, is very much a social construction, even when it comes to coffee (McCook 2013).

Matters of taste, in other words, are sometimes drivers of important historical changes. It helps, too, that they're a great way to get students interested in history. Museum curators and public historians have long known that the sensory experience of history can sometimes be more powerful and illuminating than the text-centric narrative structure that historians tend to rely upon. It's therefore worth exploring the ways in which we can apply some of these lessons to both the classroom and our own scholarly research. There's now a growing important literature on the history of the senses—with Joy Parr's *Sensing Changes* (2010) being one of the best examples in the field of environmental history—and it's up to us to use this literature in order to make the past more tangible and sensible in our own teaching, research, and writing.

References

Howes, D. (2014) *A Cultural History of the Senses in the Modern Age, 1920–2000*. London: Bloomsbury.

McCook, S. (2013) The ecology of taste: Robusta Coffee and the limits of the specialty revolution. In R.W. Thurston, J. Morris and S. Steinman (eds.) *Coffee: a Comprehensive Guide to the Bean, the Beverage, and the Industry* (248–261). Lanham, MD: Rowman & Littlefield.

Mintz, S.W. (1986) *Sweetness and Power: The Place of Sugar in Modern History*. New York: Penguin.

Mosby, I. (2014) *Food Will Win the War: The Politics, Culture, and Science of Food on Canada's Home Front*. Vancouver: UBC Press.

Parr, J. (2010) *Sensing Changes: Technologies, Environments, and the Everyday, 1953–2003*. Vancouver: UBC Press.

Smith, M.M. (2007) *Sensing the Past: Seeing, Hearing, Smelling, Tasting, and Touching in History*. Berkeley, CA: University of California Press.

Soluri, J. (2002) Accounting for taste: Export bananas, mass markets, and panama disease, *Environmental History*. 7: 386–410.

Jütte, R. (2005) *A History of the Senses : From Antiquity to Cyberspace*. Cambridge: Polity Press.

University of Guelph Special Collections. (2015) Culinary collections, www.lib.uoguelph.ca/find/find-type-resource/archival-special-collections/culinary.

14 Political effluvia

Smells, revelations, and the politicization of daily experience in Naples, Italy

Marco Armiero and Salvatore Paolo De Rosa

A sense of smell

No matter how much one tries to seal windows and doors, or to close one's eyes and ears, some things always find their way even into the most occluded lives. Naples, Italy, is the kind of city which does not leave you alone; sounds, perfumes, colors, and tastes stitch together the urban fabric as much as its architecture. Although always present in the city, garbage has become a cumbersome problem, especially since the late 1990s and early 2000s. With the streets flooded by garbage and people protesting, it is an issue that is hard to ignore. The stench of waste and the cries of people have even reached the academic world, literally entering through our office windows and awakening our senses. Solicited by this call from the world outside the university's walls, we have attempted to make 'sensing' a fundamental tool in our research. In this contribution, we present research methods that account for the relevance of senses for the politicization of people involved in the socio-environmental conflicts over waste mismanagement and illegal dumping in Naples and its surroundings. In the following pages, we first discuss the challenges of integrating sensorial experiences, and smells in particular, in both academic research and activists' practices—and sometimes at the crossroad of the two. We then delve into the specific methodologies we have devised in our research: (i) walking interviews concerned with sensations, (ii) oral sensorial histories, and (iii) "toxic tours."

Those who have visited a dump or passed through the so-called Land of Fires[1] know that it is an experience involving the whole body, where smells penetrate into the viscera, taking control of one's reactions, and impregnating clothes, hair, and skin. In Naples, the smell of waste has come to occupy the mind, becoming a daily obsession, which changes everything one sees and consumes.

As scholars, we are not trained to follow our senses, even less our emotions. Actually, for many of us the golden rule is "detachment": the further the distance between researcher and researched, the better will be the result. Some academics, maybe of the older generation, may still hear their supervisors list the dreadful consequences of becoming too involved with their object of study. Luckily enough, students do not always follow their supervisors' advice. Clearly, for those who identify as radical scholars, as we do, things are different. Radical scholars take stances and propose engagement instead of detachment. After all, it is not difficult to find inspiration and comfort in prestigious scholars who have followed that path, such as E. P. Thompson, E. Hobsbwam, H. Zinn, and D. Harvey. Today there might even be a return of the political. However, we think that the challenge of the senses goes beyond a political approach to research. The senses involve the body and its relationships with the surroundings and not only the positionality of the researcher within the geographies of political allegiances and social hierarchies.

Nonetheless, we do see a continuum between a political stance and sensorial research. We argue that the political is deeply embedded into the sensorial, even within the bodily experience of affected people. This awareness is already present in the work of Marx, whose insights on the political potential of situated sensuous experiences have been unearthed and discussed in recent years by several authors (Foster 2000). In particular, Alex Loftus (2012) has taken the sensorial engagement of humans with the socio-ecological relations making up their environments as the point of departure for thinking through an emancipatory praxis of everyday life. Following urban political ecologists, he contends that the social relations of production and the unequal distribution of power organize the concrete manifestations of the appropriation and transformation of external nature for societal needs. These processes of metabolic exchange make up the environments where we carry out our lives, and are experienced and mediated through the senses. Therefore, for Loftus, sensorial engagements shape the possibility for a politicized response to unjust metabolic relationships. Every time water does not come out of the tap, waste accumulates in the streets, air becomes unbreathable, food is contaminated, we are hit with thirst, impaired mobility, bad smell, and dangerous eating; but herein also lies the possibility for a politicization of the processes that produce specific assemblages of humans and non-humans. Senses allow for the detection of "metabolic fractures," and this consciousness becomes a precondition for rethinking the urbanization of nature and the making of the person in relation to it. Senses can even detect the penetration of capitalist relations in the body, its subsumption and its transformation into a commodity and a machine. Capitalism controls the very ecologies of the subordinate body, its cells, its cure and esthetics. In the face of the capitalist appropriation of the body, a revolutionary project of emancipation could pass through bodily awareness and sensuous engagement as the basis for engendering a transformative politics. We argue that the sensorial experience is part and parcel of the process of politicization of subaltern communities which, as Alf Gunvald Nilsen and Laurence Cox have stated, provides clues to understand the underlying structures of injustice beyond the particular features of daily oppression (Nilsen and Cox 2013, p. 73).

Especially in environmental justice struggles, the body often becomes the first place of politicization, or we may say of subjectification; it literally becomes the space where people experience the oppression of capitalist relations and the opportunity for building resisting communities. This is the case of Indigenous people affected by extractive capitalism, of workers exposed to industrial hazards, and of all the subaltern communities whose neighbourhoods become the dump sites for the well-being of elites. The centrality of the body challenges the alleged separation between economy and ecology, production and health, but it also questions the production and legitimation of knowledge. What happens to the body is a controversial issue. Experts and government agencies claim to own the body, or at least to know the language to understand it. It is not up to workers or city dwellers to say if a factory or a neighborhood is killing them. Our argument is that in challenging the expropriation of bodily knowledge, the senses play a fundamental role. In place-based struggles, people can reactivate the body in relationship to external nature, choosing to listen to its signals. Unequivocally, the sick body speaks loudly, although not always in an understandable language. However, in this essay we will not focus on the sick body, but rather on the space where the body experiences its surroundings. Sometimes, senses have been consciously mobilized in experiments of countering official monitoring programs, noses and eyes versus sophisticated technologies, to expose the contamination affecting the community. This is, for instance, the case of the Community Environmental Monitoring in Chennai, India. Here, an NGO has challenged the official truth about the contamination in the area by implementing a grassroots system of monitoring based on the bodily experience of affected people. As Shweta Naryan, the coordinator of the program, explains, the so-called bucket brigades have been instrumental in forcing the government and the corporations to address the air pollution in the industrial district of Tamil Nadu.[2] Before people organized their own independent monitoring, Naryan said, "even if the air smelled like rotten cabbage or eggs, local governments in India wouldn't respond to public protests because protestors were considered liars, not scientists" (Global Community Monitor 2013).

In *Noxious New York*, Julie Sze (2007, p. 181) has also illustrated how the bucket brigades have been instrumental in stimulating public intervention in poor neighborhoods by quantifying "bad smell" in air samples. Jason Corburn (2005, pp. 173–199) has similarly described how grassroots organizations in New York have challenged the assessments of the Environmental Protection Agency, broadening dramatically the scale of observation in order to include the direct observations and sensorial perceptions of residents.

In the above cases, the sensorial understanding of the environment has been somewhat systematized, and thereby made accessible for the work of scholars, including environmental historians and political ecologists. However, in most cases, the sensorial experience is elusive to research, especially in its historical dimension. The paucity of sources, together with an ideological repulsion for the sensorial and emotional way of knowing, has impeded research on the topic. Nonetheless, some scholars have tried to enter into the "smellscape," as Douglas Porteous (1985) has defined it. In 1982, Alain Corbin published what remains the basic text for anybody who wants to explore smells in history. Corbin came to smell from a classical social history background, proving once more the possible, though rather underdeveloped, connections linking smell to environmental history. Corbin focused mainly on the perceptions of smells and showed how smell as a field of interest and study has been progressively dismissed by the increasing authoritarianism of modern science. There was almost no room for the nose in the modernistic laboratory. Smell did not submit easily to science for two main reasons: on the one hand, smell was too subjective, almost impossible to measure and catalogue. On the other hand, smell belonged to an idea of the world that kept together internal and external ecologies, thinking about the body and the environment as being in a continuous exchange. This was a concept that was difficult for modern scientists to comprehend. For them, smell spoke the language of effluvia, not that of bacteria or viruses.

The agency of smell as a highway placing in communication the body and the environment is central in the work done by the few environmental historians who have tackled that topic. Environmental historians' meager attention to smells has gone hand-in-hand with the poor understanding of the connections between bodies' ecologies and external nature. As a matter of fact, the environmental historians who have addressed the issue of smells are generally those who have been more attentive to the metabolic relationships between environments and people. Indeed, also in the empirical case at the basis of this chapter, the awakening of the nose is linked to a context in which health and environment blend in the toxic biographies of the affected people (Chiang 2008; Nash 2006; Bolton Valenčius 2002).

Stories of odours

Since the early 1990s, Naples, the capital city of the Campania region in Italy, and its surroundings, have become the global icon of the urban ecological disaster. An area of 3,810 square kilometers—the provinces of Naples and Caserta—inhabited by approximately four million people, has turned into a huge trashcan. The synergy between government-sponsored urban waste management project and a complex network of mafia groups, industrial managers, corrupt white collar and public officials, has created a system of profiteering from the disposal of waste at the expense of local livelihoods. On the one hand, the authoritarian governance of the urban waste management for the entire region, framed by the government as an "emergency," has become an attractive business for private investors who enjoy legal derogations and optimal contractual conditions for building waste facilities based on incinerators, landfills and storage sites. On the other hand, the infamous trafficking, haphazard dumping, open-air burning and illegal disposal of hazardous byproducts, mostly from industrial production, has turned every available hole in the region into a sink of toxic scraps. Both processes have shifted the environmental costs of industrial production and waste disposal onto communities and local ecologies. This has translated into a toxic environment for the local population, affected today by threats to health, the reduction of cultivable land, and stigmatization. More than 2000 potentially contaminated sites were recorded in 2008 by the Regional Agency for Environmental Protection, and the complex links between those contaminants and the increasing cancer rates among the locals have been documented by several scientific studies.

Figure 14.1 Toxic waste ghosts in Casal di Principe

Source: Photo courtesy of Mauro Pagnano

The hazards to public health have been recognized by the central government through the insertion of three wide areas within Campania in the national record of polluted sites in need of remediation, together with 50 other places all around Italy. Nevertheless, cleaning-up works are still minimal and the authoritarian governance of environmental management and land-use planning still dominates the regional landscape.[3] Naples is a city that is literally on fire; for journalists looking for sensationalism and dark stories, Naples is the Promised Land.

Images work well in the construction of the dark tale of Naples. What is better than photographs and videos to transmit the sense of an apocalypse? But although extremely powerful, images are unable to uncover the subterranean paths at play in the Neapolitan garbage drama. Although continuously evoked in all the analyses, the mafia has stayed invisible in these visual representations. One can guess its presence in the landscape or make a metonymic connection between an object and the entire criminal organization, but it is not self-evident. The mafia is part of an underworld that can become visible only through more complex narratives. Undoubtedly, Roberto Saviano's (2006) *Gomorrah: Italy's Other Mafia* contributed dramatically to expose that reality to a wider public.

Figure 14.2 Toxic waste ghosts in Casal di Principe

Source: Photo courtesy of Mauro Pagnano.

The visual is also almost blind to the toxic waste buried in many places in the region. By definition, this is part of the underworld, not visible on the surface. In other words, there are things, and even more processes, which cannot be shown visually. The relationship of local people with waste, especially with toxic waste, is one of those processes which run under the skin of both land and humans.

Sensitive methodologies

During our fieldwork in Campania, once it became clear that the senses and physical perceptions are an important motivating factor in some people becoming activists, we began to devise specific research methodologies that could account for the sensorial as political, and that could detect within life trajectories those turning points linked to sensorial experiences of metabolic fractures. The activists had their own methodologies for understanding the radical environmental changes happening around and inside them. One of these methods, enacted by several activist groups within more than fifteen years of social mobilizations in the region, was the mapping of the countryside surrounding the towns and the cities. Armed with simple technological devices, a GPS and a camera, activists criss-crossed rural areas, suburbs and brownfield sites, documenting the places where waste was illegally discharged, abandoned, and, in some cases, set on fire. One of the most successful campaigns of this kind was promoted in 2013 by the Coordination Committee against Toxic Fires (CCF): in a single day of denunciation, activists collected hundreds of photographs and geographical locations of dumping sites, assembling a map of the state of the countryside and turning it into a collective complaint to the institutions in charge.

Toxic tours have been another crucial tool employed by activists in order to expose the contamination affecting their communities.[4] The practice of toxic tours is rather common in environmental justice struggles; it empowers local people, recognizing their knowledge, and builds connections and solidarities to support local struggles. It was through a toxic tour in 2007 that activists were able to shift the public attention from the urban trash in the streets of Naples towards toxic contamination in the outskirts of the metropolis. It was a truly Copernican revolution in the understanding of the waste crisis in Campania through which activists revealed both the ecologies and the politics of contamination in subaltern communities. This revelation occurred not through the usual tool of scholarly intervention, the written text, but through an experiential exploration of places that appeals to intellectual, sensorial, and emotional understandings. Our participation in toxic tours has had a twofold meaning for our research: it has implied the need to open up the canon of knowledge production, renouncing any pretense of monopoly, and has also challenged a positivist approach to sources. While toxic tours were evidently crucial sources of information for our research, our participation cannot be reduced to a mere extraction of information. We have been instrumental in organizing toxic tours and involving international scholars. Therefore, our sources were not just out there waiting to be mined, but we have actually contributed in creating the very sources we wished to use.

By performing these investigations several times, some of the most engaged activists developed a special sensitivity and became able to recognize different kinds of waste, inferring their hazardousness by smell. Our method of research to access the experiences in which perception was intertwined with political motivation and spatial focus has been to elaborate further the "walking interview" devised by Evans and Jones (2011). In their pilot project with the walking interview, Evans and Jones (2011, p. 858) utilized a GPS and a voice recorder to produce geo-referenced descriptions, or spatial transcripts, of local people's connections to place during their walks with interviewees that were familiar with the area. They show how mobile interviews and rigorous design generate more place-specific data than sedentary interviews and produce "a decidedly spatial and locational discourse of place, which is structured geographically rather than historically." In five interviews with Campania's activists, in which respondents were leading us on a route decided by them, we complemented the method of walking interview with sensorial data, by recording the detections of, and the reactions to, specific smells in the course of explorations in the countryside of two towns. In this way, we could highlight in the landscape zones of relevant sensuous engagement, often not recorded in official maps of contaminated sites.

Figure 14.3 Borderlines

Source: photo courtesy of Mauro Pagnano.

Andrea, a farmer in the town of Acerra, guided us in the no man's lands between old factories and cultivated fields in the middle of the countryside, and just by following his nose he uncovered a location where waste was habitually burned illegally. In his walk with us in the outskirts of a small town, Maddaloni, the memories of the past beauty of the land was continuously superimposed on the current disorder that surrounded us, offering insight into the tension felt by the locals between a past of spring waters and woods, and a present of chemical puddles and unknown materials. For him, the memory of past smells, buried today under disturbing scents, was a constant encouragement to struggle for the reclamation of "his" land. The value of this method is not only to offer a spatialized account of people's relation to place and sensorial reactions, but it could also act as a base for institutions to localize in the landscape areas in which to conduct deeper analysis and assessments of environmental monitoring.

While complementing the walking interview with sensorial detections helped us to deepen the geographical dimension of people's reactions to perceived environmental threats, an attention to smells when collecting oral histories, or rather to memories of smells, informed our research by allowing us insight into the ways experiences of environmental changes were symbolized in shared memories and acted as political triggers. In collecting oral histories, we gave particular importance to the relations linking sensuous experiences of specific places to the emotional attachment felt by the interviewees in different moments of their life.

In this way, we were able to reconstruct the links among physical environmental change (when and how the environment was changing), sensuous perceptions (the moment in time when bad smells arose), emotional reactions (the shift from happiness to anger when going to specific places), and the emergence of the political will (when and why the interviewee decided to "do something"). These oral "sensorial" histories allow for a specific focus on the space where people and their surroundings meet and are symbolized. Our role has been to create the stage for the interviewees to produce narratives of their bodies' sensitivities within specific environments, aiming to uncover the experiential base that feeds the activists' motivation.

In order to undermine the dichotomy of researcher and researched, we have also experimented with a laboratory of guerrilla narratives in which we invited a group of women to write their own biographies of contamination and politicization (Armiero 2015). Borrowing from the collective of Italian radical novelists, Wu Ming, we have argued that "stories are axes of war to be unearthed." In almost all the biographies we have collected—several also used in this chapter—sensorial, bodily experiences are crucial in mobilizing people, transforming victims into political subjects.

Our oral history project is packed with stories of people discovering the mysterious ways through which waste continuously crosses boundaries and spills into personal lives. Lucia, for instance, recollected for us the episode that transformed her from a "normal housewife," borrowing the beautiful self-definition of Lois Gibbs,[5] into a *pasionaria*.[6] Lucia tells that while studying English in a public education program, a terrible smell erupted in the classroom coming from the nearby dumps, making it almost impossible to carry on with the class. Probably, the public-sponsored English program was part of some kind of plan to pursue equal opportunity in education, but the stench coming from the dump was a reminder that environmental inequalities will always find a way to reach people, maybe entering through the window. However, that odor did not stop Lucia and her friends from studying; maybe it stopped the English course, but they started studying life cycle energy, the effects of toxic waste on health, and the recycling and disposal of garbage.

In Lucia's story, the smell performed a function of revelation, symbolically and materially. It broke through the barriers we build to protect our lives from the "external." Smell proves that no barrier is thick enough to keep the flux of toxicity outside our bodies. As Alain Corbin (1986, p. 7) has written, "The nose, as the vanguard of the sense of taste, warns us against poisonous substances. Even more important, the sense of smell locates hidden dangers in the atmosphere. Its capacity to test the properties of air is unmatched."

We are tempted to argue that the nose makes visible what stays often invisible. Of course, we do not mean to say that the piles of garbage are invisible to the people living on the edge of dumps; however, we believe that smell is an immaterial bridge able to connect the space of normality and that of the extreme otherness, acting as a political trigger that unsettles everyday life, suggesting the possibility that waste can actually enter in the human body. In her autobiographical notes, Doriana explains how the blowing of the wind changed forever the secluded paradise she and her friends believed to have built for themselves just a few kilometers from downtown Naples (Sarli 2015). The wind sweeping the beautiful house Doriana and her friends had bought on the slopes of the Astroni hill brought the sickening smell of the Pianura landfill, the oldest dump in the region, active from the 1950s, to her doorsteps. The smell uncovered the illusion of seclusion and at the same time the radical ignorance of the city elites. Apparently, Doriana and friends did not know that they bought their little corner of paradise in the neighborhood which had swallowed their garbage for decades. Overwhelmed by the stench of the landfill, Doriana began her own politicization, which would bring her to become one of the leaders of the Pianura movement.[7] Sometimes the stench does not stay in the air, and does not stop at the nose of people. Nunzia recollects the sense of dirtiness soaking her entire body when she used to patrol the lands around Naples, deeply affected by the illegal disposal of toxics:

> As I was getting nearer to the fence, my throat was burning more and more, but not only the throat, also the eyes, the face, I felt pervaded by the stench, it was not anymore only an olfactory sensation. The smell was becoming more intense, the air more thick and cloudy, suffocating. It seemed as if I was not only smelling with the nose and the mouth, but also with my skin which was soaked with that stench, with those substances. I came back home and jumped in the shower, the smell was following me everywhere, I was soaked in that stench to the bones. I was scratching my skin with the sponge, trying to erase that smell which was pervading me, but it did not go away. I looked devastated, with my face red and eyes swollen with tears and then again that smell, still with me.
>
> (Lombardi 2015, p. 31)

In the case of Nunzia, the sickness that penetrated her body seems to confirm the permeability of the human ecosystem to external agents.

In the so-called Land of Fires, odours and images figure in the making of an ecological hell. While the black smoke of the toxic fires colors the skies, the stench emanating from combustion is a familiar presence for the inhabitants. Father Maurizio Patriciello, the energetic priest who has become an environmentalist leader, described the night when the stench transformed his life:

> It was deep night. I woke up suddenly trying to breath. A disgusting stench had broken in through the window. It had invaded the room and stole all the air. Drowned in that smell, I reached the window (…) but there was no difference between the outside and the inside. (…) The stench changes your life. It is not Cogito ergo sum. Rather it is Olfacio ergo cogito. Sniff and get angry.
>
> (Demarco and Patriciello 2014, pp. 51–54)

Similarly, a life-changing sensuous experience was the one that occurred to Chiara: to prepare food for her family was not the same anymore after she got involved in her town's grassroots movement against illegal waste disposal. When they started mapping the countryside, looking for the dumping places utilized for the criminal discharging of toxic byproducts, she saw the disorder of plastics, barrels, sludge and scraps popping up amidst cultivated fields, and she smelled the revolting odor of chemicals mixed with the fresh scent of leaves and earth. Knowing that many of the vegetables she could buy in her town came directly from those fields made her suspicious of the food she was giving to her children. This fear could have become a source of despair and immobility, but luckily, it gave her the motivation to better understand the metabolism of contaminants within soil, plants and organisms, and the determination to fight back those processes turning "her" food into dangerous source of unknown threats.[8]

Conclusions

In this essay, we provide three methodological approaches that specifically address sensuous experiences and that make them an object of analysis for enriching the understanding of the links between environmental changes and social mobilizations. In the interviews we have thus far collected, smell plays two basic functions. It performs the function of the revelation in hagiography; smell is an olfactory apparition which asks for conversion almost in a literal sense. In their stories, our informants explain how the smell changed their lives, pushing them to dedicate energies and time to the cause. The smell makes the miracle to transform a passive consumer into a rebellious subject. We argue that the power of stench lays in its "ability" to establish a bodily connection between the toxic landscape and the potential toxic body. The toxicity is not anymore only in front of the observer, as, for instance, in the piles of garbage spread everywhere in the landscape, but it enters into the body through the nose and reaches the viscera, making people sick as it sticks on clothes, skin, and hair. Smell is the passe-partout which opens the door of the body, revealing that the landscape is never only out there. Of course, as in every account of dramatic conversions, also this narrative might stress too much the apparition, the ethereal event which breaks into the normality and changes it forever. Many times it is not only the smell but a more articulated set of events, relations, books, meetings and much more. In addition, the stories we have included in this chapter offer a richer explanation of the political subjectification, which can never be explained only with an olfactory epiphany. Nevertheless, we claim that the smell is more than a rhetorical tool in the making of a rebellious self-narrative. Recognizing the centrality of the nose implies a bodily understanding of politics. The space of the political is not restricted to the mind or the mouth. The nose breaks with the usual way of acting the political, reminding us that the materiality of the body is not only the terrain for governmentalizing projects but also the very agent of resistance. The nose also blends the political and the personal, overcoming a masculine practice of politicization which strongly separates feelings and sensations from political acts. By attending to smells and to other sensorial engagements with environments, researchers can integrate in their accounts and explanations a fundamental dimension of human experience often overlooked. We are aware that smells are rather intractable objects of analysis, and that they can open up complexities that are difficult to fit into a neat research design. However, we do believe that critical researchers should start taking seriously the ways ordinary people experience their surroundings and motivate their political activation. Nobody can really do science, or anything else properly for that matter, if surrounded by sickening smells. So, instead of trying to eliminate or ignore the smell, would it not be more "scientific" to start taking smell seriously and follow it? Indeed, as Arundhati Roy (2009, p. 52) has written, in order to understand history we need to "smell the smells."

Acknowledgements

Marco Armiero's research project, "Lares—Landscape of Resistance: Science, Power, and Environmental Justice in the Struggle over Garbage and Incinerators in Contemporary Naples, Italy," has been funded by the Marie Curie Intra European Fellowship under the Seventh Framework Programme for Research (FP7). Research for this paper benefitted from the People Programme (Marie Curie Actions) of the European Union's Seventh Framework Programme; under REA agreement No. 289374—"ENTITLE." The authors express their gratitude to photographer and activist Mauro Pagnano for allowing the publications of some of his pictures in this chapter. The images are part of his project on environmental injustice in Campania. For more information on his work, visit www.mauropagnano.it.

Notes

1 The term refers to the area between the northern province of Naples and the southern province of Caserta, utilized by a network of criminal organizations and business owners as an open-air dump where hazardous and nonhazardous wastes are put on fire. This definition, coined by local activists, has been picked up by all Italian major newspapers in their reports on the Campania socio-environmental situation (e.g. cf. *La Repubblica*, *Corriere della Sera* and *Il Mattino* from January to March 2014). The law approved in February 2014 as a government response to the social mobilizations of Campania people (law n.6/2014) has been labeled by politicians and media the "Land of Fires law."

2 The Bucket Brigades are self-organized teams of residents who monitor air pollution with low-tech tools, including their own senses. On the bucket brigades see Ottinge (2010).

3 On the waste crisis in Campania, see D'Alisa (2010); Armiero and D'Alisa (2012); Armiero M. (2014).

4 Phaedra Pezzullo defines toxic tours as "noncommercial expeditions organized and facilitated by people who reside in areas that are polluted by toxins, places that Bullard (1993) has named 'human sacrifice zones.' Residents of these areas guide outsiders, or tourists, through where they live, work, and play to witness their

struggle" (Pezzullo 2004, p. 236). A wonderful experiment merging activism and scholarship is Pulido, Barraclough and Cheng (2012).

5 The interview with Lois Gibbs is available online at www.youtube.com/watch?v=PrzqFPego4A (accessed on 1 November 2015).

6 The interview is in the possession of the authors.

7 Pianura is a working-class neighborhood in Naples that was the theatre of a strong anti-dump movement between December 2008 and January 2009. On the Pianura case, see the EJOLT Atlas at http://ejolt.cdca.it/conflitto/discarica-di-pianura (accessed on 1 November 2015).

8 The interview is in the possession of the authors.

References

Armiero M. (ed.) (2015). *Teresa e le altre. Storie di donne nella terra dei Fuochi*. Milano: Jacabook.

Armiero M. (2014). Garbage under the vulcano. Fighting for environmental justice in Naples, Italy, and beyond, in M. Armiero & L. Sedrez (eds.) *A History of Environmentalism: Local Struggles, Global Histories*. London and New York: Bloomsbury.

Armiero M., and D'Alisa G. (2012). Rights of resistance: The garbage struggles for environmental justice in Campania, Italy. *Capitalism Nature Socialism*, 23(4), 52–68.

Bolton Valenčius C. (2002). *The Health of the Country: How American Settlers Understood Themselves and Their Land*. New York: Basic Books.

Chiang C. Y. (2008). The nose knows: The sense of smell in American history. *The Journal of American History*, 95(2), 405–416.

Corbin A. (1986). *The Foul and the Fragrant: Odor and the French Social Imagination*. Cambridge: Harvard University Press.

Corburn J. (2005). *Street Science: Community Knowledge and Environmental Health Justice*. Cambridge, MA: The MIT Press.

D'Alisa G. et al. (2010). Conflict in Campania: Waste emergency or crisis of democracy. *Ecological Economics*, 70(2), 239–249.

Demarco M. and Patriciello M. (2014). *Non Aspettiamo L'apocalisse. La Mia Battaglia Nella Terra dei Fuochi*. Milano: Rizzoli.

Evans J. and Jones P. (2011). The walking interview: Methodology, mobility and place. *Applied Geography*, 31, 849–858.

Foster J. B. (2000). *Marx's Ecology: Materialism and Nature*. New York: Monthly Review Press.

Global Community Monitor, India, Chevron and monitoring pollution after toxic disasters, March 25 2013. www.gcmonitor.org/india-chevron-and-monitoring-pollution-after-toxic-disasters (accessed on 21 October 2015).

Loftus, A. (2012) *Everday Environmentalism: Creating an Urban Political Ecology*. Minneapolis, MN: University of Minnesota Press.

Lombardi N. (2015). Il mio nome è Nunzia, in *Teresa e le altre*, 31.

Nash L. (2006). *Inescapable Ecologies: A History of Environment, Disease, and Knowledge*. Berkeley, CA: University of California Press.

Nilsen A. G. and Cox L. (2013). What would a Marxist theory of social movements look like? In Barker C., Cox L., Krinsky J., & Nilsen A.G. *Marxism and Social Movements*. Leiden: Brill.

Ottinge G. (2010). Buckets of resistance: Standards and the effectiveness of citizen science. *Science, Technology & Human Values*, 35 (2), 244–270.

Pulido L., Barraclough L. R., and Cheng W. (2012). *A People's Guide to Los Angeles*. Berkeley, CA: University of California Press.

Pezzullo P. (2004). Toxic tours: Communicating the presence of chemical contamination. In S.P. Depoe, J.W. Delicath & M.F. Aepli Elsenbeer (eds.) *Communication and Public Participation in Environmental Decision Making* (235–254). New York: SUNY.

Porteous J. D. (1985). Smellscape. *Progress in Human Geography*, 9(3), 356–378.

Roy A. (2009). *The God of Small Things*. London: HarperCollins, 52.

Sarli D. (2015). Posillipo Pianura, solo andata. In M. Armiero *Teresa e le altre*, 102–103.

Sze J. (2007). *Noxious New York: The Racial Politics of Urban Health and Environmental Justice*. Cambridge, MA: MIT Press.

15 Minuet as method

Embodied performance in the research process

Sonja Boon

What do bodies know? And how do we integrate such knowledges into our scholarly work? In this chapter, I introduce performative inquiry as a critical approach to embodied knowledge production and then, using the minuet—the most popular eighteenth-century dance form—as a case study, I reflect on the potential of historical dance, as bodily performance practice, to illuminate our understanding of the ways that gender, class and ethnicity are both influenced by and mapped onto social spaces. I argue that our understanding of the environment is shaped by the way that we physically and physiologically move through space, and further, suggest that the way that we move is shaped by our embodiment, by the way that socially-constructed notions of sex, gender, class, ethnicity, ability, and sexuality inform our experiences of our bodily selves.

Operating from the perspectives of bodily movement and embodied knowledge—knowledge that is, as the editors of this collection observe, "felt rather than seen" (p. 5)—can open new windows into historical inquiry, windows that are otherwise inaccessible through more traditional text-based archival methods. Indeed, as Kirby has argued, "the meat of the body *is* thinking material" (Kirby 2009, p. 221). The body's materiality—its flesh, muscles, tendons, fat, bones, and guts—offers an ideal site to consider the 'nature' of humans.

Confessions

I'm not an historian. I work in the archives. I have published articles in "history journals." But I didn't study history in university. And often, I feel like a fraud. Most of my training is in music performance, and I've used historical materials—musical treatises, etiquette books, historical scores, and copies of historical instruments—to bring the music of the past to life in the present. Although musical performance is now only a very small part of my professional work, I continue to draw considerable inspiration from the act of performance itself, which requires me to, in the words of Lesa Lockford, "put my body where my reason [is]" (Lockford 2004, p. 68). This text- and performance-based background has taught me that situating the performing body as a source and site of knowledge can offer a way forward for historical research; the past can tell us stories about our bodies and further, our bodies, even living in the present day, can tell us stories about our pasts.

Musical performance and embodied knowledge

Historically informed performance (HIP), an approach to musical performance that brings to life the music of the past through close attention to the instruments, techniques and texts of the past, offers one way of bridging an otherwise insurmountable divide between the eighteenth century and our own. As a musical approach, HIP promises much: not only does it purport to open a window into eighteenth-century musical texts and textual practices, but it also offers the seductive possibility of entering into conversations about the embodied experience of eighteenth-century culture and life (Sherman 1997). Indeed, HIP promises an authentic engagement with a living history, a way to touch the past (Haynes 2007). But how might this touching happen? How might performance—as a process and product enacted in and through the body—help us to understand historical embodiment; that is, how can I use my body as a basis for understanding the bodies of those who came before me?

In musical performance, the musician balances both the rational and the intuitive. The rational musician learns the musical text: acting as a technician, she commits notes to her physical memory, slowly building a solid technical foundation. She analyzes the musical score, pulling out consonance and dissonance, noting points of climax and periods of repose. And she considers the historical context in which the work was written, situating both musical text and composer within a social, political, and cultural context. But in the spontaneity of the musical moment, the power of the intuitive takes over; the musical imagination transcends the rational—reaching beyond the human to touch the sublime. In the words of the eighteenth-century writer Mary Wollstonecraft (1987 [1796]), "my soul rose, with renewed dignity, above its cares—grasping at immortality ... I stretched out my hand to eternity, bounding over the dark speck of life to come" (Wollstonecraft, p. 153).

For Wollstonecraft, whose intensely personal epistolary narrative, *Letters Written During a Short Residence in Sweden, Norway, and Denmark*, is infused with the thoughts and feelings of a woman of sensibility lamenting the demise of an amatory relationship and equally, experiencing intimately her role as mother, the sublime is a visceral and deeply embodied experience, seemingly far removed from the abstractions of rational. In *Letters*, Wollstonecraft maps her inner journeys—which include reflections on gender roles, motherhood, social mores, and politics—onto the wild, untamed landscape of Scandinavia, with its forbidding mountains, rocky shores, rugged paths, and taciturn peoples.

Wollstonecraft, while a product of the eighteenth century, was, in many ways, a Romantic, keen to draw on the impact of the personal and the emotional in response to the extreme rationalism of the Enlightenment. We in the twenty-first century are also products of the Enlightenment. Unlike Wollstonecraft, however, we celebrate the mind/body divide and dutifully intone the Cartesian mantra, prizing the rational above all others. In academic circles, as chapter 14 also discusses, we place such value on reason that we almost negate the passionate, the intuitive, and the imaginative altogether. Instead, as Stinson observes, "[m]ost often, physicality is recognized as something that must be managed in order to obtain the best academic performance" (Stinson 2004, p. 45). Disembodiment is the rule in academic scholarship. How, then, might we approach things differently, and why might a different approach matter?

Performance as theory and process

Performative inquiry, which takes seriously the notion of embodiment as a site of knowledge production, offers one entry point into the possibilities of "thinking with the body." Performative inquiry proposes to unite the two halves of the Cartesian whole through the conception of performance as a journey. As Noble observes, "The process is the outcome, rather than a prefabricated road to some externally fixed goal" (Noble 2005, p. 52). Originally developed by creative scholars in theatre and dance, performative inquiry is a collaborative research tool that embraces the personal and the spontaneous as not only valid, but central, aspects of the research project. As Fels defines it:

> performative inquiry is a research vehicle that acknowledges performance in action and interaction as a place of learning and exploration. Its tools of inquiry, for researchers and participants, are their bodies, imaginations, experiences (shared and individual), feelings, memories, biases, fears, judgments, and prejudgments, hopes, and desires—simply, *being, becoming.*
>
> (Fels 1998, p. 29)

Through the process of performative inquiry, participants, by using the knowledges and awareness of their bodies, are more fully able to comprehend, contextualize, and transform the research they are undertaking. For Wiebe and Snowber, performative inquiry reaches to the very core of the bodily self. Knowledge, they argue, "is stored in our tissues and marrow" (Wiebe and Snowber 2011, p. 108), in the "membrane of knowing [that] resides in your cells" (ibid. p. 111).

The philosophy behind performative inquiry appears, at first glance, deceptively simple: using improvisation through the body as a way of gaining insights into the subject matter being explored. But performative inquiry draws on a wealth of research into performativity as well as critical engagement with and awareness of the body's knowledges and understandings. The body quite literally becomes the site of intellectual inquiry. As Lesa Lockford writes of her own autoethnographic research practice: "[my work] originates *from* my performing body and is methodologically informed *by* my body … through my attention to my performing body … I harvest as well as seed the epistemological ground of performance as a mode of inquiry and way of knowing" (Lockford 2004, p. 5).

Performative inquiry, then, recognises and validates the personal and the subjective as intrinsic to the research project. Noble, for example, observes that knowledge itself is an "act": "to know is to do, to create, to be, become—and I might add, to play" (Noble 2005, p. 48). Snowber, a dance scholar, concurs, noting that "the body has a huge capacity to release the imagination, uncovering places of discovery, surprise, mystery and wonder" (Snowber 2002, p. 20). Snowber argues that dance is an ideal vehicle for the generation of what she terms "body data" (ibid. p. 22) and body knowledge. She observes: "We dance the questions, we write the questions, and we go back and forth from our limbs and torsos to finding breath in our words on the page. Sensuous knowledge is our map for the journey" (Snowber 2012, p. 54).

Let's dance

Vancouver, ca. 2001. I'm in the School of Music at the University of British Columbia, where I've been hired to help direct the early music ensembles. I've got four flute students this year and we're working on baroque sonatas. A majority of mid-eighteenth century flute sonatas end with a minuet or with a stylized movement that strongly resembles one. For the contemporary flute student, whose vanity would much prefer the whiz-bang of a more traditional final movement—that is, several minutes of flash and fire that shows off just how quickly one's tongue and fingers can move—the minuet ending is anticlimactic, a letdown. Furthermore, it is confusing. Why, the thinking goes, would a composer choose to end a piece with something that, in the thoughts of many modern musicians, is so technically simple that it doesn't even really need to be "practiced" at all? After all, the minuet can be read virtually at sight. How can one approach what feels like a meek, wilted—dare I say, *boring*—ending?

This discomfiture is, perhaps, to be expected. A look at respected music history textbooks is revealing. At a general level, these texts show a distinct lack of interest in eighteenth-century dance forms other than their articulation within the emergence of eighteenth-century symphonic sonata forms. Thus, the minuet, while acknowledged as "a vestige of the ancient and honorable dance suite" (Boroff and Irvin 1976; p. 172), is nonetheless almost wholly extracted from its dance origins. Instead, scholars describe it in terms of its formal compositional elements: structure, tonality, and melody. In addition to this, they assess the minuet in relation to the larger formal structure—the symphony—of which it was often a part. Within this context, the minuet fares poorly: it is nothing more than "a form of relief" (Hughes 1974; p. 341). As Grout and Palisca state:

> The Minuet provided a relaxation, since it was shorter than either of the two preceding movements; it was written in a more popular style, and its form was easy for the listeners to follow. But the Minuet does not make a satisfactory closing movement. It is too short...and the spirit of relaxation it induces must be balanced by a further buildup of tension, climax and release.
> (Grout and Palisca 2001, p. 470)

The minuet, in other words, lacks virility.

Interestingly, in tone and approach, the commentary in these texts mirrors, almost exactly, the attitudes of my flute students. The minuet is a musical featherweight, a compositional morsel "inadequate" (Grout and Palisca 2001, p. 470) to the musical tasks of serious art music. Conversations in lessons with students usually went something like this:

> "Do you know what a minuet is?"
> "A dance in 3/4 time."
> "But what *kind* of a dance is it?"
> "Uh … like a waltz, maybe"?

And with that my students found themselves on an imagined nineteenth-century dance floor filled with couples whirling around to music by Johann Strauss, Jr. But the minuet is nothing like a waltz. It emanates from a very different cultural sphere and, as such, articulates very different values and understandings.

What is a minuet? Who danced it? Where was it danced? What stories does it tell? And why might it be important, not just for the musical performer, but for anyone interested in the eighteenth century, more broadly speaking? I talked through all of this with my students. I explained the complexities between dance steps and music and outlined the tensions between a musical movement of three beats and a dance movement of six. They, in turn, nodded obediently. And then I threatened them with dancing. Using bribery, suggested that if things hadn't improved by next lesson, we would dance. This usually struck fear in the heart of even the most intrepid student.

Minuets, for my flute students, were threats imposed upon them by a malevolent teacher whose pedagogical motives they viewed as spurious at best and positively destructive at worst. To dance—indeed, to even *think* of dancing—required these students to move into a space of radical embodiment that lay well outside their comfort zones. It meant moving their bodies in new ways, exploring things that they were not necessarily comfortable imagining. The minuet was, quite simply, a threat: to their bodily comfort, to their bodily knowledges, and to the neat tidiness of a music history textbook definition.

I would argue that as researchers, most of us are similarly resistant to our embodiment, to our bodily knowledge and our embodied stories. To dance means not only to move our bodies in new ways, but also, to move our minds in new ways. Put more simply, dancing asks us to think differently. Drawing on the body as a source of historical insight is, of course, inevitably fraught. Our bodies are shaped and styled by our integration and imbrication within our twenty-first century society. How on earth can they possibly give us insight into historical embodiment? What kinds of methodological risks might this entail? What kinds of assumptions are we at risk of undermining? And yet, if it is the case, as Simone de Beauvoir has argued, that our bodies are "the instrument of our grasp upon the world" (de Beauvoir 2009, p. 65), then we would do well to heed the call of the body, to listen to its stories, to learn from what it is saying.

Since the turn to "bodies" in scholarly research, it would seem that bodies are everywhere. Textual bodies. Visual bodies. Conceptual bodies. Anatomical bodies. Methodological bodies. Curiously, however, most of the bodies that we study—either in our research or in our classrooms—are largely *dis*embodied. They remain texts, ideas, abstractions. In addition to this, as musicians and scholars, we rarely use the bodies we do have access to—our *own* bodies—as resources, as sites of knowledge production. So, too, it was for my students, who diligently learned the abstract principles of the minuet in order to avoid the necessity of dancing. But as I think back on my years of university music teaching, I wonder why my students and I didn't just dance right away. I wonder why I only introduced the dance steps *after* our first explorations. I wonder why dance became an intellectual exercise, why it remained an imagined—rather than embodied—experience. And I wonder, too, why the suggestion of embodied learning struck fear in my students, all of whom, as musicians, were already actively engaged in what would appear—at least at first glance—to be a profoundly embodied activity.

Looking back on it, it seems to me that if the minuet remains nothing more than an abstract idea represented in text or in visual imagery, then we, as researchers, performers and pedagogues, are missing the point. And, given that the minuet was the most popular social dance of the period, we are missing a profound opportunity for understanding eighteenth-century culture and society and the texts they produced. We are also missing a unique opportunity to consider eighteenth-century politics of gender and class as they intersect with questions of space and place. After all, as music theorist, Eric J. McKee has observed, eighteenth- and nineteenth-century dance was not just about "dance"; rather it is a "multimedia art form, a form that involves the mutual interaction of physical motion, mental attitude, music, architecture, and dress … [and] involves a dynamic network of modalities (sight, sound, bodily awareness, touch, and smell)" (McKee 2012, p. 1). As such, the practice of eighteenth-century social dance—and the minuet in particular—would appear to have much to offer the scholar of eighteenth-century studies in relation to questions of identity, space, and the environments of sociability.

Social dance: Performing identity in space

Numerous memoirs written during the eighteenth century make explicit mention of the social relevance of dance, from the late seventeenth-century commentary of the Duc de St. Simon (2014 [2006]), who observes that he was "all in a sweat from the minuets and quadrilles [he] had gone through, under a very heavy coat" (chapter 25, para. 4), to the Duchess de La Tour du Pin's commentaries on the relationships between dance and women's dress (La Tour du Pin 1913, p. 9) and Princess Dashkova's musings on the relationships between dance, language and the mind: "The count's lovely daughter … dances so beautifully, and with such native grace and dignity, that her movements seemed a sort of language, which expressed the simplicity and beauty of her mind" (Dashkova 1840, p. 232). As these accounts demonstrate, dance was an ideal venue for the public display of social status: dress, manners, gestures, movements and grace were all revealed in this space (Matluck Brooks 1991, p. 64).

Dance served pedagogical, social and political functions. At a pedagogical level, for example, social dance inculcated young men and women into the finer points of bodily movement, gesture, and comportment, elements central to elite identity performance. In this way, young men and women of the elite could assimilate themselves into their cultural inheritance (Astier 1984, p. 18). Through dance, they came to understand—and locate within their body—the regulations of social comportment, from the social hierarchies that structured the dancing order to those that dictated dancing partners and the propriety of dress. The central importance of dance shaped not only social relations, but also, physical space. Matluck Brooks observes that, "The architecture of [eighteenth-century Philadelphia's] most elite taverns and hotels indicates the importance of dancing, and the formal occasions for its practice. The finest such buildings contained special dancing halls called assembly rooms where a variety of dances might be held, including private celebrations, public balls with admission by ticket only, and dancing assemblies, which were regularly occurring subscription events" (Matluck Brooks 1991, pp. 69–70). Dance, then, offered a unique window into the encounters between social and physical space, that is, between embodiment and environment.

The minuet, the most popular elite dance form of the period, would have been a regular feature of these events. Available to the contemporary scholar or dance enthusiast in dance treatises and notated choreographies, the minuet first appeared at the French court in the middle of the seventeenth century, after which it very quickly spread to other European courts (Bang Mather 1988, p. 268). By the eighteenth century, the minuet was the social dance par excellence. According to Rousseau, author of the *Encyclopédie* article on the minuet, the minuet is a "commonly-used" (Rousseau 1765, 'Minuet', para. 3) dance of "noble and elegant simplicity" (ibid. para. 1). The Reverend Mr. Walter Young (1790), concurs: "The minuet … ever has been, as is at this day accounted the most elegant and pleasing movement in music" (Young 1790, pp. 88–89).

Social dancers themselves also opined on the topic of the minuet. One young woman, of the remarkable name Ethelinda Lindamira D., observed in a 1774 letter to the *Lady's Magazine*: "I once heard an eminent dancing master say—that the minuet had been the study of his whole life, and that he had been indefatigable in the pursuit of its beauties, and at last, he could only say with Socrates, he knew nothing" (D. 1774, p. 659). Her letter continued by extolling the sensuous virtues of the dance itself. She stressed the embodied experience not only of dancing, but also of watching the dance:

> The ordinary, undulating motion of the body in common walking (as may be seen by the waving line, which the shadow of a man's head makes against a wall, as he is walking between it and the afternoon sun) is augmented in dancing, into a larger quantity of waving by means of the minuet step, which is so contrived as to raise the body by gentle degrees, somewhat higher than ordinary and sink it again in the same manner, lower in the going on of the dance. The figure of the minuet path, on the floor, is also composed of

> serpentine lines, when the parties, by means of this step, rise and fall most smoothly in time, and free from the sudden starting, and dropping, they come nearest to Shakespeare's idea of the beauty of dancing.... The other beauties belonging to this dance are the turns of the head and twist of the body, in passing each other, as also gentle bowing and presenting hands.
>
> (ibid. p. 659)

For Ethelinda Lindamira, dance was all about space, movement and patterns, that is, about the relationship between the dancing body and the space it occupied.

Dancing treatises—which taught dance steps and choreographies—asserted the social relevance of the minuet, pointing not only to its salutary health benefits, but also to its ability to reveal social graces. Nothing, writes the anonymous author (1774) of an essay on dancing:

> contributes more than motion to heighten and display the charms of beauty; and personal elegance, as it is of a fugitive nature, exists chiefly in motion; for being communicated by the principle of action that governs the person, it is found diffused through the whole body, and has no fixed residence. Hence two painters and poets always add motion as the last and highest embellishment to beauty. All the charms of an agreeable person are then in the highest exertion, every limb and feature appears in its respective grace.
>
> (Anonymous 1774, p. 522)

In this iteration, the minuet might be imagined as a species of what the eighteenth-century elite understood as gallantry, a socially acceptable site for the articulation of sensual delights and easy flirtations, to be sure, but also a space ideally designed for the embodied performance of normative gender and class roles. As Dukes observes, dancing is "an Art very conducive to the forming, in either Sex, a Genteel, free, and easy Carriage & Deportment" (Dukes 1752, p. i), points echoed by Gardiner: "the Advantages that arise from [the Art of Dancing] are many. A Gentleman or Lady cannot even enter a Room, make a genteel Bow or Courtesy, or walk graceful and polite, without being instructed in this essential part of Education" (Gardiner, 1786, p. iii). This notion of the dance as integral to gendered social relations is reflected in the structure of Gardiner's treatise itself, which is staged in the form of a gallant conversation between a "lady" and her dancing master.

Grace, charm, beauty, elegance, poetry—here we see embodied the very nature of elite sociability during this period. But such a performative embodiment of gender and class is not just a textual embodiment. It is not merely conceptual or abstract. Rather, it is located in *every* limb of the body. It is, as the previous examples have suggested, in the way that the limbs and dress reflect the flickering lights of candles in mirrored spaces. It is in the steps that are traced in patterns across the floor. It is in in the eyes of the observers, and in the pretty ankle movements of the female dancers. It is the bodily foils of masculinity and femininity: in the interactions between the muscular—and virile—male calves shaped by hours of dancing and the grace, charm and beauty of feminine body shapes and undulations as they pass each other on the dance floor. It is in the dancers' eyes as they watch and follow each other's bodies through the dance space. And it is in bodies themselves, entities that only barely touch during the course of the dance, but that are on public display for the critical eye throughout the dance movements. Finally, it is in the spatial relationships *between* dancers, and in the understandings of how the relationships between space and embodiment shape social relations, not only between the sexes, but also within the sphere of polite society itself. Elite gender, class and sexuality—all are performed in the space and time of the minuet step.

Deceptively simple, the minuet "contained the complete teaching of a dancing master and the complete training of a courtier" (Astier 1984, p. 18; see also Allanbrook and Hilton 1992, p. 144; Thorp 2003, p. 103). Measured in three in the music, the most basic dance movement of the minuet is considered in six, thus crossing the musical bar line. Where the musical gestures suggest emphases on the first beat, and less so, on the third, the *pas de minuet à deux mouvements*—the most basic minuet step—emphasizes beats two and six respectively.

The minuet is a couple's dance. The dancers begin by inscribing a serpentine or Z pattern on the floor. Starting at opposite ends of the Z pattern and facing one another, they mirror each other's movements, with three *pas de minuet* along one line of the Z, two along the diagonal and three along the other line. While the two dancers face one another at all times, they barely touch; rather, they cross paths along the diagonal. They follow each other's bodies with their eyes, or touch hands as they meet in the middle. The dance is a social affair, witnessed by many as the dancers trace their patterns across the floor (Bang Mather 1987, p. 2).

The counterpoint between musical gestures and dance steps—that tension between three and six as embodied in the rhythmic rise and fall of dancers' bodies and musicians' bows—reflects the sensual (and indeed, erotic) tensions that lie at the heart of eighteenth-century gallantry; they gesture towards a seductive ambiguity that cannot easily be contained in the regular 3/4 movement. What did the counterpoint of sexual awareness look like? How was it imagined? How was it performed? While textual explorations—in novels and memoirs, for example—can offer some insight, these questions cannot fully be answered unless we explore those sensual bodily possibilities for ourselves. In so doing, we gain new insights into the literatures and languages of love and seduction, but also, and perhaps more importantly, about the *scenes* of love and seduction; that is, how gallantry played itself out in space and place.

This rhythmic complexity, too, contributes to the elegance of the dance, an elegance that relies on the controlled sensuality of social gender relations among the elite during this period. The anonymous dancing master cited above, for example, argues that the minuet asks both dancers and observers to consider the foundational role of the body in the presentation and articulation of the social self through dance:

> There is a reciprocal effect between the sentiments of the mind, and the actions of the body. Indelicate or awkward gestures or motions are very indicative of an illiberal and unpolished mind; while contracted or gross sentiments are as evidently marked by rustic and ungraceful actions. Dancing cannot bestow generous motions or refined sentiments, but it can teach us to regulate all our exterior actions by the laws of order and decency, (in consequence of the mutual relation between sentiment and action,) it will at least tend to improve and heighten a love for grace and beauty in our ideas and sentiments.
>
> (Anonymous 1774, p. 524)

The minuet thus manifests the strict protocol of elite sociability, reminding both dancers and spectators of the accepted hierarchies of gender, class and social position. In its spatial arrangement, the minuet serves to map geographies of social identities, inscribing social relations in movement. How do individuals come together? What kinds of imagined communities are created by the minuet? What kinds of exclusions does the dance enact? How are matters of class mapped onto dance movements? And how does national identity emerge in the form of a dance? All of these questions can be answered through an engagement with dance as performance, not merely as text.

Minuets also mapped geographic identity. Sutton observes that "the distribution of the changes of weight in the *pas de menuet* varied throughout the eighteenth century, depending on the preferences of dancing masters and regional tastes" (Sutton 1985, p. 121), a point with which Tomko concurs: "Even so 'standard' a dance as the minuet received particularist interpretations in different countries" (Tomko 1999, p. 3). It was, therefore, likely that an assembled audience could read place, as well as class and gender, onto the dancers' movements. Consider for example, *Lethe, or Aesop in the Shades*, a farce by David Garrick. In this dramatic space, the minuet itself is revealed to be the farce, a species of dance designed to expose the artifices of polite society. The following passage is particularly evocative:

> Aesop: And pray, Sir, what is your business in England?
>
> Frenchman: I am arrivé dere, Sir, pour polir la nation. De Inglis, Sir, have too much a lead in deir head and too much a tought in deir head. So, Sir, if I can lighten bote, I shall make dem tout à fait François and quite anoder ting.
>
> Aesop: And pray, Sir, in what particular accomplishments does your merit consist?
>
> Frenchman: Sir. I speak de French, j'ai une bonne addresse, I dance un minuet, I sing des littel chansons, and I have—une tolerable assurance: En fin, Sir, my merit consist in one vard—I am foreignere, and entre nous, vile de Inglis be so great a fool to love de foreignere better dan demselves, de foreignere vould still be more great a fool, did dey not leave deir own counterie, vere dey have noting at all, and come to Inglande

> vere dey vant for noting at all, perdie. Cela n'est il pas vrai, Monsieur Aesope?
>
> (Garrick 1980 [1740], p. 16)

In Garrick's work, the minuet marks notions of gender, class, social propriety, and national identity at a bodily level. While an intellectual engagement with the text allows the reader entry into these concepts, it does not allow for the bodily meanings that may lie below the surface. Gender, class, social propriety, and national identity are not solely intellectual exercises. Rather, they are profoundly embodied concepts shaped by notions of physiology, poise, comportment, movement and space; that is, by embodied interactions between individuals and groups.

We can, of course, glean considerable knowledge from purely textual or visual encounters. We can also learn much from considering the body and embodiment at a conceptual level. But what of lifts, falls, proportions, balance, gesture, architecture? What of plays of light on moving dresses? What of ankles and calves? What of space, time, and movement? How do we get at these various layers without actually exploring the body itself, without actually embodying its gestures and movements? How will we learn about the landscapes of sexual desire, the topographies of gender, the environments of class, or the mappings of geography? How, finally, can we understand eighteenth-century embodiment if we don't take the time to think with our *own* bodies?

References

Allanbrook, W. J. and Hilton, W. (1992). Rhythm in Mozart's arias. *Early Music*, *20*(1), 142–9.

Anonymous. (1774). An essay on dancing, part II. *The Lady's Magazine; or Entertaining Companion for the FAIR SEX, Appropriated Solely for their Use and Amusement*, 5(10), 522–5.

Astier, R. (1984). François Morel and the art of teaching dance in the eighteenth century. *Dance Research: The Journal of the Society for Dance Research*, *2*(2), 11–23.

Beauvoir, S. de. (2009). *The Second Sex* (trans. C. Borde and S. Malovany-Chevallier). London: Jonathan Cape.

Borroff, E and and Irvin, M. (1976). *Music in Perspective*. New York: Harcourt Brace Jovanovich.

Butt, J. (2002). *Playing with History: The Historical Approach to Musical Performance*. Cambridge, UK: Cambridge University Press.

D., Ethelinda Lindamira (1774). Remarks on the minuet bowing and curtseying. *The Lady's Magazine; or Entertaining Companion for the FAIR SEX, Appropriated Solely for their Use and Amusement*, 5(12), 659–60.

Dashkova, E.R. (1840). *Memories of the Princess Daschkaw* (Vol. II). London: Henry Colburn.

Dukes, N. (1755). A concise & easy method of learning the figuring part of country dances, by way of characters. London. Retrieved from: http://find.galegroup.com.qe2a-proxy.mun.ca/ecco/infomark.do?&source=gale&prodId=ECCO&userGroupName=

sain66118&tabID=T001&docId=CW3315488767&type=multipage&contentSet=ECCOArticles&version=1.0&docLevel=FASCIMILE

Fels, L. (1998). In the wind, clothes dance on a Line: Performative inquiry – a (re)search methodology. *Journal of Curriculum Theory*, *14*(1), 27–36.

Gardiner, S. J. (1786). A Definition of minuet-dancing, rules for behavior in company &c. A dialogue between a lady, and a dancing-master (Madeley: Edmunds. 1786). Retrieved from: http://find.galegroup.com.qe2a-proxy.mun.ca/ecco/infomark.do?&source=gale&prodId=ECCO&userGroupName=sain66118&tabID=T001&docId=CW3303535672&type=multipage&contentSet=ECCOArticles&version=1.0&docLevel=FASCIMILE

Garrick D. (1980). Lethe; or Esop in the shades. A dramatic satire. In H.W. Pedicord & F.L. Bergmann (eds), *The Plays of David Garrick: Garrick's Own Plays, 1740–1766*. *1*(1–34). Carbondale, IL: Southern Illinois University Press.

Grout, D.J. & Palisca, C.V. (2001). *A History of Western Music* (6th edn). New York: W.W. Norton.

Haynes, B. (2007). *The End of Early Music: A Period Performer's History Of Music for the Twenty-first Century*. Toronto: Oxford University Press.

Hughes, D. (1974). *A History of European Music: The Art Music Tradition of Western Culture*. New York: McGraw Hill.

Kirby, V. (2008). Natural convers(at)ions: or, "What if culture was really nature all along?" In S. Alaimo & S.J. Hekman (eds), *Material Feminisms* (214–36). Bloomington, IN: Indiana University Press.

Kuijken, B. (2013). *The Notation is Not the Music: Reflections on Early Music Practice and Performance*. Bloomington, IN: Indiana University Press.

LaTour du Pin, Marquise de. (1913). *Journal d'une femme de cinquante ans: 1778–1815*. Paris, France: Chapelot.

Lawson, C. & Stowell, R. (1999). *Historical Performance of Music: An Introduction*. Cambridge: Cambridge University Press.

Lockford, L. (2004). *Performing Femininity: Rewriting Gender Identity*. Walnut Creek: AltaMira Press.

Mather, B.B. (1987). *Dance Rhythms of the French Baroque: A Handbook for Performance*. Bloomington and Indianapolis, IN: Indiana University Press.

Matluck Brooks, L. (1991). Emblem of gaiety, love, and legislation: Dance in eighteenth-century Philadelphia. *The Pennsylvania Magazine of history and Biography*, *115*(1), 63–87.

McKee, E.J. (2012). *Decorum of the Minuet, Delirium of the Waltz: A Study of Dance-Music Relations In 3/4 Time*. Bloomington and Indianapolis, IN: Indiana University Press.

Noble, S. E. (2005). Mental illness through popular theatre: Performing (in)sanely. *New Directions for Adult and Continuing Education*, *107*, 45–53. doi: 10.1002/ace.188

Rousseau, J.-J. (1765). Minuet. (P. Gay-White, trans.) In *The Encyclopedia of Diderot & d'Alembert Collaborative Translation Project*. Retrieved from http://hdl.handle.net/2027/spo.did2222.0000.256

Sherman, B.D. (1997). *Inside Early Music: Conversations with Performers*. New York: Oxford University Press.

Snowber, C. (2012). Dance as a way of knowing. *New Directions for Adult and Continuing Education*, *134*, 53–60 doi: 10.1002/ace.20017

Snowber, C. (2002). Bodydance: Enfleshing soulful inquiry through improvisation. In C. Bagley & M.B. Cancienne (eds), *Dancing the Data* (20–33). New York: Peter Lang.

St. Simon, Duc de. (2014 [2006]). *Memoirs of Louis XIV*. Retrieved from www.gutenberg.org/files/3875/3875-h/3875-h.htm

Stinson, S.W. (1995). Body of knowledge. *Educational Theory*, 45(1), 43–54. doi: 10.1111/j.1741-5446.1995.00043.x

Sutton, J. (1985). The minuet: An elegant phoenix. *Dance Chronicle*, 8(3–4), 119–52. doi: 10.1080/01472528408568908

Taruskin, R. (1995). *Text and Act: Essays on Music and Performance*. New York: Oxford University Press.

Thorp, J. (2003). In defence of danced minuets. *Early* Music, *31*(1), 100–108.

Tomko, L.J. (1999). Dance notation and cultural agency: A meditation spurred by choreographics. *Dance Research Journal*, *31*(1), 1–4. doi: 10.2307/1478304

Walls, P. (2003). *History, Imagination, and the Performance Of Music*. Woodbridge, UK and Rochester, NY: Boydell Press.

Wiebe, S. & Snowber, C. (2011). The visceral imagination: A fertile space for non-textual knowing. *Journal of Curriculum Theorizing*, *27*(2): 101–113. Retrieved from http://journal.jctonline.org/index.php/jct/article/view/352

Wollstonecraft, M. (1987 [1796]) *Letters written during a short residence in Sweden, Norway and Denmark*. London: Penguin Books, Ltd.

Young, Rev. Mr. W. (1790). An essay on rhythmical [sic] measures. *Transactions of the Royal Society of Edinburgh, Volume II: Papers of the Literary Class*, 55–110.

16 'To know the story is to love it'

Scientific mythmaking and the longing for cosmic connection

Lisa H. Sideris

Somewhere in my turbulent and apathetic teen years, I developed a fascination with evolutionary timeframes, and with the universe generally. As I write these words, they sound almost like a confession, and, in many ways, I was saved. My salvation was helped along by great storytellers like Stephen Jay Gould, whose essays on natural history I discovered in high school. Though Gould, at the end of the day, dismissed science and religion to their separate chambers, the tales he told were deeply moral and they spoke to the crusader in me. I also fell under the spell of a still-young, black turtleneck-wearing Carl Sagan whose *Cosmos* (book and TV series) I knew virtually by heart. Around age 17, I acquired a small telescope. By an incredible stroke of beginner's luck, the first time I pointed it at a patch of the night sky I vaguely knew to be traversed by the planets, I hit upon a fuzzy yellow vibrating disc. I turned the knob and the fuzzy blob resolved itself into a perfect miniature planet: a tiny golden ball with a clearly defined ring. Saturn! I laughed as I recalled Sagan's account of how early astronomers described the planet as having ears. That description was perfect. Now I was an astronomer too.

Then, as now, my science education was somewhat spotty and self-taught. I generally avoided "hard" science classes, but in my senior year of high school, I summoned the courage to take chemistry. I was completely charmed by the periodic table of elements which seemed to me so imminently reasonable and comprehensive, yet strange and fathomless. When I think back to this period, what most surprises me is that my wonder at these foreign realms of astronomy or chemistry had little connection to *nature* per se—in the sense of being out "in" nature—and even less to do with what is today my chief concern, personally and academically: the state of the natural "environment." As far as I can tell, my environmental consciousness evolved along a different path than my interest in science or the cosmos. The particular enchantment I felt in those days bore little connection to my immediate physical surroundings, yet I experienced it as a kind of sensory immersion or physical transport. What I remember vividly is the smell of my chemistry textbook when I cracked it open in the pre-dawn darkness to cram for a test. (Occasionally, I come across a book that has that exact smell and experience a wave of nostalgia.) I recall a certain headiness that came with the realization that I actually *understood* the equations. Strangely, I have no memory of chemicals or chemical reactions—their explosions, odors, or colors—though a large portion of class (my least favorite part) was devoted to lab work. I had little interest in the hands-on or shared experience of chemistry. Similarly, the feeling I recall, and often have today when gazing through a telescope, was a disembodied or disconnected awe. A telescope takes one away from Earth, often in ways that are welcome—especially, perhaps, for (all-too-embodied) teenagers. The whole cosmic encounter requires mediation by tools and is, to some extent, impossible without them. As captivating as they are, one does not really "experience" celestial objects. These early encounters, in other words, lacked the immediate sensuousness of a summer rainstorm or the dusty crunch of autumn leaves underfoot. Yet my memories of that time remain potently visceral.

Probably youth was a factor. Biologist Ursula Goodenough, who is today a vocal proponent of a grand scientific-spiritual narrative called the Epic of Evolution, recalls having had "a lot of trouble with the universe" when she was young. On a camping trip at age 20, she gazed up into a vast starry sky and experienced a kind of existential freefall so intense and frightening, she writes, that "I had to roll over and bury my face in my pillow" (Goodenough 1998, p. 9). She eventually learned to manage the "poignant nihilism" and "bleak emptiness" brought on by thoughts of the cosmos by means of an arrangement she calls her "covenant with mystery." It was okay, Goodenough decided, at last, not to have answers to 'Really Big Questions' (Goodenough 2010, n.p.).

To me, Goodenough's arrangement with the cosmos sounds more like an uneasy truce than a covenantal embrace. But then, I never had much trouble with the universe. Looking back, I suspect that part of the power I sensed during solitary moments with my telescope or chemistry textbook was my own. Perhaps turning from the cosmos to chemical elements and back, or from the history of biology to the breathtaking scale of the universe, I also experienced what Big Historian David Christian calls "the play of scales": the intellectual and aesthetic excitement and vertiginous thrill of moving from micro to macro-realities, and of sensing the possibility of their harmony. Probably my early—positive—encounter with the cosmos has relevance for my having ended up, though by a circuitous route, in religious studies (no flight from mystery here), while Goodenough went on to compose genetics textbooks and advocate for a scientific cosmology that puts the whole universe in good order.[1]

It surely has some bearing on my having spent several years of my life critically dissecting a sprawling and ambitious set of quasi-religious cosmological narratives that presume to tell the story of everything, from the Big Bang to our present, perilous moment (Sideris 2013, 2015).

Big History, the Epic of Evolution, and the Universe Story present a sweeping narrative of cosmic and evolutionary unfolding as a modern origin story for all people. These efforts to present a complete, panoramic story display elements of our Anthropocene[2] situation, for these stories narrate how we arrived at a moment where humans wield enormous influence over the biosphere, and how we might move forward without wrecking the planet or our civilization. These narratives collapse conventional boundaries between human and natural history, implicating human agency across geological or cosmic timescales. Although their emergence predates our current Anthropocene concept, it is possible to see these projects as Anthropocene narratives that evoke consciousness of ourselves as a collective entity or global agent—a *species* that emerged out of deep cosmic time (Chakrabarty 2009). Perhaps a new experience of the human species is being incubated in these cosmologies, with their religious-like mythos, rituals, and communal practices that aim to connect humans to deep time and universal consciousness. Specifically, they might be understood to create something that some commentators consider virtually impossible: a "phenomenology" of ourselves in the Anthropocene; an affective, aesthetic, or experiential identification with the human-as-species, its peculiar forms of agency, and its emergence in deep time (Chakrabarty 2009).

What insights might scholars—in a variety of disciplines—gain from attending to the parallels between grand narrative projects emerging in the discipline of history and their counterparts in religion and ethics? As a scholar trained primarily in the latter, I find much to quibble with in these grand narratives. But having articulated those concerns at length elsewhere (Sideris 2013, 2015), I will focus much of my discussion here on exploring why these projects might have popular appeal, before considering some downsides. What do people seek when they turn to these narratives—and what might they be escaping? What, for their own part, are cosmic storytellers offering their audiences, and why? As I have hinted, I suspect that the appeal of these stories lies, at least in part, in the prospect they offer of experiencing new ways of being human. Much hinges, however, on the word *experience*, for these stories address problems and objects that are not readily apprehended by humans. As historian Stephen P. Weldon (2015) puts it, "We are now struggling to come to terms with our place in a world that we have no intellectual and practical history with. We have no daily experience of the kinds of objects that we are interacting with as a species," whether those objects of inquiry are the global climate, the formation of stars, or our own geological agency.

Telling big stories

Big History is connected to a larger constellation of movements that blend deep-historical perspectives with scientific methods and discoveries to create a cosmic narrative that inspires wonder and (in some iterations) affective connection to the Earth. These movements offer a sense of belonging to a shared, coherent, and empirically *true myth*. Big History may be familiar to historians through the work of David Christian, Cynthia Stokes Brown, Fred Spier, or William McNeill, among others; my focus here is on David Christian's Big History, owing to the overlap and shared agenda of his project with the mythmaking efforts of some religion scholars and religionists.

Like Big History, The Universe Story and The Epic of Evolution endeavor to tell the whole cosmic and human story, defining humans as the universe becomes conscious of itself (Swimme and Tucker 2011; Rue 2000). These narratives often have an explicit environmental agenda, driven by the conviction that environmental ethics and values need an orienting cosmology in which to take root and flourish. The hope is that humans' dawning cosmological awareness, and the sheer awe inspired by the modern scientific story, will empower us to guide the future unfolding of cosmic and evolutionary processes into a new era—sometimes defined as a new geological or cosmic phase—in which humanity and nature will flourish together (Swimme and Berry 1992; Swimme and Tucker 2011).

The prominence of these stories in religion and ethics[3] suggests a deep-seated spiritual, affective, and aesthetic—or evolutionary—need for human connection with cosmic and deep-historical perspectives.[4] Perhaps scientific and historical narratives are beginning to perform a function traditionally assigned to religion or myth. Indeed, proponents of these movements claim that this is precisely how modern cosmology can (and should) function, and that the ascendency of these narratives is fueled by the sharply declining appeal and functionality of the traditional world faiths (Berry 1978, Rue 2004). It is instructive to place this rationale for why humans need a new story alongside more recent reflections on the Anthropocene and its attendant, but elusive, mode of species-being.

Deep time, species-level thinking

Postcolonial scholar and historian Dipesh Chakrabarty (2009) argues that the onset of the Anthropocene, and of anthropogenic climate change, entails a dramatic scaling up of the human, and a concomitant collapse of traditional boundaries separating natural and human history (and separating the disciplines from one another). These interrelated claims reflect the idea that our species has now become a planetary agent whose unprecedented power is symbolized by our capacity to alter the chemistry of our atmosphere. Humans have always interacted with their environment, but "it is no longer a question simply of man having an interactive relation with nature," Chakrabarty writes. "Now it is being claimed that humans are a force of nature in a geological sense" (2009, p. 207). This melding of human history with geological time suggests a new mode of "species thinking"- a collective self-recognition of humans as an abstract spatiotemporal entity. Yet, as the quote from Weldon, above, also suggests, this change in perspective carries with it no continuity of *experience* between humanity's past and future; as Chakrabarty says, "we never experience ourselves as a species." We have little way of emotionally identifying with the species concept, "no phenomenology of us as a species" (2009, p. 220). Chakrabarty suggests that species thinking may nevertheless act as a "placeholder for an emergent, new universal history of humans that flashes up in the moment of the danger that is climate change" (2009, p. 221). At times, and in a move that aligns him with thinkers such as biologist E.O. Wilson, whom he cites with some frequency, Chakrabarty appears to suggest that deep-time, species-consciousness is *necessary* for coping with the climate crisis (2009, p. 213). Emotional, phenomenological identification with concepts beyond our grasp may be critical to our survival.

This suggestion puts Chakrabarty's diagnosis of the Anthropocene into conversation with emerging cosmological narratives. The turn to (or "return of"[5]) universal history and a universal human subject as key to confronting global challenges, environmental or otherwise, is central to these stories. A case in point is *Journey of the Universe*, a multimedia project spearheaded by "evolutionary philosopher" Brian Swimme and religion scholar Mary Evelyn Tucker. Both were students of the late cultural historian and "geologian" Thomas Berry. As early as 1978, Berry issued a call for a "New Story," citing the traditional faiths (notably, Western religions) as outdated and ill-suited to modern environmental and existential crises (Berry 1978). Swimme and Tucker appeal to universal, collective experience and species-level consciousness, claiming that there is "little validity to the idea that humans are isolated individuals, for each of us arises out of an ocean of experience and understanding acquired by our species as a whole" (Swimme and Tucker 2011, pp. 90–91). We are urged to become "not just nation-state people, but universe people," a form of human that is "natural to the universe" (Swimme and Tucker 2011, p. 113). Swimme and Berry (1992) propose that humans will never come to grips with a dawning era of enhanced human interaction with the planet, a period they call the "Ecozoic,"[6] until we think of ourselves as a species. "We have thought about the human as nations, as cultures, as ethnic groups, as international organizations, even as the global human community," they write, "but none of these articulates the present human-Earth issue so precisely as thinking about the human as a species among species" (Swimme and Berry 1992, pp. 259–260) The human story must be seen as "integral with the larger life story and the larger Earth story" (Swimme and Berry 1992, p. 260).

All this suggests a merger of natural/geological history and human history. Epic of Evolution proponents Loyal Rue and Ursula Goodenough (2009) urge the adoption of a "consilient"[7] college curriculum that narrates the human and cosmic story as a coherent, compelling evolutionary tale around which the entire curriculum should be organized: "One world calls for one story" (Rue and Goodenough 2009, p. 181). Writing several years before Chakrabarty heralded the collapse of the natural/human history binary, William McNeill introduces readers to David Christian's *Maps of Time* (2004) as a project that "unites natural history and human history in a single grand and intelligible narrative" (McNeill in Christian 2004, p. xv). Christian similarly describes Big History as erasing the "borderline between history and natural sciences" (Christian 2010, p. 7). In all of these narratives, the supreme achievement of a combined human and cosmic story, the daunting and daring task of assembling a comprehensive whole, is *itself* held up as an object of awe and reverence—a product of wondrous scholarly creativity that renders order from chaos, like the universe itself. The wonder-rhetoric frequently tacks back and forth between wonder at the emergence and existence of the universe and wonder at the emergence and existence of the comprehensive narrative (and by extension, those bold enough to attempt its creation).

Making the cosmic connection

Why is Big History needed? Like those offered for The Universe Story or The Epic of Evolution, rationales for Big History range from practical and pedagogical, to ethical and aesthetic, to spiritual and metaphysical. On the practical side, Big Historians frequently maintain that students are confused and uninspired by the fragmented knowledge they receive about human origins and history—a bit of foreign language here, some math there, a smattering of state or national history. Putting history into epic form arouses their enthusiasm and excitement for learning. Simply put, Big History gets results in the classroom. Moreover, proponents often maintain that a global society needs global, generalized knowledge and that "globalization itself becomes a vehicle for Big History's expansion of awareness" (Rodrigue et al. 2015, p. 9). Regarding ethical justifications, the claim is made that in the absence of an overarching, integrated narrative, we lose our moral compass. These claims often begin to sound circular: to wit, without a unified story, we cannot have unity. Unity and harmony are marshalled as self-evidently good and universally desired and desirable. Without Big History we are "fated to live within a fragmented, endlessly shifting intellectual universe, deprived of the philosophical and ethical anchors of a more unified vision of how things came to be" (Rodrigue et al. 2015, p. 10). Big Historians are collectively "in love with the idea of such integration" (Rodrigue et al. 2015, p. iv–x). As the language of love and deprivation suggests, many who are drawn to this sprawling but coherent vision display a kind of erotic longing, a "tormenting desire for unity"—unity of knowledge, unity of the disciplines, unity of experience (Smocovitis 1999). Again, the underlying assumption is that this unificatory drive is itself a universal feature of the human species: "We believe that the aspiration for a syncretic worldview, for the comprehension of the whole of human history of this immense globe, was (and always will be) inherent in humans" (Rodrigue et al. 2015, p. 9). Big History aspires to "explain existence" by constructing bridges between fragmented areas of knowledge (Rodrigue et al. 2015, p. 11).

Following Emile Durkheim, David Christian invokes our modern disorientation and lack of meaning as instances of "anomie," while some religion scholars refer to "amythia," the absence of a plausible, comprehensive myth (Rue 2004). "The absence of a unifying story," Christian argues, "may be an important ingredient in the pervasive quality of loss and disorientation in modern life" (Christian 2005, p. 29). Note that the claim being made here is not just that humans rely on narratives (of some sort) to make sense of their individual lives, or to negotiate their role within a given social group, as some social theorists have argued. The claim is that humanity as a whole needs a common story that encompasses *everything* in cosmic and human history and that gives us a place within the universe. The driving assumption is that humans were once guided by appealing, authoritative macro-narratives that served these functions. Above all, Christian believes that creation stories once operated as trustworthy, up-to-date maps—"attempts to use the *best available knowledge* to place society within a large, often cosmological context" (Christian 2010, p. 10, my emphasis).

Religion scholars may scoff at the highly *literalist* interpretation of creation myths—*best available knowledge*—that motivates enthusiasts like Christian to proffer a scientific-historical narrative in place of (say) Genesis creation myths; after all, only strict Biblical literalists or dedicated creation scientists engage mythic narratives as empirically accurate, propositional statements about life's origin and history. Nevertheless, modern myths like Big History appear to be addressing a lack that is keenly felt by some (certainly by those actively promoting these narratives). Loyal Rue (2000) embraces the Epic of Evolution as a story that informs us of *how things are* and *which things matter*. Prior to his conversion to the Epic, he felt his own values to be disconcertingly unmoored, like a "moral imperative just floating loose. … I was in search of a cosmology" (Rue quoted in Barlow 1997, p. 71). His turn to biology and other sciences led him to grasp the mythic potential of science as a full-blown, comprehensive cosmology and solid ground in which to anchor moral commitments. "All of a sudden, whoomp! It goes down like a taproot. Now it's grounded in the cosmos" (Rue quoted in Barlow 1997, p. 71). For some adherents, then, these stories provide, or restore, a connection between existent values and a larger framework of meaning.

But these narratives also aim to provide, or restore, something more profound and nebulous: religious-like experiences of wonder and awe. Cosmic storytellers often speak of awe as a reaction to the beauty and vastness of "nature," as when beholding the starry sky or a towering forest. But their awe is also (or ultimately) awe at the abstract "immensity of scale" of cosmic space and time. Macrohistory, Christian writes, "is *about* scale. It is about what musicians call 'diapason': the contrasts, juxtapositions and insights that can be achieved by moving through the complete range of available scales" (Christian 2005, p. 26). But what does it mean to wonder at scale itself? Is this encounter compelling enough to attract adherents? Can it turn them into dedicated environmentalists?

In making his "case for awe," Neal Wolfe (2014) writes that Big History tells the "profound evidence-based story of how the universe as we know it began, developed, led to our emergence as a species, culminates in the present moment, and continues unfolding into the future" (Wolfe 2014, p. 339). Yet we never experience the *beginning of the universe* or our *emergence as a species*, or the *continued unfolding into the future* of the cosmos. We directly experience a star-filled sky, but not "how stars are formed, how far away they are," or that we are "looking millions or billions of years into the past" (Wolfe 2014, p. 334). Similarly, Wolfe celebrates awe at the "realization that if all the strands of DNA in a human body were strung end to end, the resulting thread would stretch to the sun and back as many as seventy times!" (Wolfe 2014, p. 337).

This brand of gee-whiz science may well provoke amazement at what Wolfe calls a *realization*, and yet we—especially, we without an electron microscope—have no phenomenology of DNA. This leads one devotee of Big History to ask: in what sense do those who read or write grand cosmic narratives "actually live the familiar concepts" of biology, physics, or other sciences that drive the storyline? (Blundell 2015, p. 2). The intellectual dimensions of Big History need to go hand-in-hand with "the lived-experience," so that one can somehow sense the storyline "as opposed to only *knowing* it" (Blundell 2013, p. 1). The experiential dimension is sometimes characterized as a mystical perception of "relational unity" with the universe, a salutary diminishment of our "habitual sense of individuation" (Wolfe 2014, p. 337). But how to create these experiences?

Some cosmic storytellers personalize and *phenomenologize* the story by drawing from the template of religious ritual and practice. To invoke a metaphor suggested by David Christian's diapason, or "play of scales," they endeavor to supply not only the written score but the music as well. Epic of Evolution evangelists Connie Barlow and Michael Dowd preside over a variety of evolutionary rituals, in religious and secular settings. We might see these performances as another kind of play—ritual or symbolic play that allows participants to engage imaginatively and bodily with what is typically remote, unseen, or incomprehensible. These rituals include a Cosmic Communion that invokes Carl Sagan; the recitation of evolutionary parables and songs; ritual use of Great Story beads that mimic rosary beads; and a candlelit Cosmic Walk that symbolizes the unfolding universe (Barlow 2013).

In Cosmic Communion rituals performed in liberal churches or spiritual retreat centers, participants are "anointed with 'stardust' (glitter) to signify, as Carl Sagan pointed out in the 1980s, that we are quite literally 'made of stardust'" (Barlow 2013, n.p.) They may string together Beads of the Cosmic Rosary intended to represent key moments in evolution. The shape and color of beads are suggestive of cosmic events like supernovae, or signify the evolution of particular life forms such as dinosaurs, birds, or flowers. Cosmic beads can be personalized, much like a charm bracelet.

For the Cosmic Walk, a rope is placed on the ground in a spiral shape, symbolizing 14 billion years of cosmic unfolding; candlelit stations around the spiral represent cosmic and evolutionary events, such as the Big Bang, the death of dinosaurs, and the emergence of early humans (Barlow 2013). Interestingly, markers near the end of the Cosmic Walk symbolize major discoveries by twentieth century scientists, including the discovery of the expanding universe—the very discovery that "revealed" the storied structure of our cosmos. In this way, the ritual becomes self-reflexive, visually evoking the central insight that humans are the universe becoming conscious of itself. Human knowledge of the cosmos is thus ritualized as an event of cosmic proportions and significance.

Proponents of scientific cosmologies have at times debated whether aesthetic accoutrements can facilitate stronger connection with the larger story. In a forum discussion in the *Epic of Evolution Quarterly*, Goodenough posed the question, "Does the Epic Need Art?" One respondent denied needing any artsy props—paintings or poems—to evoke awe at the subatomic world or interstellar space. "Just the sheer knowledge, partial though it may be as brought to me by the various sciences, leaves me stunned and breathless" (Goodenough 1998, p. 18). Another doubted that traditional art forms could ever evoke the vast temporal dimensions of the story. "A spatiotemporal canvas is needed to capture the meaning of the scientific story of creation" (Goodenough 1998, 19).

Creating a spatiotemporal canvas for the unfolding cosmic story seems to be a goal of cinematic projects, such as Swimme and Tucker's film *Journey of the Universe*, billed as "big science, big history, big story" (Synopsis 2015, n.p.). With Swimme as our quasi-omniscient guide, the film takes viewers through the 14-billion-year "journey" of cosmogenesis. We travel from a small Greek island to the vast expanses of the universe, or from the birth of the universe to the realm of genome science, evoking the play of scales—microcosm/macrocosm contrast—that excites Big Historians. Our journey concludes on a distinctly Anthropocene note, highlighting humanity's recent "epic impact" on Earth's biosphere (Synopsis 2015, n.p.). The upshot is that knowledge of the vast cosmos—better yet, knowledge enhanced by cinematic immersion—will draw us into an intimate and responsible relationship with Earth.

Cosmopolitan environmentalism

This move from cosmology to ethics—zooming out in order to zoom back in with intensified ethical and emotional connection to nature—is symptomatic of our Anthropocene moment, though it has deeper roots.[8] For it is only with the perspective afforded by deep space and time, we are told, that humans apprehend the essence of our relationship to Earth and the radical nature of our present transformation of it. Christian, for example, claims that a "planetary scale" of 4.6 billion years is most appropriate for exploring "the human relationship with the environment and the biosphere," for it highlights the integration of *our* identity with the whole, interrelated organism that is Gaia (Christian 2005, p. 34). Only on this scale, moreover, can we "see" the human impact on the biosphere and compare it with other transformative impacts, such as asteroids (Christian 2005, p. 35). A similar move is made by a certain celebrity astrophysicist who, like Swimme, acts as cosmic tour guide. In the reprised version of *Cosmos*, Neil DeGrasse Tyson boards his Ship of the Imagination and ferries us through cosmic space and time. Occasionally, his narrative focuses on Earth's tumultuous climate history, thereby highlighting the relative climatic *stability* of the Holocene period during which our civilization arose and on which our survival utterly depends.

Not infrequently, Tyson ventures a more tendentious claim about the connection between space exploration and care for planet Earth, suggesting that the environmental movement *as a whole* owes its origins to the space program. On Earth Day 2015, I participated in a conference that featured Tyson as keynote speaker. In the course of his energetic and entertaining lecture, Tyson took pains to demolish the standard narrative, namely, that Rachel Carson and *Silent Spring* sparked the environmental movement. It was not images, conjured by Carson's prose, of DDT-ravaged eggshells or squirrels gnashing their teeth in the death throes of chemical poisoning that provoked an outpouring of empathy and action. It was the image of Earth from space. Without that iconic image, we would never have begun to think "globally" (picture *Whole Earth Catalog*, Tyson suggests). We have NASA to thank not only for our environmental legislation, but also for organizations like Doctors Without Borders, Tyson maintains, because the image of earthrise allowed us to envision our planet without boundaries. As with David Christian's invocation of Gaian unity as a normative principle, Tyson contends that Earth as a single, unified whole—a whole with which humanity could bond emotionally—gave birth to unified, global environmentalism.

Disorder and its discontents

Unity, identity, universality. These are the organizing principles of cosmic mythmaking. Humans require a common, comprehensive story, and a global environmental movement, we are told. To complete this vision, however, we must also achieve unity among the disciplines. E.O. Wilson refers to such unity as consilience. The quest for disciplinary unity—a seamless body of knowledge, or unifying paradigm—makes frequent appearances in these Anthropocene narratives. Christian writes that scholars who engage in work that crosses disciplinary boundaries and scales "will soon find they are taking part in the larger project that E.O. Wilson called consilience" (Christian 2005, p. 25). He promises that much will be gained from consilient "cooperation," though his descriptions of interdisciplinarity evoke militaristic invasion or mutiny rather than cooperative sharing: "Macrohistory will seem like a series of booty raids into neighboring disciplines," he predicts. "There is a lot of conceptual and methodological loot out there for historians willing to do the work and take the risks …" (Christian 2005, p. 36). He writes admiringly of boundary-crossing historians like Alfred Crosby, praising their "swashbuckling style … suggestive of a booty raid" (Christian 2005, p. 37).

A sense of daring and excitement—bold transgression rewarded with untold glory—animates these endeavors that spurn disciplinary boundaries and move breathlessly across disparate scales. Yet, swashbuckling flamboyance seems out of place with the decentered, non-individuated, mystically unbounded self that we might expect to emerge from habituation to the cosmic perspective or the *longue durée*. A paradoxical mix of humility and hubris is ever-present in these stories. Not only is cosmogenesis mythologized and consecrated; so too is our species' heroic quest to comprehend and narrate it. The adventure of mapping the universe and uniting all knowledge conjures mythic figures like Prometheus, the bringer of fire; or Icarus, who flew too near the sun. "Let us see how high we can fly before the sun melts the wax in our wings" (Wilson 1998, p. 7). Some decidedly unscientific rhetoric—appeals to *faith*, *hope* or *deep conviction*—permeates discussions of consilience and universality. Wilson confesses a metaphysical longing for consilience—a "conviction, far deeper than a mere working proposition, that the world is orderly and can be explained by a small number of natural laws" (Wilson 1998, p. 5). Christian has faith that universal history will reveal "profound orderliness" (Christian 2010, p. 20). A more innocently aspirational mood is suggested by Christian's characterization of Big History as part "genuine attempt at prediction," part "letter to Santa" (Christian 2010, p. 7). The quest for universal history and consilient unity seems buoyed by something almost *make-believe*—a fervent wish, a refusal to attend to sober naysayers and their unlovely visions. However it is expressed, the desire for unity, totality, and law-like order has a dreamlike quality for these starry-eyed seekers.

The wishful tone of these projects and their feel-good appeals to unity and harmony may mask their less egalitarian aspects. Big History and associated projects prompt us to reflect more deeply on both methodological and ethical questions that are also at the heart of Anthropocene debates. For example, scholars who engage in bridging the disparate disciplines and timescales by means of consilient passageways may find some partners to be more genuinely collaborative than others. Consilience is not necessarily a two-way street. "Scholars convinced that their method is consistently superior and universal," environmental historian John M. Meyer notes, "are most likely to pursue [multidisciplinary] endeavors by reducing all others to their singular preferred method. The result is holism manifest as reductionism" (Meyer 2014, p. 17) Christian's language of looting and booty raids, as I have said, suggests something other than mutual sharing of disciplinary riches—even while he speaks of historians raiding the *scientists'* bounty, rather than vice versa. From the standpoint of consilient integration, however, it is often the humanities that are regarded, paradoxically, as both the poorer cousin and the primary object of conquest.

Also of great concern to some environmental scholars—and I count myself among them—is that the scaled-up *species* vision of humanity obscures important differences among human cultures and their varied contributions to the global crisis in terms of fossil fuel consumption; it also downplays the critical role played by capitalism in fueling the Anthropocene. A given individual's "imprint on the atmosphere varies tremendously depending on where she is born," as human ecologist Andreas Malm points out. Humanity as a whole is thus "far too slender an abstraction to carry the burden of culpability" (Malm 2015, n.p.). Nor, I would argue, should this abstraction be put in the service of a global mythmaking enterprise that presumes to tell "everybody's story."[9] In the context of the Anthropocene and the present global crisis, there can be no depoliticized story of "all of us."

There are many different stories we can and ought to tell about what it means to be human. Depending on where we choose to look, we may derive a variety of meaning and messages, many of them incommensurable. Big Historians tout the empirical and scientific basis of their mode of storytelling. Yet interrogating even just one branch of the sciences—say, biology—can produce radically different interpretations of the human and our place in planetary unfolding. The species-level view of humans as a collective agent and discrete entity might be recognizable to paleobiologists, but a microbiologist would reject this image in favor of one that regards each of us as a conglomerate of multiple species, akin to a coral reef. Where paleobiology may seem to underwrite species oneness and solidarity—the ethic at the heart of Big History or The Universe Story and The Epic of Evolution—the microbiologist's perspective makes it difficult to talk about a human "we" at all. "While 99.9 percent of our human DNA is shared," historian Julia Adeney Thomas notes, "our microbial cells may have as little as 50 percent of their genetic profile in common." For scholars and laypeople who look to science to inspire a narrative of species solidarity, she notes, "this finding is disturbing" (Thomas 2014, p. 1595). Those who turn to science to craft a coherent story about who *we* are or what it means to be human will (if they are truly paying attention) come away perplexed and bewildered. On the other hand, equipped with the proper attitude toward this array of images and meanings—one that does not seek erasure of difference for the sake of narrative coherence and closure—we may come away with a genuine sense of awe and wonder at this diversity of disciplinary insights into who we are and where we fit in. For practical and ethical purposes, we may choose to attend to particular scales more than others, without rejecting the multiplicity of options and scales available to us from the disciplines. Where we choose to focus will necessarily reflect our values. If a deeper sense of affective and sensory connection to the natural world and its myriad inhabitants is what we hope to foster, I would argue that it seems most appropriate to attend to something far more local than the cosmic scale, for as David Abram notes, the "sensuous world is always local, and it is never merely a human world" (Abrams 2000, n.p.). The cosmic scale can be awe-inspiring, but it is not the scale on which we find or make meaning in our day-to-day lives.

Ultimately, the vision that inspires dreamers of Big History is one of greater certainty and simplicity, a universe in which everything falls into place and each of us knows, on good authority, who and where we are. Diapason, it is worth noting, implies not just range, but emergent harmony from the play of scales. However recent the science, this harmonious vision of the cosmos is ancient. The word cosmos, Sagan liked to remind us, means order. Its antipode is chaos. These storytellers, as I have suggested, assume that their own profound longing for cosmic identity, unity, order, and harmony is itself a species-wide trait. Whether that cosmic dream—with or without a robust phenomenology—will have broad intellectual and emotional appeal remains to be seen. I, for one, will continue to register my dissent from this orderly project and its assumptions. I do not feel its power and allure, nor can I revel in its carefully managed, empirically verifiable, law-abiding forms of wonder. Whether or not a cosmic blueprint, or a complete user's guide to the planet, can inspire deeper connection and concern for nature is another question. On that score, too, I have doubts and misgivings. Time alone will tell.

Notes

1 In addition to publicly advocating for the Epic of Evolution, Goodenough teaches a course on it at Washington University in St. Louis.

2 The Anthropocene is the proposed but as yet unofficial name for the present geological epoch in which humans have become a dominant force on the Earth system.

3 Their popularity in the subfield of religion and ecology is especially pronounced.

4 Some Epic of Evolution proponents are drawn to sociobiological or evolutionary psychological explanations of the origins and adaptive function of religion or myth.

5 Christian often refers to a *return* of universal history, suggesting that it once held sway (see Christian 2010).

6 The Ecozoic bears some resemblance to what some call a "good" Anthropocene: an era in which humans become wise planetary managers to the benefit of Earth and humanity (see Sideris 2015).

7 Consilience is E.O. Wilson's term for the unity of knowledge, the conviction that the disciplines are converging on a single, totalizing explanation of life, the mind, and the universe (Wilson 1998).
8 A number of scholars have explored the connections between the space program and the rise of environmentalism. For a work that argues for the similar goals of environmentalism and space exploration see Cockell (2007).
9 The expression "everybody's story" is Loyal Rue's from his book by the same title (Rue 2000) but is frequently used by proponents of cosmic storytelling.

References

Abram, D., Jensen, D. (2000). David Abram interviewed by Derrick Jensen. Alliance for Wild Earth. http://wildethics.org/awe/essay/david-abram-interviewed-by-derrick-jensen. Accessed 7 October 2015.

Barlow, C. (1997). *Green Space, Green Time: The Way of Science*. New York: Springer.

Barlow, C. (2013). Ritualizing big history. www.metanexus.net/blog/ritualizing-big-history. Accessed 15 October 2015.

Berry, T. (1978). The new story. *Teilhard Studies* (Number 1, Winter). New York: American Teilhard Association.

Blundell, R. (forthcoming 2016). Experience and big history. *The Ways that Big History Works: Cosmos, Life, Society, and Our Future. (From Big Bang to Galactic Civilizations: A Big History Anthology*. Volume 3). Delhi: Primus Books.

Chakrabarty, D. (2009). The climate of history: Four theses. *Critical Inquiry*, 35 (Winter). 197–222.

Christian, D. (2004). *Maps of Time: An Introduction to Big History*. Berkeley/Los Angeles, CA: University of California Press.

Christian, D. (2005). Macrohistory: The play of scales. *Social Evolution & History*, 4 (1). 22–59.

Christian, D. (2010). The return of universal history. *History and Theory*, Theme Issue 49, 6–27.

Cockell, C. (2007). *Space on Earth: Saving our World by Seeking Others*. New York: Macmillan.

Goodenough, U. (1998). Does the epic need art? An IRASnet dialogue. *Epic of Evolution Quarterly*. Fall, 18–19. www.thegreatstory.org/Epic-Evol-Journal.html. Accessed 15 October 2015.

Goodenough, U. (2010). My covenant with mystery. www.npr.org/blogs/13.7/2010/08/27/129471676/my-covenant-with-mystery. Accessed 15 October 2015.

Kennard, D, and Northcutt, P. (Dirs) (2011). *Journey of the Universe*. DVD. Mill Valley, CA: InCA Productions.

Malm, A. (2015). "The Anthropocene Myth." *Jacobin*. March 20. www.jacobinmag.com/2015/03/anthropocene-capitalism-climate-change. Accessed 16 September 2015.

Meyer, J. (2014). Less is more. In Emmett, Rob, and Frank Zelko (eds), "Minding the Gap: Working Across Disciplines in Environmental Studies." *RCC Perspectives* 2014, no. 2. www.environmentandsociety.org/sites/default/files/2014_i2_web.pdf. Accessed 1 September 2015.

Rodrigue, B., Brinin, L., Korotayev, A. (eds). (2015). Preface: The sphere of big history, and introduction: From big bang to galactic civilization. *Our Place in the Universe: An Introduction to Big History (From Big Bang to Galactic Civilizations: A Big History Anthology)*. Volume 1. Delhi: Primus Books.

Rue, L. (2004). *Amythia: Crisis in the Natural History of Western Culture*. Tuscaloosa, AL: University of Alabama Press.
Rue, L. (2000). *Everybody's Story: Wising Up to the Epic Of Evolution*. Albany, NY: SUNY Press.
Rue, L. Goodenough, U. (2009). A consilient curriculum. In C. Genet et al. (eds). *The Evolutionary Epic: Science's Story snd Humanity's Response*. Santa Margarita, CA: Collins Foundation Press.
Sideris, L.H. (2013). Science as sacred and myth? Ecospirituality in the Anthropocene age. In Rozzi et al. (eds). (2013). *Linking Ecology and Ethics for a Changing World: Values, Philosophy, and Action*. New York: Springer, 147–162.
Sideris, L.H. (2015). The confines of consecration: A reply to critics. *Journal for the Study of Religion, Nature, and Culture*, 9 (2), 221–239.
Smocovitis, V.B. (1999). Essay review: The tormenting desire for unity. *Journal of the History of Biology*, 32, 385–394.
Swimme, B. and Berry, T. (1992). *The Universe Story: From the Primordial Flaring Forth to the Ecozoic Age—A Celebration of the Unfolding Cosmos*. New York: Harper Collins.
Swimme, B. and Tucker, M.E. (2011). *Journey of the Universe*. New Haven: Yale University Press.
Synopsis of film. http://journeyoftheuniverse.org/synopsis. Accessed 15 October 2015.
Thomas, J.A. (2014). History and Biology in the Anthropocene: Problems of Scale, Problems of Value. *American Historical Review*. Roundtable: History Meets Biology, Vol. 119, No. 5 (December), 1587–1607.
Tyson, N.D. (2015). Science as a way of knowing. Keynote address. Ninth Annual Nelson Institute Earth Day Conference. University of Wisconsin. April 20.
Weldon, S. P. (2015). Big history, deep history, and the problem of scale. www.inhabitingtheanthropocene.com/2015/01/12/bigdeephistory. Accessed 29 September 2015.
Wilson, E.O. (1998). *Consilience: The Unity of Knowledge*. New York: Vintage.
Wolfe, N. (2014).The case for awe in teaching big history. In R.B. Sime, M. Behmand, T. Burke (eds). *Teaching big history*. Berkeley/Los Angeles, CA: University of California Press, 336–342.

17 The cycling historian

Exploring environmental history on two wheels

Stephen Bocking

At the junction of a road and an abandoned railroad just east of the village of Keene, the bicyclist faces a choice. Continue on the railroad, now a trail, as it gradually inclines, gently curving around the hills ahead—an easy ride. Or follow the road as it extends straight, then turns at a right angle, then continues straight again and almost straight up (it seems) a steep hill—a serious workout. Whatever choice is taken, the difference between rail and road becomes immediately evident—and not just in the bicyclist's muscles, but in the divergent opportunities they provide to experience the natural and cultural features of this region of southern Ontario, Canada.

These choices, and the experiences they present, are familiar to any bicyclist traveling in this or similar landscapes. The distinctive character of these experiences is especially evident when compared with travel by automobile: "[l]andscapes that disappear behind the window glass of a speeding car envelope the cyclist. They smell, they speak, they make you suffer and sing" (Barber 2014). Bicycles enable new experiences, even in familiar landscapes. Indeed, this quality appears in many personal histories: bicycles are the device by which countless children first constructed their own environment, exploring their neighbourhoods beyond the reach of parents. In the history of environmental experience, bicycles are inescapable.

Nevertheless, while appreciated as means of recreation and transportation, bicycles have received less attention from historians. As Graham Robb once noted, even in the country where the Tour de France is a metonym of national identity, the "effect of the bicycle on daily life is now drastically underestimated by many historians, who tend to see it as an instrument of self-inflicted torture" (Robb 2007, p. 341). This neglect is especially striking in environmental history. Environmental historians often insist on the use of not just traditional forms of evidence such as archival materials, printed texts, visual records and oral testimonies, but study of landscapes themselves, to identify the evidence of changes in nature and in human activities that will reward the careful observer. (As Donald Worster has explained, environmental history sometimes requires a "ramble into fields, woods, and the open air. It is time we bought a good set of walking shoes, and we cannot avoid getting some mud on them" (Worster, 1988, p. 289)). To these scholars, bicycling offers particular advantages, and not only as a means of observing landscapes. It provides an entry into understanding how historical actors have experienced the landscape, whether by bicycling or through other forms of transportation that operated at a similar pace and scale, particularly in the era before the automobile. Yet this significance is rarely remarked on by environmental historians.

This paper will discuss the methodological possibilities that bicycling presents for environmental history. Some of these have been evident in the historical experience of bicycling itself; accordingly, the paper begins with that history. The discussion will then turn to the distinctive benefits of bicycling for historical method. These benefits will be illustrated through a case study of a landscape especially well-suited to historical analysis on two wheels: south central Ontario. This is a predominately rural region, highly accessible to the bicycle, and possessing a rich environmental history.

Bicycles in history

Much of the scholarly and popular study of the history of bicycling has focused on technological development: tracing the sequence of technical and stylistic innovations, from the "boneshaker" to the high wheeler to the safety bicycle—the precursor of today's conventional bicycle.[1] Historians of technology, most notably Wiebe Bijker, have also explained how innovations in bicycles have not occurred in isolation, but in the context of larger trends in technological, economic, social and cultural change (Bijker 1995).

Understanding the history of bicycles in relation to their larger social contexts becomes essential when considering the "bike craze" that briefly took hold in Canada and elsewhere in the 1890s. Technological development was the immediate factor igniting wider interest in bicycling: the safety bicycle—easier and safer to ride by older and younger people, including women—made bicycling far more inclusive. High wheelers, in contrast, had been both expensive and intimidating, and so were mainly the preserve of privileged young men. Other technical innovations, including pneumatic tires and lighter frames, also attracted new riders. Numbers tell the story: by 1895, 90 stores in Toronto alone were selling bicycles (Norcliffe 2001, p. 202). They appeared in every city, in rural areas, even in remote regions, including the Klondike, where they served as a significant form of transport during the gold rush (Cole 1990; Humber 1986, p. 22; Norcliffe 2001, pp. 216–222). A similar phenomenon occurred in the United States and Europe; internationally, world tours by bicycle became objects of public fascination.

Bicycles meant mobility. Adopted across diverse classes and occupations (including the social "elites"), they provided a liberating personal mobility, free from the limits of both walking and railroad timetables—in the words of historian Eugen Weber, they were an "incomparable harbinger of progress and emancipation" (quoted in Norcliffe 2001, p. 7). (In France, bicycles have even been credited with increasing the average height of the population by reducing the number of marriages between blood relations (Robb 2007, p. 342)). This new mobility was especially significant to women. As American social reformer and feminist Susan B. Anthony proclaimed in 1896: "Let me tell you what I think of bicycling. I think it has done more for the emancipation of women than anything else in the world" (Bathurst 2011, p. 101). Women adopting the bicycle saw it not just as practical transportation or recreation, but as an emancipatory technology that affirmed their dignity and equality (Herlihy 2004, p. 266). (Emancipation had limits, however: in 1894 the League of American Wheelmen, the premier bicycling organization in the United States, voted to exclude black people from membership.) Overall, bicycles were central to a brief but striking period in technological and social history. Epitomizing progress and the expansion of personal geographies, they marked, and even encouraged, the transition from the railway age to the age of the private automobile. Bicycles represented modernity. Displacing the horse as a means of personal transportation, they were considered among the most significant technological innovations of the era, alongside the steamship, railroad, telegraph and telephone (Herlihy 2004, p. 2).

Among their consequences were new ways of experiencing and understanding environments. Sometimes this experience was vicarious, the product of watching, not doing. For example, while the Tour de France became one of the world's leading sport spectacles, it also delivered to the French a better understanding of their own national territory: it "gave millions of people their first true sense of the shape and size of France, but it also proved beyond doubt that the land of a thousand little pays was still alive" (Robb 2007, p. 343). But of even greater significance to the environmental history of bicycling was its role in giving ordinary people a novel way of gaining access to the countryside. Numerous observers celebrated city dwellers being able to "escape" to the country, and to value what they found there: as one commented in 1895, the "bicycle has done more to foster love of out-of-doors than any other invention" (Herlihy 2004, p. 264). The consequences were felt in many areas of life, including the study of nature. As David Allen explained in his history of British naturalists, the "most conspicuous feature of the nineties was the arrival of bicycles in strength. For the first time a cheap and highly flexible form of personal transport came within reach of the majority, and pursuits like natural history, traditionally a matter of short-term forays into ever remoter districts, were among the prime beneficiaries" (Allen 1976, p. 224). One of the advantages of bicycling, Allen noted, was its almost total silence—wild birds would allow a bicycle to pass within just a few feet.

These consequences exemplified wider changes in how people were experiencing rural areas by redefining the relation between city and country, shifting the edges separating suburban and rural environments, and changing the meaning itself of the term "countryside" (McCullough 2015). In this newly accessible rural terrain, working landscapes became tourist sites, defined by a new recreational geography, with hotels for bicyclists on weekend excursions, and tearooms where they could refresh themselves before returning to the city (Norcliffe 2001, p. 19).

But these excursions could also often be ordeals, as bicyclists struggled along rutted, muddy, or rocky roads. The challenges they encountered on rudimentary rural roads encouraged many to take action. They assembled a substantial body of geographical literature for their own guidance, pioneered the use of road signs, and built a network of bicycle paths. They also became policy advocates. A major focus, particularly in the United States, was advocacy for better roads. In the 1890s the League of American Wheelmen adopted a leading role in the Good Roads lobby. Although there was a similar roads lobby in Canada, bicyclists were apparently less prominent in it (Wells 2012, pp. 25–26; Norcliffe 2001, p. 149). Bicyclists also urged that existing passageways such as trolley rights-of-way and canal towpaths be adapted for bicycling (McCullough 2015). This advocacy had some consequences: encouraging road improvements, and influencing city design in subtle ways. But its significance went beyond improved surfaces. The notion itself of roads as a policy issue was an early manifestation of the idea, central to the emergence of environmental politics, that common spaces are matters of not only private initiative but public responsibility, and so require the organized attention of the state.

Other political implications of bicycling also began to emerge in the 1890s. These included greater interest in the preservation and improvement of scenic and rural landscapes, underpinned by a view of the countryside as existing in a state of settled harmony between nature and culture. Illustrated materials produced by bicyclists strengthened the ethical ties between bicycling and conservation, extending that ethic to cultural as well as natural resources (McCullough 2015). Bicyclists also imagined how their machines could bring some of the quiet and cleanliness they found in the country back into the city. By replacing horses (and their manure and clattering hoofs), bicycles would help create a more pleasant urban environment—the same hopes that, somewhat later, became attached to automobiles (Friss 2015; Herlihy 2004, p. 259; Tarr, 1996). Bicycling thus illustrated how a distinctive mode of experiencing the environment through technology could bring wider public attention to issues of scenic and landscape preservation—an outcome that anticipated the better-known story of how, in the United States, knowing nature through the automobile encouraged the establishment of national parks (Louter 2010).

Bicycles also had other indirect environmental and social consequences, particularly through technological innovations. The novel mechanical contrivances and modes of industrial organization fostered by bicycle production would become essential to other forms of technology. Airplanes and automobiles adopted innovations developed for bicycles, including pneumatic tires, wire spokes, steel tubing, ball bearings, and chain and shaft drives. The Wright brothers owed their talents as airplane builders to their work as bike mechanics—an experience they shared with Henry Ford and other early car makers. Bicyclists' advocacy for better roads also smoothed the way for drivers. These new technologies, of course, would have consequences far beyond those that could be traced to bicycling. But perhaps most significant, as David Herlihy has argued, is that the bicycle helped redefine the nature of transportation itself. It created a demand for independent and private travel that the railroad, with its fixed schedules and communal accommodations, could not meet, and that would itself only eventually be satisfied by automobiles (Herlihy 2004, pp. 299–300). The ironies, of course, are obvious: although bicyclists literally helped pave the way for automobiles, they have since had to fight to maintain their presence among them—an eco-friendly technology thus enabled its far more polluting replacement.

By 1897 the bicycling industry, facing declining sales, had collapsed in a mountain of debt and unsold bicycles. Sales nevertheless recovered after 1900, as bicycles became more affordable and reliable (thanks to systems of mass production), and so spread across the social strata. They also began to be used not just for recreation and exercise, but in a variety of occupations. Even after the boom of the 1890s, the bicycle was still widely accepted as a practical and versatile machine, and was now within the reach of much of the population (Herlihy 2004, p. 297). But it had lost the sense of excitement and prestige, as influential professional and business classes became, like Mr. Toad in *The Wind in the Willows*, more excited about the next novel machine—the automobile. No longer were bicycles felt to epitomize modernity (Norcliffe 2001, p. 187). So while in some industrialized countries, particularly in Europe, bicycles came to be widely used for practical transportation and recreation, in North America (except occasionally, such as during the Depression and in response to gasoline rationing during the Second World War), they came to be seen as mainly for children. In 1933, while Americans owned 17 cars for every bike, Europeans had seven bikes for every car (Herlihy 2004, p. 328). Postwar suburban development and the great commuting distances it required hardwired car-dependency and the marginal role of bicycles in America. While cars also became far more important in postwar Europe, cultural and physical factors preserved a place for bicycles as an option for adults.

Revival came in the late 1960s, with a renewed enthusiasm for bicycling. Novel bicycle designs, especially the 10-speed, fed demand, as did interest in personal health and exercise and concerns regarding environmental pollution and energy conservation. A variety of other ties between bicycles and environmental politics also emerged. Visions of future societies guided by environmental values often placed bicycles in a central role; bicycles also featured in counterculture movements as a form of resistance to modern technology (nicely reversing their role epitomizing modernity during the 1890s) (Perry 1995). More recently, interest in recreational bicycle paths, including the conversion of abandoned railroad lines, has encouraged new definitions and spatial configurations of protected areas. Concerns regarding climate change and urban congestion have led many cities to invest in bike lanes and bike-sharing programs, while encouraging wider debates about who belongs on the road. The emergence in California and elsewhere of mountain bikes since the 1970s also inserted bicycling into debates about access to wilderness. The essence of mountain biking technology, of enabling bicycles to go where they don't normally go, raised objections among traditional trail users concerned about the consequences of greater access to and mobility within undisturbed landscapes. In interesting ways, therefore, bicycling became tied to the emergence of a new politics of energy and the environment, urban planning and transportation, and the wilderness.

The history of bicycling is also an aspect of the larger history of the relations between technology and perceptions of landscapes. Several scholars have examined the perceptual implications of technologies such as the automobile (Barnett 2004), the steam engine (Roberts 2004), and the airplane (Bocking 2009); together, they illustrate the diverse ways in which technology helps to define places. The political implications of the view from the automobile have also been considered in studies of wilderness protection (e.g. Louter 2010). Often the automobile has been criticized for its role in the loss of an intimate connection to the landscape. A representative argument is Thomas McIlwraith's: "[f]ast motoring transformed our view of the countryside between towns and farmsteads from that of a pleasant and interesting space ... to that of a barrier separating the settlements, to be conquered or at least endured" (McIlwraith 1997, p. 15).

In contrast, slower, self-powered forms of mobility can enable other ways of perceiving landscapes. For example, Joy Parr has explored the significance of walking to the ordinary experience of living in a place: it is, she explains, a "spatial practice" that "coagulates time, expands distance and makes places dense and prickly with details and complexities" (Parr 2010, p. 82). Perceptions of the environment through the body become essential: sensory experiences of sight, sound, touch, taste and smell that precede the construction of meaning through thought and language, and that often cannot be conveyed in words. Perceptions, including those other than sight, are thus central to the experience of everyday life; finding ways to access them should be, as Parr has argued, an essential element of the practice of environmental history (Parr 2010, p. 9). This requires going beyond the study of written or visual records.

Environmental history on a bicycle

Just as bicycling has been, as we have seen, an historically significant way of seeing, understanding and acting in landscapes, so too does it provide ways of experiencing and interpreting an environment historically, enabling one to move through it while keeping in mind the essential historical question: "How did things get to be this way?" (Cronon 2003, p. 171). This contribution of bicyclists to historical practice is itself historically situated. Bicyclists have often been among the most discerning observers of landscapes. As Robert McCullough has described recently, American bicyclists in the late-nineteenth century did not just explore their countryside in unprecedented numbers, they were also keen observers: "[p]ossessing a heightened awareness of their surroundings and stirred by the compositional elements inherent in those landscapes, these ... wheelmen had become discerning landseeërs" (McCullough 2015, p. x). They also produced copious records of their journeys: periodical articles, books and diaries, and road books prepared by state divisions of the League of American Wheelmen. Rural life and the relations between urban and rural landscapes were of particular interest. For example, Archibald Lampman, an Ottawa civil servant, bicycled around southern Ontario, reproducing in poems such as "Heat" the atmosphere and sensibilities of this region (Humber 1986, p. 22). Other especially energetic bicyclists provided more extensive, if less lyrical, records of roads and landscapes. Among these was Lyman Hotchkiss Bagg, who in the early 1880s set out on a series of long-distance rides in the United States and Canada, producing along the way a detailed record of road conditions and other features of rural landscapes. His book, *Ten Thousand Miles on a Bicycle* (1887) has been said to be the best surviving record of the state of roads in Canada during this period (Bagg 1887; Norcliffe 2001, pp. 157–168). His report of other cyclists traveling in 1885 through southern Ontario captures what roads were like in the era before uniform standards: "Through the town [Cobourg] the road was all right, but ½ m. [mile] out they came to a steep hill that forced a number to dismount, and then came a stretch of soft sand, followed by a mile of loose stones of assorted sizes, and then a stretch of ruts, and so on throughout most of the day's ride" (Bagg 1887, p. 320). These records are now of special historical value because they offer glimpses of landscapes that would soon be transformed by the automobile and by the redefinition of road improvement as a public responsibility.

These records also testify to the distinctive ways of perceiving landscapes available from the perspective of a bicycle. They therefore hint of the value today of bicycling as an historical method. This is partly a matter of speed, and how it shapes opportunities to observe, reflect on, and make sense of the many layers of history and the subtle features that form landscapes. As landscape architect John Stilgoe has argued, the car "insulates the motorist from every sort of nuance. [It] moves too fast for its driver to notice much, and when it slows, it obstructs then jams traffic". Nor can a car venture into narrow roads or trails. The automobile also lacks "the cloak of invisibility bicyclists and walkers quickly take for granted" (both quotes in Stilgoe 1998, pp. 9–10). So while drivers speed along, indifferent to topography, focused on the road in front rather than the countryside around, bicyclists can move at a slower pace, observing details, stopping as necessary for closer examination. While walking is often invoked as an ideal method for observing historical details and landscapes (see, for example, Cronon et al. 2008), a bicyclist can observe with a similar level of care, but is also able to cover more ground, more quickly—thereby benefiting from both the practical advantage of observing a significant area in a reasonable period of time, and the potential to detect the larger patterns of regional landscapes. Bagg himself noted in 1887 these advantages of bicycling, suggesting that while travel by train would be too fast and travel by foot too slow to appreciate the country, bicycling, which he referred to as "walking large," set the right pace (Bagg 1887, p. 303). McCullough agrees, explaining that the speed of bicycling "endows cyclists with the ability to record mental images of various land features, recalling them and connecting them in creative ways and forming visual prospects that remain invisible to others… today, bicycling's landseeërs can spot the traces of history that others disregard in our built and cultural environments and can offer imaginative and fitting visions for reclaiming those forgotten places" (McCullough 2015, p. x). He adds that, having the ability "to move at distinctive speeds, a cyclist can lengthen a sense of place in both distance and time, allowing sensory perception to linger in thought or memory and providing opportunities to revisit places imaginatively from a distance" (McCullough 2015, p. 11).

Bicycle-based environmental history study has greater potential in some landscapes than others. Huge regions, such as the Great Plains, or those with few roads, such as northern Canada, are obviously poorly suited to study on two wheels. Where this potential is most evident is in those landscapes that combine appropriate scale and access. Settled regions, such as the rural areas of eastern Canada and the United States, provide such landscapes. Situated in the "middle" ground between wilderness and city, with ample roads and mosaics of land use that testify to complex histories of interaction between humans and nature, these places present many opportunities to use bicycles to explore their pasts.

Of particular interest to me is one region within southern Ontario: the area east of Toronto, centered on Peterborough. The home of my university, it is also a place where I have devoted considerable time over the years to exploring back roads and former railroad lines. The region encompasses diverse landscapes. In the south, glacial forces formed fertile plains, rolling hills, and lakes and rivers that would become essential to transportation, settlement, and industry. In the north, the Canadian Shield lies beneath forests and lakes, its granite prohibiting cultivation over most of the terrain. Throughout their history these landscapes have been transformed by human activities, even as these activities have themselves been shaped by natural features. The region also possesses several sites essential to the history of Canadian bicycling: pioneering bicycle factories, some of the country's first bike races, even the route that Bagg followed (and described in meticulous detail) back in the early 1880s (Humber 1986). In the following discussion, the practices and outcomes of bicycle-based environmental history study will be explored using this region as a case study. The objective will be to demonstrate how bicycling, in combination with other forms of historical method and evidence, can contribute to understanding several aspects of the region's environmental history, including land surveys, settlement, forest clearing and agriculture, the formation of transportation networks, industrial expansion and tourist development.

We can begin with the natural features of this landscape. These bear countless signs of past transformations, evident to any observers who have both awareness and the time to reflect on what they see and feel as they move across this terrain. This region is the product of diverse geological forces, but especially glaciation. (The Wisconsinan ice sheet of the last Ice Age retreated northwards about 12,000 years ago.) These forces produced a distinctive and varied landscape of drumlins, moraines, eskers and glacial spillways, as well as lakes, wetlands, and rivers (Ecclestone and Cogley 2009). Geology and physiography have, in turn, guided how people have lived in the region, shaping transportation patterns (especially by water) and, by determining soil quality, slope, and other variables, influencing agriculture and other land use activities. Suitability for agriculture generally declines as one travels north. Even in the south, although there is much fertile land, the land is often rough and stony, better suited to pasture than crops. Drumlins and other glacial remains are interspersed with poorly drained plains underlain by clay, creating areas of marsh and bog that have posed obstacles to agriculture. On the Shield thin soils discouraged permanent settlement, encouraging a distinctive cultural landscape.

To understand these patterns one can draw on the varied sources available to the historian, such as geological maps, historical series of topographic maps, aerial photos, and surveys of the physiography of the region. But bicycling can also aid in the study of the relations between these natural features and human choices relating to the environment. In fact, these relations can be best appreciated at the slow speed of bicycling. For example, the distribution of rock types, such as limestone beds overlying Shield granite, can be observed in road cuttings. This pace can also enable observation of how woodlots are often situated on land that is poorly drained, or that is otherwise unsuitable for cultivation, and how the steep slopes of drumlins and the rocky soils of moraines are typically reserved as pastureland. Through physical exertion, one can get a stronger sense of the role of glacial forces in shaping the region's landforms, whether this is experienced by labouring up the sides of drumlins, or, as a welcome break, speeding across the flat farmland and wetlands that once formed the floor of glacial Lakes Peterborough and Jackson. If, as environmental historians often insist, features of nonhuman nature are also historical actors, there is value in a mode of study that involves experiencing them directly, gaining thereby a visceral sense of their materiality and their capacity to influence human choices.

A second aspect of regional environmental history accessible to the bicycling historian are local ways of making things such as buildings, fences, and other structures out of and in the environment. As McIlwraith (1997, p. 9) notes, these "structures will speak reliably, if we know the language" (and, he might have added, if we pass by slowly enough to hear what they have to say). They testify to how, as Stilgoe has argued, the most ordinary features of a landscape often contain traces of the activities and consequences of previous generations; they also form links between the material world accessible to our senses and the concepts that we use to think about how we organize society and our relations with nature (Stilgoe 1998, p. 6). Consider, for example, fences, which are ubiquitous in this landscape. Seemingly fragile and easily destroyed, many stretches of old fences have managed to survive the transformation of rural roads from narrow dirt tracks to wide paved highways, thanks to the generous width (one chain—66 feet, or 20 metres) of the original surveyed road allowances (McIlwraith 1997, pp. 267–268). When examined with careful attention to local variations and regional patterns, their distribution and mode of construction can tell us about the influence of traditions from settlers' home regions, the development of the agricultural economy, property relations, and relations with local resources, including land and timber. Fences began to be important in this region in the mid-nineteenth century, as settlers finished clearing their lots, thereby making property boundaries more important. The evolving rural economy, including a shift later that century towards rotations of crops and livestock characteristic of mixed farming, encouraged the proliferation of fences not only along farm boundaries but within farms, separating fields of typically about four hectares (McIlwraith 1997, p. 41).

The diverse fence types apparent to bicyclists today also tell stories. Snake-rail fences, set at angles, were among the earliest type: they took more land, but required less labour, and thus reflect the relative abundance of each resource during that era. In contrast, later straight fences of wood or wire used land more economically, reflecting its increasing value. Tripod fences that do not require the digging of postholes reflect adaptation to thin, stony soils. Fences made of upended stumps may be byproducts of farm mechanization in the second half of the nineteenth century: for the first time tractors required farmers to remove stumps from fields. In subtle ways, therefore, fence types mirror the relative availability of land, labour, and capital, technological change, and adaptations to local environmental conditions.

The styles and materials of built structures offer other clues to changing rural economies and resource uses. Like fences, these are apparent to bicyclists moving slowly enough to perceive their subtle patterns. Early in the white settlement era log barns lacking foundations were built in this region to store wheat for export. Some examples still survive, providing hints of the geography of this aspect of the early rural economy. By the 1870s the region had been forced out of the wheat market by cheaper American production; farmers responded by taking up mixed farming, and so lifted their barns onto new foundations, or built new barns with space below for livestock (McIlwraith 1997, pp. 8, 41). The use of local resources as building materials can also be traced through careful observation. For example, houses made of logs (likely dating to the first generation of settlement) can be identified, even if covered by a veneer, by small, gable-end windows squeezed up under the eaves (McIlwraith 1997, p. 72). Houses placed in open land, often on hilltops, testify to settlers' desire—said to be distinctive to North America—for light and wind on all four sides (Stilgoe 1998, p. 128). Bicycling can also provide views of the subtle but long-term effects of earth moving: barn ramps and leveled fields, railway and road cuttings, embankments and wayside pits—illustrating how the environmental history of a region includes a reshaping of the earth itself.

Traces of other built forms may also survive, invisible to all except those able to pass by more slowly, but with much to say about regional environmental history. For example, mapping the distribution of saw mills and grist mills can contribute to understanding geographical patterns of settlement, industry and agriculture, the ties between the regional economy and the local environment, including the transformation of local sources of energy and resources into commodities (Steinberg 1991), and the history of impacts on stream hydrological regimes and on stream life, including salmon (formerly widely abundant in this region, but largely extirpated by the late 1800s). Their locations can be identified through various historical records, including old maps, county atlases, and fire insurance plans. But the landscape itself should also be studied, as sometimes abandoned mills and their associated dams are hidden near bridges or under foliage, where they can be more readily noticed by a bicyclist than by motorists speeding by. Tiny communities that once served their agricultural hinterlands, but whose role has been eliminated by the automobile, and so testify to the changing geographies of daily life, can similarly be readily noticed and explored by bicycle. Other signs of evolving geographies of mobility are evident in the remains of buildings lining abandoned railroad lines, including warehouses, granaries, and stations. These tell of the links that once existed between city and country formed through the movement of commodities. These signs can be found by bicycling along the many kilometres of railroad lines in this region that have been converted to trails. Some railroads, however, were abandoned so long ago that only fragments remain, and so are impassable by bicycle. For example, the Peterborough-Cobourg railroad, abandoned by 1861, is now only evident in overgrown hedges that arc across fields, as well as in a few buildings, now private homes, that once served as railroad hotels or stations.

Other possibilities for historical interpretation concern the evolving relations between human and natural landscapes as represented, for example, by land surveys and road networks. In the early 1800s, the Colonial Office sought to impose an orderly system of land allocation, applying a uniform grid survey system that embodied ideas about order and rationality. This system subsequently guided road construction and the allocation of property rights, with a host of consequences for people's relations to the environment. Bicycling along the roads produced by this system provides many opportunities to explore the practical relations between this rational grid and the "irrational" local landscape. In this region, road cuttings commonly ascend the steep-sided drumlins that run in rows that correspond to the direction of ice movement, but that are diagonal to the imposed grid; the exertion these drumlins demand from the bicyclist testifies to the disharmony between grid and landscape. Elsewhere, streams and roads bend in mutual deference, illustrating compromises between rationality and local realities, as do occasional jogs in the road pattern that compensate for surveyors' mistakes.

This grid and accompanying transformations rendered nearly invisible, even to bicyclists, another dimension of this landscape: thousands of years of Indigenous occupation. A few traces of this history can still be seen: the Alderville Black Oak Savanna (grasslands maintained today through controlled burns that reproduce once widespread Indigenous management practices), as well as sacred places such as the Serpent Mounds burial sites north of Rice Lake and the Petroglyphs (ancient images carved in rock) northeast of Peterborough. There are also more subtle traces of Indigenous history—such as the roads that follow ancient paths or portage routes, and that now cut across the grid; and place names, including the Otonabee River (from Odoonabii-ziibi, Ojibwe for "the river that beats like a heart"—a reference to its rapids, many now drowned by the reservoirs of the Trent-Severn Waterway). But bicycling offers no special insights into these traces.

Other relations between human activities and natural environments that can be observed by bicycle include the role of these environments in shaping settlement and land uses. Bagg observed these relations as he bicycled eastward across southern Ontario in 1883:

> the further I proceeded, the poorer grew the soil, and, similarly, the appearance of the people and their habitations. The roads likewise deteriorated and became rougher and stonier as the country grew more bleak and barren. Its rocky and sterile nature impressed me most on the final day, when the houses were fewer in number and meaner in quality than ever before,—some of them being nothing else than rudely-built log-cabins.
>
> (Bagg 1887, p. 302)

A difficult history of human-environment relations is similarly evident in the landscape in the north of this region, on the edge of the Shield. Bicycling along the narrow winding roads in this area reveals many traces of the difficulties encountered when settlers from elsewhere attempted to impose ideas of agricultural progress on this difficult terrain. Census figures tell of how the local agricultural population peaked in about 1881, before experiencing decades of depopulation, with numbers only now recovering to their earlier level. Observations today give texture and meaning to these numbers: small fields of poor soil carved from second-growth woods; log barns and houses, many abandoned or never modified or replaced by the second generation frame houses evident in more productive regions; and snake-rail fences that were never improved into straight fences. These traces testify to the unsuccessful effort after 1860 to extend northwards small farms after the better land to the south had been settled (Wynn 1979). This effort itself had been motivated by a belief in agricultural expansion as the basis for colonial development. Today, proliferating cottages reflect the displacement of this struggling agricultural economy by urban "settlers" that attach more value to scenic and recreational possibilities than to soil. In these and other ways the relations between economic activity and the landscape, including the gradual realization by settlers of its (and their own) poor prospects, are retained in local buildings and settlement patterns, evident to the careful observer.

Much can also be learned from close study of the spatial patterns of land uses: how land, water, shelter, transportation and other features are arranged. At the speed of a bicycle details can be noted that might otherwise be missed: the quality and apparent viability of farms, the state of the fields (cultivated or abandoned, grazed or plowed), the state of the houses, barns and other outbuildings (well-maintained or dilapidated)—all of which are clues to the evolution of the rural economy. Close observation of abandoned rural industries and farm depots, "ghost" villages, and even the recent appearance of solar power facilities can also testify to this evolution.

Bicycling also presents opportunities to examine evolving networks of mobility. The ability to travel along both roads and former railroad lines enables study of the contrasting relations between topography, property, labour and capital that these represent. Most obviously, abandoned railroads testify to the twentieth-century triumph of the automobile. But they tell other stories as well: about the linked consequences of the physical requirements of railroading (gentle inclines and curves, engineered rail beds, and causeways and bridges able to withstand heavy loads), the railroads' legal capacity to extend their rights-of-way across land and property lines, and the costliness of labour in relation to land in nineteenth century Canada, which required railroads to plan routes that would not require a large workforce for their construction.[2] The contrast with roads is instructive: while railroads are respectful of topography but heedless of property, roads are precisely the reverse. This has had ample consequences for travelers' divergent experiences of the landscape: unlike railroad passengers, the road traveler, constrained by the grid and property rights, encounters head-on hills and other topographic features (Parr 2010, p. 18). The difference can be experienced and recorded not only in topographic maps and photos, but in a bicyclist's muscles.

Bicycling along abandoned railroads can even be seen as a kind of time travel. These once-busy corridors, now almost deserted, provide an experience of the landscape somewhat akin to that of railroad passengers (not entirely, of course: the bicyclist experiences neither the railroads' precisely regimented timetables—so novel in a landscape once regulated by the sun and planting seasons—nor the sociable experience of being a passenger). This illustrates how bicycling can also provide insights into the more intangible dimensions of the environmental history of a region, including people's perceptions of landscapes. As an embodied

Figure 17.1 Trail on former railroad line – curving in response to topography

Source: photo credit: Stephen Bocking.

Figure 17.2 Country road – straight line across the landscape

Source: photo credit: Stephen Bocking.

approach to the study of history, bicycling gives access to sensory histories, enabling one to approach people's everyday experience of past environments (Parr 2010).

Consider, for example, Catharine Parr Traill's observations as she and her family journeyed in 1832 through this region:

> About three miles above Peterborough the road winds along the brow of a steep ridge, the bottom of which has every appearance of having been formerly the bed of a lateral branch of the present river, or perhaps some small lake, which has been diverted from its channel, and merged in the Otanabee. On either side of this ridge there is a steep descent; on the right the Otanabee breaks upon you, rushing with great velocity over its rocky bed, forming rapids in miniature resembling those of the St. Laurence; its dark, frowning woods of sombre pine give a grandeur to the scenery that is very impressive. On the left lies below you a sweet secluded dell of evergreens, cedar, hemlock, and pine, enlivened by a few deciduous trees.

> Through this dell there is a road-track leading to a fine cleared farm, the green pastures of which were rendered more pleasing by the absence of the odious stumps that disfigure the clearings in this part of the country. A pretty bright stream flows through the low meadow that lies at the foot of the hill, which you descend suddenly close by a small grist-mill that is worked by the waters, just where they meet the rapids of the river.
>
> (Traill 1836, Letter VII, 25 October 1832)

Traill's writings, including *The Backwoods of Canada*, are well-known literary records of the experience of early settlers in Upper Canada. As this passage attests, they are also of interest to environmental historians. Her observations provide a valuable visual record, of, for example, the "dark, frowning woods of sombre pine" (now vanished), or the rapids on a stretch of the "Otanabee" (Otonabee) River that is now, thanks to the locks and dams of the Trent-Severn Waterway, a series of quiet lakes. Her commentaries – such as her remarks about a "fine cleared farm" and "odious" stumps—illustrate early settlers' aspirations regarding transformation of this landscape. Further, her account conveys the distinctive perceptions of scale and distance, and of the difficulties presented by local topography, that were part of the experience of movement during this era. A bicyclist moving at a pace akin to that of movement during an earlier, pre-mechanized era presents the possibility of at least partly reproducing the sensory and physical experience of travel at that time: not just the exertion required, but the experience of, say, the gradual appearance of a new view as one moves around a hill or approaches its crest.

By reproducing (in part) the experience of being in and moving within an environment in earlier eras, bicycling can also aid in understanding how certain places became valued in relation to social priorities and interests. For example, as a few historians have explained, certain landscapes have been appreciated as having healthful qualities, and so became sites of recreation and recuperation (Mitman 2007). Certain areas within this region of Ontario have also been valued in these terms: in the early 1800s the landscape around Rice Lake appealed to many seeking leisure; several decades later, lakes further to the north along the edge of the Shield began attracting urban dwellers seeking respite from the stress of city life. The re-creation through bicycling of past experiences of landscapes can help in understanding how and why sensory perceptions, together with ideas about health and the environment, led to certain places becoming valued in these terms.

While visual evidence is so often privileged as evidence of environmental change, bicycling can provide opportunities to access other senses: to understand what a place sounds, smells, or feels like, which can then lead to questions about the origins of these sensations. Bicyclists relying on their muscles can also gain insight into the implications of experiencing a place through one's body, and so how human endurance shaped the experience of landscapes in the past: for example, the placement of inns in relation to steep hills and at intervals appropriate for tired horses. More generally, bicycling provides many opportunities to be reminded of the importance of muscle power in environmental history, not just in modes of travel, but in, for example, the need to husband scarce labour in settler economies.

Conclusion

A distinctive feature of bicycling is that it is done at a pace that permits careful observation and slow reflection. It therefore provides many opportunities to assemble diverse observations and forms of evidence and experience in order to understand the complex mosaic of a rural landscape. In doing so it enables access to essential stories of regional environmental history: the extension of land surveys; the expansion of agricultural settlement; the rise and decline of resource industries; the evolution, at several scales, of trading and transportation networks; shifting perceptions of the landscape; and the relations between land, resources, labour and capital. To deduce these stories one must draw on the forms of evidence conventionally associated with historical research. But it also requires getting out and experiencing the environment: treating the ground as an historical document, identifying the multiple meanings embedded in it, and thus the links between environmental change and human ideas and interests. Bicycling is one essential but often overlooked way of obtaining this experience. Just as it has been part of many episodes in environmental history, so can bicycling be applied to exploring this history.

Notes

1 For example, David Herlihy (2004) structured his comprehensive history of the bicycle in terms of these and other distinctive styles.
2 In contrast, nineteenth-century British railroads tended to be straighter: this required less land (which was expensive), but more engineering works, including tunnels, that drew on large numbers of lowly paid workers.

References

Allen, D. E. 1976. *The Naturalist in Britain: A Social History*. New York: Allen Lane.

Bagg, L. H. [a.k.a. Karl Kron]. 1887. *Ten Thousand Miles on a Bicycle*. Self-published. The University Building, Washington Square, New York.

Barber, J. 2014. Massive bike route to showcase Ontario's green space. *Toronto Star*. www.greenbelt.ca/toronto_star_series_gbbikeroute_2014.

Barnett, G. R. 2004. Drive-by viewing: Visual consciousness and forest preservation in the automobile age. *Technology and Culture*, 45(1), 30–54.

Bathurst, B. 2011. *The Bicycle Book*. London: Harper Press.

Bijker, W. E. 1995. *Of Bicycles, Bakelites, and Bulbs: Toward a Theory of Sociotechnical Change*. Cambridge, MA: MIT Press.

Bocking, S. 2009. A disciplined geography: Aviation, science, and the Cold War in Northern Canada, 1945–1960. *Technology and Culture*, 50(2), 265–290.

Cole, T. 1990. *Wheels on Ice: Bicycling in Alaska, 1898–1908*. Alaska Northwest Publishing Company.

Cronon, W. 2003. *Changes in the Land: Indians, Colonists, and the Ecology of New England.* Afterword. New York: Hill and Wang.
Cronon, W. and students. 2008. Learning to do historical research: Sources: How to read a landscape. www.williamcronon.net/researching/landscapes.htm.
Ecclestone, M. and G. Cogley. 2009. The physical landscape. In P. Adams and C. Taylor, (eds.) *Peterborough and the Kawarthas*, 3rd edn (19–40). Trent University Geography Department.
Friss, E. 2015. *The Cycling City: Bicycles and Urban America in the 1890s*. Chicago: University of Chicago Press.
Herlihy, D. V. 2004. *Bicycle: The History*. New Haven, CT: Yale University Press.
Humber, W. 1986. *Freewheeling: The Story of Bicycling in Canada.* Erin, ON: Boston Mills Press.
Louter, D. 2010. *Windshield Wilderness: Cars, Roads, and Nature in Washington's National Parks*. Seattle, WA: University of Washington Press.
McCullough, R. L. 2015. *Old Wheelways: Traces of Bicycle History on the Land.* Cambridge, MA: MIT Press.
McIlwraith, T. F. 1997. *Looking for Old Ontario: Two Centuries of Landscape Change.* Toronto: University of Toronto Press.
Mitman, G. 2007. *Breathing Space: How Allergies Shape Our Lives and Landscapes.* New Haven, CT: Yale University Press.
Norcliffe, G. 2001. *The Ride to Modernity: The Bicycle in Canada, 1869–1900.* Toronto: University of Toronto Press.
Parr, J. 2010. *Sensing Changes: Technologies, Environments, and the Everyday, 1953–2003.* Vancouver: University of British Columbia Press.
Perry, D. B. 1995. *Bike Cult: The Ultimate Guide to Human-Powered Vehicles.* New York: Four Walls Eight Windows.
Robb, G. 2007. *The Discovery of France: A Historical Geography from the Revolution to the First World War.* New York: W. W. Norton and Company.
Roberts. L. L. 2004. An Arcadian apparatus: The introduction of the steam engine into the Dutch landscape. *Technology and Culture*, 45(2), 251–276.
Steinberg, T. 1991. *Nature Incorporated: Industrialization and the Waters of New England.* Amherst, MA: University of Massachusetts Press.
Stilgoe, J. R. 1998. *Outside Lies Magic: Regaining History and Awareness in Everyday Places.* Walker and Company.
Tarr, J. A. 1996. The horse—polluter of the city. In J. A. Tarr, *The Search for the Ultimate Sink: Urban Pollution in Historical Perspective* (323–333). University of Akron Press.
Traill, C. P. 1989 (1836). *The Backwoods of Canada.* Toronto: McClelland & Stewart.
Wells, C. W. 2012. *Car Country: An Environmental History.* Seattle, WA: University of Washington Press.
Worster, D. 1988. Doing environmental history. In D. Worster (ed.). *The Ends of the Earth: Perspectives on Modern Environmental History.* Cambridge University Press.
Wynn, G. 1979. Notes on society and environment in old Ontario. *Journal of Social History*, 13(1), 49–65.

Part IV

Digital research

18 Online digital communication, networking, and environmental history

Sean Kheraj and K. Jan Oosthoek

On January 23, 1996, Professor Dennis Williams from Southern Nazarene University declared, "Today begins a new epoch [for] all of us who have been discussing environmental history electronically for the past few years." This was the first message sent to the H-ASEH (later renamed H-Environment) email listserv, now more than twenty years ago. H-ASEH was the successor to the previous ASEH-L email listserv that Williams started in 1991. While some environmental historians had previously communicated in smaller, closed computer networks in the past, the advent of email listservs and the H-Net consortium was the beginning of widespread online digital communication and networking among environmental history scholars.

In the two decades since Williams posted that first message, online digital technologies have opened up new avenues for the communication of research findings and the development of research networks among environmental historians. In fact, environmental historians have been leaders in the use of online digital communications technologies in the environmental humanities. As Cheryl Lousley (2015, p. 3) recently wrote, "Among environmental humanities scholars, it is the environmental historians who have been most adept at reconfiguring scholarly research and communication in light of emerging digital possibilities." This chapter examines the changing uses of online digital technologies for communication and networking in the environmental history research community. It offers a survey of the history of online environmental history activities and a snapshot of the contemporary uses of such technologies. These technologies have influenced scholarship in the field of environmental history in a number of important ways. First, they have facilitated the development and growth of regional, national, and international scholarly networks. Second, they have extended the reach of environmental history research findings, making this research accessible to communities beyond the academy, including educators, policy makers, journalists, and public history audiences. Finally, environmental historians have begun to make use of online digital technologies for the development of new forms of scholarly publication.

Creating online scholarly networks

From the early-1990s to the mid-2000s, environmental historians primarily used online digital technologies for the development of scholarly research networks. This was precisely the purpose of ASEH-L and H-ASEH. These email listservs provided an online discussion forum for members of the American Society for Environmental History and other users around the world. ASEH-L began in the early-1990s with 300 mostly American subscribers and was hosted by Texas Tech University before migrating to the H-Net system. Just as today, these email listservs created an international online forum for scholars in environmental history to discuss common issues, post announcements about conferences and publications, and share other news.

In the first month, H-ASEH users posted sixty-one messages on a range of topics. One of the earliest topics for discussion was a general conversation about the state of the field of environmental history and how historians defined this relatively new area of study. As editor of *Environmental History*, Hal Rothman enthusiastically described the field as "one of the most exciting new disciplines to emerge in the past two decades." He wrote an early introductory statement about environmental history, the ASEH, and the journal itself. "We intend for the list to be the location of dialogue and debate about the field of environmental history," wrote Rothman in January 1996, "as well as a place where books are reviewed, new publications announced, and other matters of interest to the members discussed and debated" (Rothman 1996). For the most part, this was how environmental historians used the listserv.

Conferences, workshops, symposia, and other events filled the early message threads on H-ASEH as scholars made use of the listserv as a platform to advertise various activities and meetings. Within a day, Charlotte Zoe Walker (January 24, 1996) from SUNY Oneonta posted the first conference announcement for an event on multicultural perspectives on environmental writing. Ray Bromley (January 29, 1996) followed up with an announcement about a one-day conference at the Lewis Mumford Center at the University of Albany.

Teaching issues arose early in H-ASEH discussion with historians seeking advice on course development. For example, Robert Schwartz (January 29, 1996) sought help with developing a new course on environmental thought in Europe from 1600 to the present. Joel Tarr's first message (January 29, 1996) on the listserv was a request for videos relating to Native American attitudes toward nature. Within a month, Sarah Elkind was already working on creating an online repository for syllabi and other teaching resources (Williams February 28, 1996).

While H-ASEH began as an online tool for members of the ASEH and much of the early discussion focused on US topics, scholars studying topics and working outside the US made use of the listserv to develop scholarly networks. For instance, Mats Widgren (January 25, 1996), a geographer from Stockholm University, reached out to find other scholars working on "the historical and social contexts of intensive farming." Richard Stuart (January 25, 1996) from Parks Canada sought information about environmental historians working on topics relating to Canada. Laxman D. Satya (January 30, 1996) started an extended discussion of sources relating to the environmental impacts of British colonialism in nineteenth-century India. And Horacio Capel from Universidad de Barcelona shared a number of references to scholarship on conservation history in Spain. The global reach of environmental history was evident in the early use of this email listserv in the mid-1990s.

To reflect the increasing number of non-American subscribers of H-ASEH, 20 per cent of the 1200 members in early 2000, H-ASEH changed its name to H-Environment. This change became particularly urgent with the creation of the European Society for Environmental History (ESEH) the prior year. ESEH would become one of the supporting professional organisations of H-Environment, which was reflected in the creation of a new editorial board that included a representative of ESEH and the appointment of a European list editor. Through ESEH, the number of non-American subscribers of H-Environment continued to expand (H-Environment editors, April 6, 2000).

Early discussions on H-ASEH differed somewhat from the way that environmental historians use email listservs today. For the most part, discussion on the current H-Environment listserv is dominated by announcements without much back and forth conversation. In the first months of H-ASEH, however, it was common for environmental historians to pose open-ended questions and generate some discussion among users. This most often took the form of seeking resources for particular topics or issues in environmental history. On 5 February 1996, Tina Roberts, a graduate student at University of Maine, started a discussion thread seeking resources on the role of women in the environmental movement. This generated eighteen responses over the remainder of the month with suggestions for further reading and references to particular women in the environmental movement.

While most of the direct online interaction among environmental historians in the 1990s took place on H-ASEH, scholarly associations also made use of websites to share information and resources with their members. "Understanding the past for its impact on the future," read the header of the Forest History Society website in early 1998. As one of the earliest environmental history associations to establish a presence on the Web, Forest History Society led the way by providing information about environmental history for its members and general readers. Duke University started hosting the website in the mid-1990s. In addition to providing information on membership, relevant archives, its oral history program, publications, and awards, the Forest History Society website included an extensive bibliography: an online version of a computer database that the society had kept since the early 1980s. This quickly became one of the most influential online resources for environmental history scholarship.

ASEH and, later, ESEH both established websites to share information with members and provide resources about the emerging field for a general readership (Figure 18.1). ASEH first started with a website directly tied to the H-ASEH/H-Environment email listserv but eventually fully migrated to its own domain at aseh.net in 2002. This website focused specifically on membership issues, the society's annual meeting, and the journal *Environmental History*. In 2007, ASEH began publishing a digital version of its newsletter. ESEH first established its website in 2000, constructing a similar site to that of ASEH and other historical scholarly associations. It included membership information, conference details, and links to additional resources.

The case of ESEH, in particular, illustrates the power of the Web to create networks. An earlier attempt to create a scholarly environmental history association, the European Association for Environmental History (EAEH), never really got off the ground. The last print newsletter was published in 1995 and then it fell silent. Without a ubiquitous communication tool, such as the Web, it proved almost impossible to create an international network in the culturally and linguistically diverse continent that is Europe. From the start, the website was a central part of ESEH's communication strategy. Here, scholars could sign up as members and find information about the organisation, conferences, and publications (Winiwarter, March 31, 2001). A combined use with H-Environment proved very effective and by the time of the first ESEH conference in September 2001 about 300 scholars had registered as members mainly through the website. In recent years the website has diversified and is now linked to a social media presence and H-Environment has mostly lost its function as a communication channel for ESEH members.

Up to the early-2000s environmental history focused on issues in North America and Europe, with most of the online content appearing in English. More recently, however, an ever-growing number of researchers from Asia, the South Pacific, Africa, and Latin America have been joining what has become a global environmental history community (Reinaldo 2008). This resulted in the creation of the Latin American and Caribbean Society of Environmental History (SOLCHA in Spanish) in 2004 and the Association of East Asian Environmental History in 2009. Both societies had a web presence from the beginning but the sites are not yet as thoroughly developed as those of the ESEH and ASEH. The sites present basic information about the organisations, membership, conferences, and news in the field. The sites play an important role in communicating and organizing the biannual conferences of both organisations.

By the early 2000s, ASEH, ESEH and H-Environment had established an online presence that had helped to create international scholarly networks in environmental history. But in large countries with relatively small and scattered populations, in particular in Canada and Australia, it was felt that a more regional focus was needed in order to connect scholars. This led to new online initiatives of network building and online innovation during the first decade of the twenty-first century. In 1997, the *Environmental History Network* was established at the Australian National University with the aim to facilitate communications among scholars working across a range of university departments, government agencies, businesses, and cultural and scientific institutions. A few years later the network established an online presence and encompassed all states of Australia and New Zealand, reflected in the new name of the network: *The Australian and New Zealand Environmental History Network*. The

Figure 18.1 Three examples of early environmental history sites: ASEH (2002), ESEH (2000), and EH website (1998). Note the simple structures and the lack of interactive features.

Source: Images courtesy of ASEH, ESEH and Jan Oosthoek.

network became a successful bulletin board around which a diverse group of researchers from around Australia and New Zealand coalesced. The site of the *Australian and New Zealand Environmental History Network* is closely linked to the Centre for Environmental History at the Australian National University which administers the site. The latter also has an active web presence and complements the network with news and stories related to members of the Centre and the wider research community in Australia and New Zealand.

In 2004, Alan MacEachern (University of Western Ontario) and a team of environmental historians and historical geographers founded the Network in Canadian History and Environment (NiCHE), an organization focused on scholarship in environmental history in Canada. This team won funding from the Social Sciences and Humanities Research Council of Canada to develop a national research network and disseminate environmental history scholarship to academic and non-academic audiences. In doing so, NiCHE leveraged online digital tools in order to connect researchers across Canada. The organization began with a traditional static website, but soon made use of online collaboration software by Groove Networks to facilitate the development of collaborative research papers and to lead real-time online discussions of issues and themes in Canadian environmental history. In 2007, NiCHE relaunched its website using the Drupal content management system. This allowed members to write and submit their own content for the site. The commenting feature of the new site also allowed members to respond to one another and generate discussion and conversation. This use of user-generated content and commenting tools blended the information sharing features of early environmental history websites with the online engagement and interactivity of email listservs. In 2008, NiCHE started publishing an audio podcast called, *Nature's Past*, which further extended the reach of the network to communities beyond the academy.

Reaching wider audiences

While the development of scholarly networks dominated environmental historians' early use of online digital technologies, some researchers also saw the potential to use the internet as a communications tool to reach broader public audiences. With environmental issues capturing greater public attention in the 1990s, some historians began to envision a public role for environmental history, one that could be facilitated by online technologies. Dissemination of scholarly work has always been important. Only a few decades ago this was mainly done through scholarly journals and, in the humanities, through books. Some better-known scholars contributed to public debate through newspapers, popular magazines, and, in some cases, television and radio programs. Overall, dissemination of scholarly material was conducted through a few media with a limited reach. In 2004, Catherine Christen and Lisa Mighetto co-edited a special issue of *The Public Historian* to coincide with the first joint meeting of the ASEH and the National Council on Public History. In that issue, they made note of the degree to which environmental historians addressed "the 'usefulness' and uses of environmental history in a variety of forums, including ASEH conferences, and in recent years also on the H-Environment web site and e-mail discussion network" (Christen and Mighetto 2004, p. 10). Based on those online discussions, they hoped to use the joint meeting of these two groups as an occasion to ask how environmental history might benefit from the adoption of public history's approach to the dissemination of historical knowledge. Online digital technologies would come to play a significant role in that effort.

Outside of the activities of the main scholarly associations, a few individual researchers began to develop websites as resources for other researchers and general audiences with an interest in environmental topics. These websites served as some of the earliest online resources for the dissemination of environmental history to audiences beyond the academy. In 1996 Bill Kovarik published the first version of his *Environmental History Timeline* on the Web. This timeline was originally published earlier that same year in a book entitled *Mass Media and Environmental Conflict* (Neuzil and Kovarik 1996). Over time the website and timeline evolved and expanded to include more international topics and take the timeline back to prehistory. At present the site also contains a blog with posts commenting on current environmental events, putting them into historical context, and providing short biographies and material for students. The site stands very much on its own and is not embedded in current online social media platforms which would extend the reach of the site even further.

A site that is currently firmly embedded in a network of online social media is *Environmental History Resources*. In 1998 this site began its life as a personal website of Jan Oosthoek, then an environmental history graduate student based in Scotland. The site was initially hosted by Tripod.co.uk and called "Environmental History Website" and was intended as a general resource for a global audience. Within months the site was moved to the servers of University of Stirling and renamed "Oosthoek's Environmental History Homesite." In 2005 the site was closed down by the University of Stirling and in early 2006 Oosthoek relaunched it as *Environmental History Resources*. Here Oosthoek continued to focus on developing publicly accessible environmental history resources that were intended not just for a scholarly audience, but a broader public history audience. The site is used as a platform to explain the field to non-experts, especially students new to the field, and earliest content included a series of introductory essays. In the spring of 2006, Oosthoek launched *Exploring Environmental History*, the first environmental history audio podcast, which was a significant departure from previous uses of online audio. Instead of recorded lectures, Oosthoek produced a regular, edited, episodic series, which continues to this day. More recently *Environmental History Resources* has added a dedicated YouTube Channel with short introductory videos and the site is closely linked to a network of online social media, including Twitter, Google Plus and the scholarly social network ResearchGate.

From the mid-2000s, a new type of online academic presence emerged in the form of blogging. A number of individuals (mainly graduate students and early-career scholars) embraced blogging to showcase their work to a wider audience.

Blogging among environmental historians developed not just as a form of self-publication and promotion. Instead, it added to the academic processes of research, writing, and debate. Blogging emerged as a productive scholarly activity as scholars turned research, a talk, a topic they were teaching, and even commentary on current affairs into blog posts.

Since the late 2000s, several environmental historians have published regular blogs about their research. For instance, in late 2009, Katherine Knight, an early-career scholar, launched Envirohistory NZ, a blog about her own research and conservation issues in New Zealand. The aim of the site is to "to stimulate discussion and exploration of the way we relate to our environment through history." Many of the topics on this blog have ended up in a monograph about the environmental history of the Manawatu region of the North Island of New Zealand. Also in 2009, Sean Kheraj (York University) started an academic blog about his work in Canadian environmental history, featuring short articles about ongoing research, teaching, and current affairs. In 2013, Dolly Jørgensen (Luleå University of Technology), president of the ESEH and contributor to this volume, started a research blog with a focus on her project on the history of species re-introductions. This blog stands out as an explicit example of a project-focused blog created with the intention of helping Jørgensen develop her ideas and present preliminary findings from her broader research project (see Chapter 19 in this volume).

Environmental historians have also collaborated on group blog initiatives, multi-authored regular blogs with editorial teams. The Forest History Society launched one of the earliest environmental history group blogs in 2008 called *Peeling Back the Bark*. This blog highlights the archives and photo collection of the Forest History Society through numerous articles and features (Lehmen, August 8, 2008). In 2011, Jim Clifford, then the project manager for the Network in Canadian History and Environment, launched *The Otter~La loutre* (originally called *Nature's Chroniclers*), a group blog focused on Canadian environmental history. Based on his experience co-founding and editing the web publication, *ActiveHistory.ca*, Clifford developed a multi-authored group blog model that saw the publication of hundreds of short articles each year. These articles included a mix of research findings, opinion, historiographical debate, and current events, offering a rich, new form of scholarly communication and the fostering of ongoing discussion among environmental historians in Canada. Similar blogs have since emerged, including *Edge Effects*, a digital magazine edited and published by a graduate student team from the Center for Culture, History, and Environment at University of Wisconsin, Madison.

While many of these blog posts may never contribute directly to academic articles, many others have ended up in published papers, book chapters, and even books. As a result, blogging has become part of the online academic public landscape. According to digital historian Tim Hitchcock (July 28, 2014), blogging has become "a way of thinking in public and revising one's work, to make it better, in public." He argues that knowing that there is an audience forces scholars to think a little harder about the reader, the standards of record keeping and attribution that underpins research. Indeed, as Jørgensen found in the case of her own research blog, which she discusses in the next chapter, "With a little time commitment, a love of writing, and a desire to share insights, having a research blog can be an integrated part of the research process" (Jørgensen, September 6, 2013).

Online social networking

Within the past few years, new online social networking technologies have emerged, assuming some of the former scholarly networking features of email listservs and early website development. Online social networks represent a great opportunity to create communities that go well beyond the traditional limits of academic networks to include not only professional historians and practitioners of other disciplines, but also policy makers, other educators, and journalists. Environmental historians tend to use Twitter most actively for online social networking, but they can also be found on Facebook and Google Plus. Users share links to articles and other media relevant to the field. They also debate and discuss key issues and themes. Some scholars use online social networks to organize panels and workshops for annual conferences and other meetings. On Twitter, they also adopt specific hashtags for conferences in order to share and discuss conference papers in real time. Social networks are good for creating communities, sharing ideas, debating issues, finding collaborators, and disseminating research.

One of the largest online social networking groups in environmental history can be found on #envhist, the most common hashtag used among environmental historians for general discussion on Twitter. In 2009, during a conversation with Adam Crymble (website administrator for the Network in Canadian History and Environment), Twitter user @SmithMillCreek first proposed using a single hashtag to help organize the tweets of environmental historians. Since then, the #envhist hashtag has been used to coordinate discussion among hundreds of Twitter users, many of whom are environmental history scholars. According to Wilko Hardenberg's (March 29, 2012) analysis, the hashtag gained popularity beginning in April 2011 and users posted more than 2,300 tweets with the #envhist hashtag within a year. Unfortunately, technical limitations on accessing historical data from Twitter make it difficult to determine the total use of this hashtag today. However, environmental historians, journalists, policy makers, and the broader public continue to use #envhist and Twitter as tools for scholarly networking.

One significant limitation to the use of Twitter is the limited number of characters that can be used (up to 140) and its short memory. Blogs get around these limitations but these do not have the flexibility of building large interactive communities that share information easily. Other online social networks, including Facebook and Google Plus incorporate the networking capabilities of Twitter and some of the features of blogging. These platforms, however, do not allow users to organise a network around a dedicated hashtag like the use of #envhist on Twitter. However, the recent update of Google Plus, which puts an emphasis on its so-called communities allows for the creation of online social networks that connect people interested in a specific topic, making it possible for them to discuss the topic as well as to disseminate new research, publications and announcements. Communities are often actively managed and moderated so that only relevant items are posted and shared among members and the use of tags as well as topics simplifies searches. Google Plus currently hosts three profiles related to environmental history and one community.

A most impressive example of how an environmental history community could be run on Google Plus is the Biocultural Landscapes and Seascapes (BCLS) community. It was started by Jean-Baptiste Pichancourt, a research ecologist at the Commonwealth Scientific and Industrial Research Organisation (CSIRO) in Brisbane, Australia, in December 2012. The BCLS community is an interesting online space for environmental historians and has a wide interdisciplinary membership. Pichancourt and others designed the community for people interested in issues at the intersection between biodiversity and cultural diversity. In January of 2016 the BCLS community reached over 4,000 members and was renamed "Biocultural Diversity" to reflect the diverse membership of the network that encompasses scientists, social scientists, and humanists as well as policy makers, land managers, NGOs, and other stakeholders from civil society (BioCultural Diversity, January 14, 2016).

New platforms for publication

In addition to using online digital technologies to facilitate public engagement and networking, environmental historians have also developed new online platforms for scholarly publishing. As Peter Roberts wrote in 1999, "I believe the potential advantages of moving toward electronic publication for scholarly work far outweigh any possible disadvantages associated with such a move." Environmental historians have indeed seen many of those advantages in recent years. All of the major journals in the field now publish online electronic editions and many of the university presses publish e-book copies of their monographs. But environmental history scholars have begun to use online digital technologies to produce new forms of scholarly publication beyond the traditional print journal and monograph.

For many years, H-Environment has published digital book reviews, once the exclusive purview of scholarly journals. H-Net began its book reviews project in the 1990s and H-Environment joined the effort in 1997. Dennis Williams served as the first book review editor and posted the first call for contributors on January 28, 1997, just one year after launching the H-ASEH listserv. Within a few months, Charles C. Kolb (1997) published the first H-Environment book review. As an online and freely-accessible digital publication, H-Environment book reviews reach a wide and growing audience of readers.

In recent years, however, H-Environment has broadened its approach to reviews, exploring new forms made possible by online digital technologies. In particular, H-Environment Roundtable Reviews stands as an example of how online digital technologies offer the potential to explore new forms of publication. In January 2011, Jacob Darwin Hamblin announced the publication of a new feature intended to complement the existing book reviews project. H-Environment Roundtable Reviews assembles a group of three to four scholars to review a single book in the field of environmental history. H-Environment published the first issue of this new feature on January 31, 2011. H-Environment Roundtable Reviews, now in its sixth volume, has published forty issues. While the format is not entirely novel, it was only on the rare occasion that a traditional scholarly journal would assemble a roundtable review of a single monograph or essay. The online digital platform allows H-Environment to devote an entire publication to the roundtable format.

Within the past few years, digital technologies have facilitated a flourishing of new open-access online journals in the fields of environmental history and environmental humanities. Indeed, the dissemination of freely available online journals through open-access publishing has been one of the most significant consequences of the adoption of online digital technologies in academia and it has been influential in the development of new online journals in environmental history and environmental humanities. Michael Geist recently wrote, "The shift toward open access becoming the default form of disseminating research in many fields is a remarkable change given that conventional publishing in expensive subscription-based journals was the standard in many areas of research as recently as ten years ago" (Geist 2016, p. 30). Open-access online publication has offered these journals the opportunity to reach global audiences and publish research that may not have fit well in traditional print publications. In 2012, a collaboration of scholars from universities in Australia, Canada, the United States, and Sweden founded *Environmental Humanities* and published its first issue. This international peer-reviewed journal publishes exclusively in an open-access online digital format. *Environmental Humanities* is part of the Directory of Open Access Journals, joining a number of new online journals that have adopted an open-access publishing model. Since 2010, the Rachel Carson Center for Environment and Society has published its own open-access online journal called, *RCC Perspectives*, a publication focused on less formal articles in environmental humanities. Most recently, the Centre for Environmental History at Australian National University launched *International Review of Environmental History*, its own open-access journal (available in print and online). The appearance of these new journals suggests a trend toward open-access online journal publication in environmental humanities, although the most influential journals in the field of environmental history remain subscription-based print journals.

We see other experiments in online digital publication that push the boundaries of how environmental historians communicate and disseminate their research. The Arcadia Project is one such digital environmental history project that attempts to publish research findings in novel ways. A project of the Rachel Carson Center for Environment and Society and the ESEH, the Arcadia Project is a unique web publication of original peer-reviewed environmental histories. The project is organized into collections that examine key themes in environmental history with case studies from around the world. Each case study is authored by a leading scholar on the topic and includes a short essay, a bibliography for further reading, and links to relevant online resources. Moreover, the collections are mapped and situated on a timeline so that users can explore global environmental histories spatially and temporally. As a so-called "born digital" project, Arcadia makes full use of rich media, hypertext linking, and non-linear structure in its presentation of environmental history scholarship. The short text and emphasis on references to additional resources position this project as a public history initiative as well, offering users an introduction to a wide range of topics in environmental history. Arcadia and other new online publications suggest rich possibilities for the development of new forms of scholarly publication for environmental history.

Conclusion

From email listservs in the 1990s to the development of international online social networks of scholars, policymakers, educators, and journalists, the past two decades of the use of online digital technologies for networking and communication clearly demonstrate the ways in which environmental history has been at the forefront of broader changes in digital scholarship. This survey suggests a number of significant conclusions about the influence of digital technologies on environmental history.

First, the field of environmental history has developed alongside the growth and the use of online digital technologies in academia. To understand the development of the field from a historiographical perspective, it is important to consider the role that online digital technologies played in fostering communities, facilitating debate, linking researchers around the globe, and promoting the dissemination of research findings. The launch of H-ASEH in 1996 took place concurrently with an international flourishing of the field, especially in regions outside of the United States. In many ways, the growth of the field in parts of the world where geographic distance or cultural barriers isolated researchers from one another can be attributed, in part, to the exploitation of online digital technologies for networking and communication.

Because online digital technologies played such a significant role in the development of the field of environmental history around the world, the archiving and preservation of digital communication is especially urgent. Historical records of online social networks, including Twitter, Google Plus, and Facebook are difficult, if not impossible, to access, and many early websites in environmental history have not been adequately archived. Future research on the historiography of the field will be impeded by these technological limitations and failures to preserve this digital content.

Second, while environmental historians have had a long history of using online digital technologies for networking and communication and they may have been particularly prominent in this regard when compared to other sub-disciplines in history, it may also be true that environmental history's use of online digital technologies is symptomatic of broader changes in the humanities and social sciences over the past twenty years and a reflection of how the Internet and the Web have shaped academia more broadly. Certainly in the case of open-access online publishing, environmental historians have not been exceptional. Many other fields of academic scholarship have embraced open-access publishing more thoroughly. Still, the particular case study of environmental history is worth considering in detail as an example of how the Internet and the Web have shaped academia in the late-twentieth and early-twenty-first centuries.

Third, the use of blogging and social networks, such as Twitter, have started to influence the way that a younger generation of scholars conducts the processes of research and writing. For example, blogging has become a tool for testing new ideas, refining writing, and honing in on new research questions. The immediacy of the Web has sped up feedback processes compared to older print publications. In addition, the Web has the potential to increase the visibility and impact of research, something that universities and funding agencies increasingly emphasize (Biswas and Kirchherr 2015).

Finally, while environmental historians have been particularly adept at using online digital technologies for communication and networking, they have not yet widely adopted digital technologies for research analysis. This is still an emerging area in the field, but there are current projects that demonstrate the future directions for digital scholarship in environmental history. In particular, in the areas of historical geographic information systems (HGIS) and text mining, we see new, exciting environmental history research. The use of "big data" and the development of infrastructure for executing large collaborative research projects is still in its infancy in the field of environmental history. However, projects such as Trading Consequences and Sustainable Farm Systems stand as models for new digital scholarship in the field. The work of scholars at the Spatial History Project at Stanford University also represents future possibilities for the further integration of HGIS into environmental history scholarship. The use of online digital technologies in environmental history will likely continue to be significant for networking and communication, but will take on an even greater role in research analysis in the future.

References

@SmithMillCreek. (2009, December 15). Message posted to Twitter, https://twitter.com/SmithMillCreek/status/6696754705.

BioCultural Diversity. (2016, January 14). Retrieved from https://plus.google.com/+BioculturalDiversityLandSeascapesCommunity/posts/5qsx3jFUQ 5f

Biswas, Asit and Julian Kirchherr. (2015, April 9). Citations are not enough: Academic promotion panels must take into account a scholar's presence in popular media. Retrieved from http://blogs.lse.ac.uk/impactofsocialsciences/2015/04/09/academic-promotion-scholars-popular-media/

Bromley, Ray. (1996, January 29). Message posted to H-ASEH, http://h-net.msu.edu/cgi-bin/logbrowse.pl?trx=vx&list=h-environment&month=9601&week=e&msg=ugGOGsI3N0TlJQCsiOUC1A&user=&pw=.

Christen, Catherine A. and Lisa Mighetto. (2004). Introduction: Environmental history as public history. *The Public Historian*, 26(1), 9–20.
Clifford, Jim. (2011, January 29). New NiCHE group blog: Nature's chroniclers. Retrieved from http://niche-canada.org/2011/02/10/new-niche-group-blog-natures-chroniclers.
Geist, Michael. (2016). Fair access: Strikes the right balance on education and copyright. *Academic Matters*, January.
H-Environment editors. (2000, April 6). Message posted to H-ASEH, http://h-net.msu.edu/cgi-bin/logbrowse.pl?trx=vx&list=h-environment&month=0004&week=a&msg=rDeNTyXsqcYm9c8FYuxQXw&user=&pw=.
H-Environment. (2011). Review of McNeill, J.R., *Mosquito Empires: Ecology and War in the Greater Caribbean, 1620–1914*. H-Environment Roundtable Reviews, 1 (1). https://networks.h-net.org/j-r-mcneill-mosquito-empires-ecology-and-war-greater-caribbean-1620-1914-roundtable-review-vol-1-no.
Hamblin, Jacob Darwin. (2011, January 28). Message posted to H-Environment, http://h-net.msu.edu/cgi-bin/logbrowse.pl?trx=vx&list=h-environment&month=1101&week=d&msg=cw0P5rrwD63V5snvvSurEg&user=&pw=.
Hardenberg, Wilko. (2012, March 29). #envhist: Social media and community-building in environmental history. Retrieved from www.antspiderbee.net/2012/03/29/envhist-social-media-and-community-building-in-environmental-history.
Hitchcock, Tim. (2014, July 28). Twitter and blogs are not just add-ons to academic research, but a simple reflection of the passion underpinning it. Retrieved from http://blogs.lse.ac.uk/impactofsocialsciences/2014/07/28/twitter-and-blogs-academic-public-sphere.
Jørgensen, Dolly. (2013, January 1). Launch of research blog. Retreived from http://dolly.jorgensenweb.net/nordicnature/?p=1.
Jørgensen, Dolly. (2013, September 6). Why are there not more research blogs? Retrieved from http://blogg.umu.se/forskarbloggen/2013/09/why-are-there-not-more-research-blogs.
Kheraj, Sean. (2009, May 29). Welcome, now let's begin. Retrieved from www.seankheraj.com/?p=5.
Knight, Katherine (2009). About. Retrieved from https://envirohistorynz.com/about/
Kolb, Charles C. (1997, June). [Review of Potts, D. T., Mesopotamian Civilization: The Material Foundations]. H-Environment, H-Net Reviews. Retrieved from www.h-net.org/reviews/showrev.php?id=1058.
Lehmen, Eben. (2008, August 8). Welcome. Retrieved from https://fhsarchives.wordpress.com/2008/08/08/another.
Lousley, Cheryl (2015). Canadian editor's introduction: Political ecologies, public humanities—a report from Canada. *Resilience: A Journal of the Environmental Humanities*, 2(2), 1–5.
Neuzil, Mark, and William Kovarik. (1996). *Mass Media & Environmental Conflict: America's Green Crusades*. Thousand Oaks, CA: Sage Publications.
Oosthoek, Jan. (2006). Environmental History Resources. Retrieved from https://www.ehresources.org/
Reinaldo, Funes Monzote. (2008). The Latin American and Caribbean society of environmental history. *Global Environment*, 1, 244–49.
Roberts, Peter. (1999). Scholarly publishing, peer review and the Internet. *First Monday: Peer-Reviewed Journal on the Internet*. 4(4). Retrieved from http://firstmonday.org/ojs/index.php/fm/article/view/661/576

Roberts, Tina. (1996, January 26). Message posted to H-ASEH, http://h-net.msu.edu/cgi-bin/logbrowse.pl?trx=vx&list=h-environment&month=9602&week=a&msg=PBgf4Uht54tgKmKJRgbSBg&user=&pw=.

Rothman, Hal. (1996, January 26). Message posted to H-ASEH, http://h-net.msu.edu/cgi-bin/logbrowse.pl?trx=vx&list=h-environment&month=9601&week=d&msg=OT9/ogFQPvRhvzv5RlfIuQ&user=&pw=.

Satya, Laxman D. (1996, January 30). Message posted to H-ASEH, http://h-net.msu.edu/cgi-bin/logbrowse.pl?trx=vx&list=h-environment&month=9601&week=e&msg=pebR8FjysucjeVIsCBMm/g&user=&pw=.

Schwartz, Robert. (1996, January 29). Message posted to H-ASEH, http://h-net.msu.edu/cgi-bin/logbrowse.pl?trx=vx&list=h-environment&month=9601&week=e&msg=aoItjQXrIjM%2bSJbz6F/GfQ&user=&pw=.

Stuart, Richard. (1996, January 25). Message posted to H-ASEH, http://h-net.msu.edu/cgi-bin/logbrowse.pl?trx=vx&list=h-environment&month=9601&week=d&msg=Z%2bX44fJcVpo7049N8Qs9mA&user=&pw=.

Tarr. Joel. (1996, January 29). Message posted to H-ASEH, http://h-net.msu.edu/cgi-bin/logbrowse.pl?trx=vx&list=h-environment&month=9601&week=e&msg=xdc6JOo%2bj7umq1j2emjnbw&user=&pw=.

Walker, Charlotte Zoe. (1996, January 24). Message posted to H-ASEH, http://h-net.msu.edu/cgi-bin/logbrowse.pl?trx=vx&list=h-environment&month=9601&week=d&msg=pr6rr54mFDCEobqt2XsxlA&user=&pw=.

Widgren, Mats. (1996, January 25). Message posted to H-ASEH, http://h-net.msu.edu/cgi-bin/logbrowse.pl?trx=vx&list=h-environment&month=9601&week=d&msg=JmToqwGr%2bfrqhDQsMGcX7A&user=&pw=.

Williams, Dennis. (1996, January 23). Message posted to H-ASEH, http://h-net.msu.edu/cgi-bin/logbrowse.pl?trx=vx&list=h-environment&month=9601&week=d&msg=x63e6QkShUWX35uf4TIe0Q&user=&pw=.

Williams, Dennis. (1997, January 28). Message posted to H-ASEH, http://h-net.msu.edu/cgi-bin/logbrowse.pl?trx=vx&list=h-environment&month=9701&week=e&msg=R7uBHjF42azdeEb9U4cBLw&user=&pw=.

Winiwarter, Verena. (2001, March 31). Message posted to H-Environment, http://h-net.msu.edu/cgi-bin/logbrowse.pl?trx=vx&list=h-environment&month=0003&week=e&msg=SBUcP0ApodpoXw1Soa85oQ&user=&pw=.

19 A new place for stories

Blogging as an environmental history research tool

Dolly Jørgensen

Digital history has been understood as writing the history of computing, relying on digitally available source material, using digital tools for analysis or presentation of data, or disseminating research through digital media. At first glance, blogging, or more properly 'web-logging', the practice of writing discrete entries called 'posts' on a website which are then displayed in reverse chronological order, would seem to be just another digital mode of research dissemination. Within the academic sphere, research blogs are primarily assumed to be dissemination vehicles for either one's own research or discussion of others' more traditionally published research (e.g. Powell et al. 2012; Schema et al. 2012). But over the course of three years of blogging about my research project 'The Return of Native Nordic Fauna,' I have found that committing to a research blog fundamentally altered my research process and my scholarly publications.

My project is a cultural history of animal reintroduction focused on the Scandinavian case studies of European beaver and muskox, but as I have worked on the project it has expanded to encompass re-wilding and de-extinction as contemporary ideas for bringing back nature. My work is situated as an environmental history in that it seeks to better understand the complex relationship between humans and the non-human natural world. Because I started this project while working in an ecology and environmental science department, the scholarly publications that have resulted from my research reach across disciplinary boundaries with some appearing in natural sciences journals and others in humanities/social science venues.

There is nothing inherent in 'The Return of Native Nordic Fauna' that would predispose it toward digital media. Unlike the earliest scholarly analysts of blogging, Torill Mortensen and Jill Walker (2002), who worked on online gaming, my project is not about online material, Internet communities, or digital data. The project is an environmental history, not in any way a digital humanities endeavor. The decision to use a blog was based on the idea that it would be a knowledge dissemination tool. I had never thought of my blog as a community-building tool or history discussion forum, which has been touted as a major reason for blogging research (such as Bonetta 2007 and Wilkins 2008). Indeed, my blog does not generate a large amount of discussion through the comment feature, although there has been some discussion on other social media networks such as Twitter and Facebook. I was not anticipating that it would modify my research process and certainly would have agreed with Mortensen and Walker (2002, p. 254) who claim that "we are not positing that writing a weblog will change the articles we publish in scholarly journals," even while recognizing blogs as potential influences on research. Now I think differently.

In this chapter I present an experiment: five pseudo-posts about blogging. I call them pseudo-posts because, unlike the true online format of a blog, clickable links and embedded visuals are not possible. Yet I've tried to simulate the reading of a blog through indicated links (they are underlined and you can find the web addresses in the footnotes) and writing style. Each pseudo-post uses the title of one of my 2013 posts as a launching point and explores how blogging changed the process of my environmental history research.

Post #1, 1 January 2013: Launch of research blog

To kick off the New Year 2013, I launched my research blog.[1] My project, funded by the Swedish research council Formas, was just beginning. I had written in the proposal that I would make a research blog as a form of research dissemination, but I had never written one before, and so I wasn't entirely clear what it would look like. I made no mention in my first post about how much I would blog, but I had a personal commitment to post twice a week. I did that the first year, then changed to once a week in 2014. Although that amount of writing may sound substantial, keep in mind that I was working 100 percent on research at the time.

Because I made that (internal) commitment to write a post twice a week, I forced myself to work through the research material immediately. In my prior research projects, I had operated in a more divided 'phase' mode. I collected material, then I analyzed it, then I wrote about it. Sure, I didn't hold everything until the end without looking at, but I didn't make a concerted effort at analysis until after a fair chunk of material was in hand. For 'Return of Native Nordic Fauna', the process was different because the steps had to be much more integrated and happen simultaneously. A frequent blogging schedule forced me to do analytical and writing work often and early.

At the very beginning, I did not have enough primary source material to write about my particular cases, and so I tried to place my future work into contemporary debates about wildness, conservation policy, and the historian's work. By the end of the first month, I had gathered enough material to tell short histories of beaver skeletons,[2] hunters as godfathers,[3] and roaming muskox.[4] It was not always easy to figure out what to write about twice a week that first year, but material always seemed to pop up that was interesting, since I was continually gathering, analyzing, and writing.

Frequent writing had two unanticipated effects on my research. First, I found a new voice. I did not want my blog to be comprised of long, involved research articles, but I also wanted to avoid the micro-post of a little thought. Instead, I focused on pieces that would be in the 500 to 1000 word range—something I could write in a few hours, including looking up material. While short by

Figure 19.1 A postcard dated 24 September 1919 sent to the director of the Skansen zoo offering to sell the family's pet badger to the zoo. A gem I found in the Nordiska museet archive and blogged about (http://dolly.jorgensenweb.net/nordicnature/?p=151)

Source: Photograph by author.

academic publication standards, these were by no means "half-thought, naked ideas" (Mortensen and Walker 2002, p. 267) or even "fieldnotes-in-public" (Wakeford and Cohen 2008, p. 307). Fast and short(ish) writing meant that I would 'talk' the post in my head. I wrote them basically in the same way that I would speak out loud. This opened up a whole new world of historical writing for me because I didn't feel I had to wordsmith every sentence. I did not have to have exactly the right phrase or the most erudite word choice. I could just be me. This has made my writing in all situations, even academic ones, flow better and feel more natural.

Second, I got in the habit of looking for good stories. My research blog is a collection of stories. Some are more narrative than others, but all of them tell a tale. This has carried over to the way I write now in more traditional publications. I am more interested in finding a good story and telling it in an interesting way than I used to be. As someone trained originally in engineering, I tended to be factual rather than compelling in my writing. I can see that blogging is helping shift my writing toward the narrative, making it more engaging to read.

Post #30, 4 April 2013: The hidden reintroduction

I was looking at a scan of the letterhead used by Peder Jensen-Tveit, the Norwegian supplier of beavers for the reintroduction efforts in Sweden, which had been sent to the Natural History Museum (London), when I noticed something odd. Under Jensen-Tveit's name and address, there was a list of items that he could provide, including living and stuffed beavers, crania, and skeletons—all things I would expect—but also on the list was "bæverbiller (*Platypsyllis castoris*)." I had no idea what that was, except that obviously by the name it was related to beavers. Doing some quick online searches, I discovered that the "beaver beetle" is a louse found only on beavers.

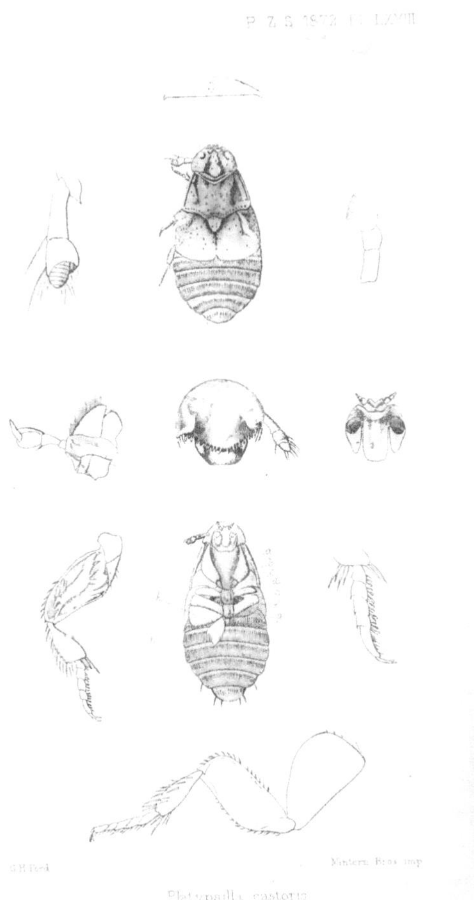

Figure 19.2 The earliest published drawing of *Platypsyllus castoris*

Source: Reproduced from J. Le Conte, On Platypsyllidæ, a new family of Coleoptera, *Proceedings of the Zoological Society of London* (1872), 799–804

If I had been thinking only about traditional scholarly publications, there would have been no reason to look further into the matter. But because I was also writing a blog, I decided that it might be interesting to write a post about "beaver beetles," since Jensen-Tveit had sold them (or tried to sell them). Why would Jensen-Tveit think someone would so much want to buy a louse that he would list it on his letterhead?

I dug deep into the history of the louse and it turned out that this little critter was a controversial discovery of the nineteenth century. Much scholarly ink had been spilled about who found it first, which beavers it had been found on, and whether it was a beetle, a louse, or something else. In this search I also noticed that there had been a number of papers published in the last 20 years claiming finds of 'the first' beaver beetle in a particular area. As a historian working on beaver reintroduction, I could see a connection: beavers had been absent in much of Europe for hundreds of years and so of course there were no beaver beetles until reintroduction projects brought the beavers back. The lice had hitched a ride on the beavers.

I wrote up a blog post with the information I'd gathered, calling the beaver beetle story a hidden reintroduction.[5] The story, however, nagged at me because it showed how recent scientific finds of the beaver beetle had failed to account for environmental history. I decided to work up an article about the louse and its history. This meant significantly more research into entomology practice of the nineteenth century and even a site visit to the Naturhistoriska riksmuseet in Stockholm to see their *Platypsyllus castoris* specimens (of course, I blogged about the visit[6]). After back-and-forth negotiations with the journal *Conservation Biology*, I ended up publishing a broad review of parasite co-reintroduction from a historical standpoint rather than an article focusing only on the beaver beetle case study (Jørgensen 2015b).

I wrote about the beaver beetle because a blog gives the researcher the freedom—the intellectual room—to explore topics that are slightly off-topic without feeling that it is a waste of time. Then, because I opted to pursue the beaver beetle story for my blog, I ended up writing a scholarly article in a top-notch journal. A traditional article had its start in making a hidden story visible on my blog.

Post #64, 31 July 2013: On the time I drank castoreum

In July 2013 I had the opportunity to try a special liquor made by infusing alcohol with the castoreum glands from a beaver. When I saw the beverage BVR HJT,[7] which is the only commercially available version of this drink, and is short for bäverhojt (literally 'beaver shout' in English), for sale as a shot in a restaurant in Gällivare, Sweden, I knew I had to try it.

In my research on beaver reintroduction, it had been unavoidable to think about the beaver's near extinction in Europe, and one of the primary products of the beaver hunt was castoreum. Historically castoreum was infused in alcohol and taken as a medicine for a wide variety of ailments, from headaches to epilepsy. I didn't know much about 'beaver gall' (as it is called in Sweden) but when I saw the BVR HJT, I recognized it as the modern version of medicinal castoreum, albeit for a non-medical use. I bought a shot and drank it.

In the days before blogging, that would likely have been the end of the story. It would be an experience to mention over dinner with friends or perhaps to talk to other environmental historians about at a conference. Drinking castoreum was not immediately applicable to my scholarly writing on reintroduction, and so it might have faded into oblivion. But in the blogging era, I got to write about it.

In my post,[8] I not only described the experience of drinking castoreum, but I also set it into historical context. I had to do some quick reading about the history of this particular medicine, turning up references to it in ancient and medieval literature as well as in nineteenth-century medical practice. While I had read a few things about castoreum before deciding to write the post—mainly from my own primary sources about beaver reintroduction in Sweden—most of what I wrote about was new to me. I had to do new research, research I'd previously not thought I would do, in order to write the post.

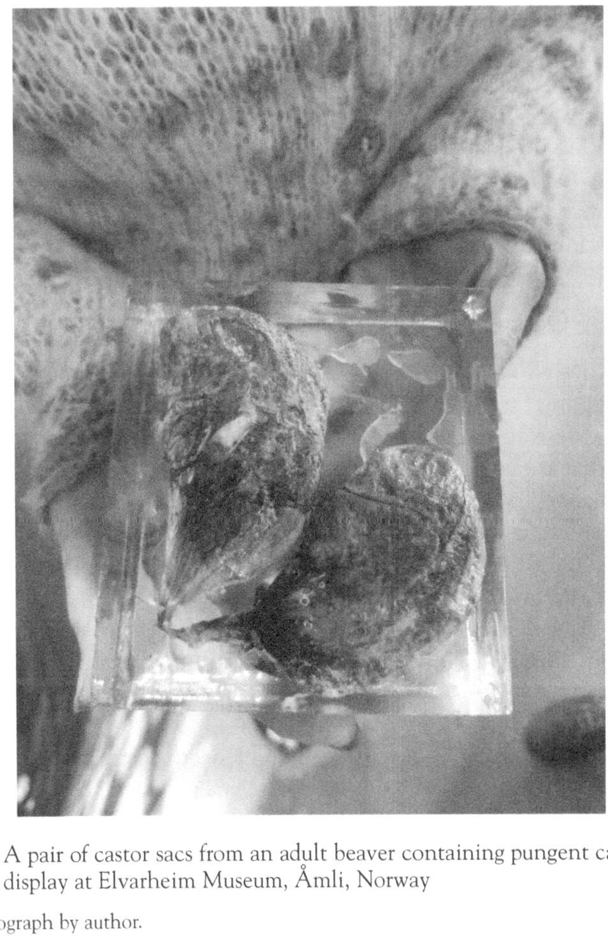

Figure 19.3 A pair of castor sacs from an adult beaver containing pungent castoreum on display at Elvarheim Museum, Åmli, Norway

Source: Photograph by author.

Since then I've written several posts about castoreum, including one focused on an impromptu search of newspaper advertisements[9] and another pointing out the difference between castoreum and castor oil.[10] These teasers have gotten me interested in writing a cultural history of castoreum at some point, but I do not know if it will ever make it into a print publication. By pursuing the story of castoreum on my blog, my research agenda has expanded.

During historical research, historians often run into side stories, things off the main track. We may collect some document scans or photographs, but they often end up as dust-collecting bytes on the computer with nowhere to go in a 'proper' publication. The research blog, however, gives the historian a place for those stories.

What's particularly fascinating about my post on drinking castoreum is that it has been the most popular post on my blog. It had been viewed 1,994 times as of 5 October 2015. A big chunk of those—nearly 500—came in March 2014 when an article on National Public Radio (NPR) titled 'Does beaver tush flavor your strawberry shortcake? We go myth busting'[11] linked to my post because it discussed bäverhojt. I continue to get 50 to 60 hits a month on my post, often through the NPR link. NPR did not interview me for their article, yet my work as an environmental historian had a mark on the presentation of a news story. Some NPR readers know more about castoreum and its history through my post than they ever anticipated. Side note: There is a similar story with my post Wild boar woes[12] which comes in as my second-most read post because it was linked to in a Reddit thread about a photograph of a wicked-looking Norwegian wild boar. You never know in what context your research will show up.

Post #80, 7 October 2013: Museum menageries

My project design included examining how reintroduced animals are presented to the public in zoos and museum exhibits. I was interested in finding out how animal histories are told in those settings. While text is important in those settings—what is said, or often not said, about the species' history and relationship to humans—the visual is just as key. The visual is read like a text.

Luckily, a blog is a visual space. Unlike the standard academic article or book which might have one image for every 4000 or 5000 words, my posts have at least one for every 500 words. Part of the reason for that is that it is easy to use images in the digital blogging platforms, but more important is that every post should have a 'highlighted image,' which is a standard functionality of most blogging layouts and shows up in social media contexts, and posts should be visually interesting to readers.

Having the ability, and indeed the requirement, to post pictures along with my text on the blog has changed the way I collect photographs. When visiting a myriad of museums,[13] I now take photographs of the displays from all kinds of angles. I want to try to see the stuffed beavers or muskox as representations of landscape, as representations of a species, and as individual animals who lived and died. Watching my daughter take photos of the muskox family on display at the Polar Museum in Tromsø, Norway, reiterated how the visual matters in the understanding of the story. The photographs I collect are so much more than documentation of a particular exhibit's layout. They try to capture the visual story.

On my blog, I have analyzed some of the stories I encountered from my museum and zoo visits, from the muskox who didn't belong in a Svalbard diorama,[14] to Bruno the bear as transgressor,[15] to the darkness of a room of extinct animals.[16] In each case, the visuals matter as much if not more than the text. The need to have images on my blog has made me more sensitive to the power of the image. It has reoriented my research to incorporate the form, not just the substance, of the reintroduction stories. In other words, how the reintroduction story is told matters.

My first significant field work expedition for the project took place in June 2013. As part of that trip I stopped by the Vitenskapsmuseet in Trondheim, Norway, to see if they had any muskoxen on display. The muskox I encountered on the second floor moved me and I took several photos of the stuffed, still animal, including a selfie. One photo I took of it was awarded a prize in the A

Figure 19.4 A muskox displayed in a diorama of the Svalbard archipelago, even though the population introduced there died out in the early 1980s. Frösö Zoo, Sweden

Source: Photograph by author.

Snapshot of the Field 2013 competition of the Network in Canadian History and Environment (NiCHE).[17] I reflected on the muskox encounter on my blog:

> Within a dusty exhibit titled 'Natur-miljø' (Nature environment) on a dimly lit balcony-style floor, I saw this secluded muskox standing in the corner. It is a descendent of muskox caught in Greenland and reintroduced to Norway in the 20th century. It seemed to cry out for attention to break its dreadful loneliness with the words poignantly written by curators on the ground: 'Just touch me!' (*Bare ta på meg!*) And indeed it appears to have been touched many times over the years, as almost all of its qiviut, the muskox's softest and most valuable wool, was missing from the middle of its back. Like many other visitors to the museum, I touched the muskox, but in addition to hair, I could feel the cold plastic form over which the skin had been stretched to make the mounted specimen standing before me. In my tactile encounter in the corner, reaching out and touching the muskox exposed how unnatural nature often is.

Even only six months into blogging, I had already become attuned to the visual as a key element of reintroduction stories. My data collection strategy reflected the search for a something to share on my blog. I had encountered this muskox and its hair because I was thinking about the form of the story. Blogging changed my way of seeing.

Post #88, 13 December 2013: Migrant muskox

In December 2013 I presented my second talk about the reintroduction of muskoxen in Scandinavia. The first had been a standard 20-minute academic conference paper given in August, but this was a longer colloquium presentation at the Rachel Carson Center[18] in Munich. The presentation, titled "Naturalised national identities: Migrant muskox in northern nature,"[19] offered a reconceptualization of muskox reintroduction as an environmental migration. I wanted to find out what would happen if I treated the muskoxen like the subjects of studies in human migration. I built up the talk in large part based on material I had blogged.

The first part of the presentation focused on the capture and forced relocation of muskox calves from Greenland to Sweden and Norway. I had explored much of its material in two previous posts: one about the context of the great Greenland hunt[20] to capture muskoxen and the other on the specifics of the expeditions of 1899 and 1900[21] by Alfred Gabriel Nathorst and Gustaf Kolthoff, respectively, to capture calves.

The second part placed the reintroduced Norwegian muskoxen in the social landscape as dangerous immigrants. I included verses from a poem written by a woman who had decided she could no longer walk in the mountains because she feared a muskox attack. It was a poem about loss[22] that I had first published in its entirety (in my own translation from Norwegian) on my blog in October 2013.

Figure 19.5 "A shot muskox bull" from East Greenland

Source: Reproduced from Gustaf Kolthoff (1901) *Till Spetsbergen och Nordöstra Grönland år 1900* (Stockholm: Skoglunds förlag), p. 192.

The talk continued with stories of muskoxen who actually attacked people, although always because they felt threatened by humans who approached too close. The material was drawn extensively from my post "When curiosity kills"[23] from June that year, which had told the story of the 1964 death of Ola Stølen, the first victim of an attack by reintroduced muskox.

The final section examined the reaction to the Norwegian muskoxen which crossed the national border into Sweden in 1971. I re-appropriated information which I had uncovered on the Swedish parliamentary debates over muskox when I wrote a post about the April 2013 release of a female muskox[24] to supplement the Swedish herd.

My academic presentation in Munich was heavily indebted to online text I had previously written. For each post, I had worked through a set of primary materials, reaching out and gathering necessary sources. When it came time to write the presentation, I wove together these stories into one narrative. The work did not stop there either: I expanded upon and modified my presentation to become an academic article that was published in the edited collection *The Historical Animal* (Jørgensen 2015a). My blog does not exist as a stand-alone entity. It is integrated into my project as one of many forms of thinking through, writing through, and delivering the project as a knowledge product.

Closing thoughts about blogging as research tool

What I hope these five vignettes about my research blog experience show is how a blog can be an integrated tool in research. Although blogs may provide space for research dissemination, discussion, or community building, I have found that the greatest effect of my blogging is a shift in my scholarly practice by embracing two central aspects of research blogging: writing often and sharing stories.

First, a blogging commitment means writing often. I have dedicated time to this writing—time which some might think could have been better spent on traditional academic publications, but I disagree. Writing often has improved the way I write, so that I have a much more narrative and talkative style than I did before. It modified the project's progression in that sources were continuously collected, analyzed, and written about. It has meant putting ideas out there before they are fully polished, and before they are published formally. Many people are afraid to do that, either for fear of changing their minds, being criticized, or having their ideas stolen. I have never experienced any of those things. Instead, writing on a blog has made my environmental histories more lively and better thought out. Doing historical research in public by showing the research process through a blog does not hinder traditional publication. It rather encourages and enhances it.

Second, a blog provides a place for stories. Like Cronon (1992), I believe environmental history needs to embrace storytelling. I have been attuned to the draw of a good narrative and the visual because of blogging. Both have affected how I collect data, as well as how I write about it. Many of the academic publications coming out of my project were first shared in short versions as stories. In addition to writing posts about my central research questions, I have found liberating the ability to write a post of under 1000 words about an off-topic but interesting subject. I've felt more free to pursue the unusual, strange, or fascinating than I had when I was publishing only traditional academic works. This has had direct implications for my scholarship, as these 'side stories' have turned into important academic outputs of the project. The stories I tell on my blog may or may not attract readers—it's often hard to predict how a post will be received. My experience does show, however, that once you put your research online, it takes on a life of its own. Others will link to it in posts, news stories, tweets, Facebook posts, and a myriad of other places that I had not expected. That too has to be embraced as part of sharing stories.

After writing 184 posts and nearly 137,000 words on my blog, I can say that it has been time well spent. I intend to write even more. Historians, who depend on writing and telling stories, would do well to think about integrating research blogs into their practices. Like me, they may find that they are better historians for it.

Notes

1 http://dolly.jorgensenweb.net/nordicnature
2 http://dolly.jorgensenweb.net/nordicnature/?p=70
3 http://dolly.jorgensenweb.net/nordicnature/?p=106

4 http://dolly.jorgensenweb.net/nordicnature/?p=129
5 http://dolly.jorgensenweb.net/nordicnature/?p=429
6 http://dolly.jorgensenweb.net/nordicnature/?p=1542
7 http://losmith.se/sprit/bvr-hjt/
8 http://dolly.jorgensenweb.net/nordicnature/?p=1015
9 http://dolly.jorgensenweb.net/nordicnature/?p=1062
10 http://dolly.jorgensenweb.net/nordicnature/?p=2360
11 www.npr.org/sections/thesalt/2014/03/26/293406191/does-beaver-tush-flavor-your-strawberry-shortcake-we-go-myth-busting
12 http://dolly.jorgensenweb.net/nordicnature/?p=234
13 http://dolly.jorgensenweb.net/nordicnature/?p=1186
14 http://dolly.jorgensenweb.net/nordicnature/?p=850
15 http://dolly.jorgensenweb.net/nordicnature/?p=1087
16 http://dolly.jorgensenweb.net/nordicnature/?p=2269
17 http://niche-canada.org/
18 www.carsoncenter.uni-muenchen.de/index.html
19 www.youtube.com/watch?v=3NvbmikWnGo
20 http://dolly.jorgensenweb.net/nordicnature/?p=1034
21 http://dolly.jorgensenweb.net/nordicnature/?p=664
22 http://dolly.jorgensenweb.net/nordicnature/?p=1235
23 http://dolly.jorgensenweb.net/nordicnature/?p=747
24 http://dolly.jorgensenweb.net/nordicnature/?p=652

References

Bonetta, L. (2007) Scientists enter the blogosphere. *Cell*, 129, 443–445.

Cronon, W. (1992) A place for stories: Nature, history, and narrative. *Journal of American History*, 78, 1347–1376.

Jørgensen, D. (2015a) Migrant muskoxen and the naturalization of national identity in Scandinavia. In Nance, S. (ed.) *The Historical Animal*, 184–201. Syracuse: Syracuse University Press.

Jørgensen, D. (2015b) The conservation implications of parasite co-reintroduction. *Conservation Biology*, 29, 602–605.

Mortensen, T. and Walker, J. (2002) Blogging thoughts: personal publication as an online research tool. In Morrison, A. (ed.) *Researching ICTs in Context*, InterMedia Report 3/2002, 249–279. Oslo: Unipub.

Powell, D. A., Jacob, C. J., and Chapman, B.J. (2012) Using blogs and new media in academic practice: potential roles in research, teaching, learning, and extension. *Innovative Higher Education*, 37, 271–282.

Shema, H., Bar-Ilan, J. and Thelwall, M. (2012) Research blogs and the discussion of scholarly information. *PLOSOne*. DOI: 10.1371/journal.pone.0035869

Wakeford, N. and Cohen, K. (2008) Fieldnotes in public: Using blogs for research. In Fielding, N., Lee, R. M. and Blank, G. (eds) *The Handbook of Online Research Methods*, 307–326. London: Sage Publications.

Wilkins, J. S. (2008) The roles, reasons and restrictions of science blogs. *Trends in Ecology & Evolution*, 23, 411–413.

20 Cultivating the spirit of the commons in environmental history

Digital communities and collections

Kimberly Coulter and Wilko Graf von Hardenberg

This is a call for more and better collaboration. Not just among individuals, but between whole sub-disciplines of history. Environmental and digital history, as Andrew Murphie suggests in a roundtable published by the journal *Environmental Humanities*, have astounding similarities when it comes to collaboration, sharing, and improving the ways heritage is preserved (Murphie, Jørgensen, Gibbs, and Hardenberg 2013). Both disciplines can only gain from adopting (and adapting) methods, tools, and ideas from each other.

We are all part of a "scholarly ecosystem" (Underwood 2015), a metaphor that seems particularly apt for explaining the links between digital and environmental history and how the two disciplines are connected. In this essay, we call on environmental historians to examine not only their methodologies and epistemologies, but also the role of the digital in the political economy/ecology of environmental history knowledge production and distribution. In particular, we ask them to consider how digital developments affect—and might improve—their own practices and the practices of their community. Moreover, in environmental history and related fields, more needs to be done to connect disparate research efforts, to reach a broader public, and to ensure that the past informs the future. By cultivating a "spirit of the commons," or what others call a "culture of togetherness" (König 2015), digital tools present a wealth of opportunities to increase transparency, communication, and democracy, expand data, and uncover new sources, leading to new perspectives and new research questions.

A young field with roots in environmentalism and in an understanding of the environment as culturally constructed, environmental history is inherently dynamic and diverse. The field has strong regional societies and an energetic, collegial community that includes colleagues in places where it is not recognized as a field. Those who identify as environmental historians have a variety of views on research priorities and frameworks, what lessons environmental history holds for practitioners and policymakers, and how activism fits in this spectrum. In spite of these differences, most agree that communication—both scholarly and public—is essential to a field that "aims to contribute positively to human understanding and decisions" (Coulter and Mauch 2011, p. 7). You don't have to be a digital native to understand the advantages of reaching broad groups of people with common interests. For environmental historians, as is explained in this volume by Kheraj and Oosthoek (see Chapter 18), subscribing to H-Environment, a subgroup of H-Net (Humanities and Social Sciences Online), is a main way of keeping informed about developments in the field.

Yet environmental history, like the rest of historical scholarship, is still very much bound to the classic interpretation of research work as an enterprise in which solitude grants freedom. As Anthony Grafton (2011) reminds us in a piece published a few years ago in *Perspectives on History*, while historians are expected to perform "unremitting, solitary intellectual work," this vision is however, at least partly, mythical: "Every realized work of scholarship, even one that bears the name of a single author, is the product of many individuals working together in different ways." Environmental history, an inherently transdisciplinary field in which research depends on both cooperation with practitioners in other fields and the ability to read and interpret research outputs from a variety of disciplines, is a pre-eminent example of the need to fully embrace such a point of view. The same is even truer for digital history.

By "digital tools" we mean the technologies (such as databases, search engines, application programming interfaces, mapping software, and social media) that help us gather, organize, filter, discover, connect, visualize, contribute, and share information. Others' engagement with this information contributes to its transformation into knowledge. For example, the Smithsonian's Bumblebee Project (https://transcription.si.edu/project/6780) shows volunteers images of digital photographs of bees, and lets them input the information found on each bee's tag, including the location where the bee was collected, directly into a new database. Information that used to be viewable only in person and by appointment, will become discoverable and comparable, accessible to everyone. This will transform the kinds of questions researchers can ask, and may even lead to new knowledge about the reasons for pollinator decline. In addition to facilitating the production of new knowledge, digital tools help researchers more easily discuss, cite, and share their findings beyond traditionally small circles of colleagues. Such digital tools—already too prevalent to be ignored—have indeed the potential to disrupt habits and structures in a positive way and make the cooperative nature of research more evident and widely accepted. There is hope for increased acceptance of these tools as proper products of scholarship: in June 2015 the American Historical Association approved its *Guidelines for the Professional Evaluation of Digital Scholarship by Historians*, recognizing the burgeoning number of formats adopted by professional historians in the twenty-first century and the need to give credit for all contributions to historical knowledge independently from the medium (Denbo 2015). This essay, in which we introduce an absolutely not-exhaustive sample of digital developments, practices, and collections that present opportunities to move toward a new, more collaborative, disciplinary ecology, aims to fuel a discussion on this topic and attract new scholars to experiment on the border of the digital and the environmental. We start by explaining what we believe could be the role of a "spirit of the commons" in digital environmental history, then provide a few examples of running projects, and finally discuss obstacles and opportunities of such a venture.

The new "spirit of the commons" is digital

The word 'commons', to an environmental historian, evokes shared resources and the necessity of user-informed systems (as opposed to the state or the market) for their sustainable management. Discussing human overpopulation by referring to William Forster Lloyd on the overgrazing of common land (1834), Garrett Hardin (1968) used the term "tragedy of the commons" to explain how, if individuals exploit a shared resource according to self interest, they will ultimately deplete and destroy the common resource. Yet privatization or top-down systems of management have other negative consequences: when private profit drives the system, resources can collapse and disappear, according to Elinor Ostrom, who in 2009 won the Nobel Prize in Economics for her analysis of economic governance of common property by groups using it. Ostrom insists there is no top-down panacea, no one right way of managing the commons. She argues for the need to draw on local knowledge of ecologies to understand their institutional differences, and for building enough variety to cope with the diversity in the world (Stockholm Resilience Centre TV, 2009). Her arguments inform not only studies of natural resource management, but have influenced studies and practices related to the management of cultural resources as well. Studies of information and intellectual property increasingly identify sets of "constructed cultural commons," or "pools of information resources," that serve common functions, generally "promoting valuable spillovers of information and knowledge distribution" (Madison, Frischmann, and Strandburg 2010).

Digital tools, technologies, and systems have already radically transformed how we value and maintain knowledge. Online lectures, massive open online courses (MOOCs), open-access repositories, digital catalogues, shared data and archival sources, more open peer review, open-source code, and community-driven acquisitions have changed the ways we work, the materials we examine, and the people with whom we engage. Posthumus and Sinclair (2014, p. 258) highlight the discrete nature of digital information, and its role in allowing the effective processing of data and enabling "one of the conceptually simplest and most powerful affordances (capabilities) of digital text: search." The digital, however, as they say as well, is not only central in having completely changed what a search is, but appears to provide a bounty of opportunities also in facilitating what John Unsworth (2000) has termed the *scholarly primitives* of the humanities (e.g. discovery, annotation, reference, etc.). A completely new array of tools is available to the historian to understand and interpret environmental change and human/nature relations. Digital environments show potential to allow improved understanding of natural ones; the virtual appears more and more necessary to understand the real.

As Gerben Zaagsma (2013) points out in his essay "On Digital History," not only is the digital not new, but also taking it seriously is not optional. He encourages reflection on how it affects and modifies our methods and epistemologies. This reflection is urgently needed. But we should not stop there. Applying, if in a limited way, this environmental metaphor to *knowledge* as resource focuses our attention on its political economy. Who can produce and access knowledge? On what structures does this production and access depend? Who manages these structures? Who profits from them? As academics, it is our responsibility to cultivate conditions for maximal knowledge spillover. Such a spirit of the commons is central to the academic enterprise, for environmental history perhaps even more so. But how can we do this, and do it in a way that allows for user-informed systems that sustain our resources?

Maintaining the spirit of the commons in the digital humanities means to promote the valuation and maintenance of digital resources as public goods. New processes for evaluation, distribution, and discovery are transforming academic systems. Already, scholarship can bypass some gatekeepers and offer alternative ways of quantifying impact. We are almost in front of the realization of William Wulf's *collaboratory*, a "center without walls, in which the nation's researchers can perform their research without regard to physical location, interacting with colleagues, accessing instrumentation, sharing data and computational resources, [and] accessing information in digital libraries" (Wulf 1989, p. 19). Possibly the only difference between today's digital academic infrastructure and Wulf's vision is that we are not limited to the national dimension now.

We are arguably transitioning to a completely new scholarly communication system in which online technologies will greatly facilitate research dissemination. In a comment in *Nature*, Jason Priem holds that, as the journal article radically innovated the ways in which the final products of scholarship are assessed and spread, the Web will allow tearing down the boundary between process and product (Priem 2013). Research work will gradually become open and public, and new algorithmic methods of judgment and filtering will then replace peer review. Here it is extremely important to establish and promote standard best practices, in order to ensure that we agree to build projects with compatible moving parts—such as common metadata standards—that will allow projects to "speak the same language" and thus keep open their future prospects to remain accessible and relevant. In his *PLoS* blog response to Priem, Martin Fenner (2013) cautions however that interoperability is paramount, since "both scholarly communication and the web in general have a tendency for centralization." The risk is thus that, in front of a scientific infrastructure lacking interoperability, instead of an increase in the number of actors, ecosystemic openness, and a revival of academic self-determination, we may face an oligopolistic concentration in the hands of a few tech companies. Open access and effective interoperability standards are particularly relevant to environmental humanities scholarship, as Fred Gibbs observes, because they have the potential to greatly facilitate the challenges posed by the need to aggregate studies at different geographical and chronological scales: "The significance and viability of collected, created, and organized (and hopefully standardized) data and research becomes manifest only as far as they remain visible and available for reuse and re-appropriation–both within the environmental humanities as well as for other design and planning efforts" (Gibbs, in Murphie et al. 2013).

Yet building successful interoperable scientific infrastructures depends on many things: state research funding, agreed-upon standards of professional societies (e.g. library organizations), and the good will of open-source (and commercial) software developers. For historical scholarship, a valuation and maintenance of its knowledge as a public good without reliance on digital tools is unthinkable. In fact, using these tools in an informed and constructive way is becoming essential to its effectiveness. It is not about washing existing research with some out-of-the-box digital varnish. Rather, people design tools for particular purposes. We attribute metadata, contextualize information, evaluate it, and determine paths to its retrieval. Failing to recognize this means relinquishing the reigns of our work to someone else. Environmental historians, in particular, need to begin a conversation about how digital tools inform their ways of knowing, and also about what role they want digital tools to play in the political economy of their knowledge production and distribution. Below we introduce a few examples of digital developments, communities, and collections that present opportunities for environmental historians to move toward more collaborative knowledge production and distribution.

Digital developments, communities, and collections for environmental history

Free and open access to research results, data, and primary sources are central to the unimpeded development of knowledge. These are not new features of the digital era: libraries and archives have long provided free access to resources, while journals have disseminated the results of research, more or less broadly, for over a century, and conferences have been, more or less successfully, spaces of communitarian engagement. What is new is the scale and scope that the digital allows the concepts of "free" and "open" to reach: from the virtually unlimited number of people who can be reached to the radical reduction of the cost (and possibly environmental footprint) of sharing in the post-print era, for example. Projects aimed at opening up academic communication range thus from small, voluntary community initiatives to major state-sponsored plans and projects furthering and funding open-access goals, requirements, and standards.

One example of an open-access gateway supported by existing academic infrastructures that specializes in resources about human participation in, and understanding of, the environment is the Environment & Society Portal (www.environmentandsociety.org). The digital publication platform and archive of the Rachel Carson Center for Environment and Society (RCC, www.carsoncenter.uni-muenchen.de/index.html), a nonprofit joint initiative of the University of Munich (LMU) and the Deutsches Museum, the Portal addresses the environmental humanities scholarly community as well as the interested public. The fellows of the RCC are actively involved in contributing to the Portal and curating its content; in this way, a temporarily *in situ*, dynamic roster of international scholars brings its expertise and resources to broader audiences.

Online knowledge portals like the Environment & Society Portal aim to connect diverse users and cultivate public interest in humanities topics, in some ways similar to magazines, museums, and libraries. Knowledge portals may also provide different ways of locating, classifying, or disseminating content. They may serve users by making content freely and openly accessible and by allowing them to find content most relevant to their needs. Users may also be able to create or contribute to content, analyse or share it, connect with other users, and subscribe to alerts. Such portals are methodologically important for the humanities, in which finding content "relevant to the user's needs" may not mean precise search results, but results that encourage discovery. In other words, new perspectives and knowledge may be stimulated not by providing what one is already looking for, but by suggesting unexpected alternative content, considering the context within which the information was produced or hinting at new perspectives. Unexpected yet relevant results may be generated not only by suggesting results based on user-supplied metadata, but also by finding metadata that user inputs have in common and linking them. Metadata may thus be linked and connected independently from the users' search strategies, leading to a wider palette of possible results. By attributing metadata to each item, and ideally by following standardized taxonomies, items loaded onto the Portal become more findable and let people make more specific and powerful searches. As of now the Environment & Society Portal's most popular discovery tool is its keyword explorer, which allows users to follow the connections between items according to conceptual similitudes and allegiances. To a certain extent it allows users to recreate (and further develop) what Cunningham et al. (2013) refer to as the feeling of physically browsing the library shelves.

Public, private, and nonprofit enterprises all over the world offer a broad range of non-specialist open repositories that show great potential for the research needs of environmental historians. The most well known of these free and open-access repositories are probably The Internet Archive (www.archive.org) and Project Gutenberg (http://gutenberg.org). Others, like Flickr Commons (http://flickr.com/commons), offer access to visual resources provided by a huge number of independent research facilities, including the Library of Congress and the British Library. Finally, projects like the Biodiversity Heritage Library (www.biodiversitylibrary.org) and the Global Plants database (http://plants.jstor.org), with their breadth of freely readable sources, not only improve current research methodologies in biodiversity and botanical research by making historical literature openly available, but have also become essential tools for environmental historians and historians of science (Svensson 2015). Other repositories, such as HathiTrust (http://hathitrust.org) or JSTOR (www.jstor.org), work along the lines of mixed models, allowing almost global access to sources in the public domain (usually, according to US legislation, works published before 1923), while limiting access to other material to partner institutions and members.

The availability of an ever-increasing number of online sources may radically change the way environmental history is practiced, allowing for completely new dimensions of comparative work and the development of research questions that make full use of data sets that have been until now considered too disparate.

Besides an increase of the primary sources made available in electronic format, environmental history has much to gain also from a concerted effort to aggregate relevant born-digital material. Our blog, *Ant Spider Bee* (http://antspiderbee.net), is an attempt at setting up a community of people specifically dedicated to fostering the interaction and mingling of the digital and environmental humanities. For example, environmental historian Andrew Stuhl shares a successful class experiment to produce born-digital entries describing environmentally significant places and events (2015). The hope is that digital tools positively affect our methods, perhaps especially concerning pedagogy and global history.

Human curation of information remains more essential than ever, and semi-automating aggregation can afford us more time for this work. Thanks to a generous grant from the Alfred P. Sloan Foundation, *Ant Spider Bee* has partnered with the Roy Rosenzweig Center for History and New Media as a PressForward (http://pressforward.org) pilot partner. Many websites, blogs, and news sources syndicate their content for free for subscribers. If you know what feeds may produce content of interest, a web aggregation and curation tool like PressForward makes it easy for you (or volunteers) to review and republish the items. As the blog's title refers to Francis Bacon's metaphor for different types of knowledge work, we call the resulting curated stream of posts "the Ant," for its gathering function. Our experience with this tool has shaped our understanding of the topic as well: in attempting to aggregate "digital environmental humanities" material, we realized there is no such thing per se. Casting a broad net, we saw results as fitting three categories: environmental humanities items that employ digital technology; digital humanities items that may have environmental relevance; or environmental technologies with as-yet-untapped significance for humanities scholars. Also the *Environmental Humanities Now* (http://environmentalhumanitiesnow.org) web aggregator, a now discontinued experiment set up by John Christensen and Ursula Heise, was powered by PressForward. This site had the rather broader and more ambitious aim to showcase scholarship and news of relevance to the environmental humanities community, spanning the disciplinary divides and implementing a post-publication editorial review system.

A further example of a community of environmental historians making great use of digital tools and methods to promote its aims and foster research is the Network in Canadian History & Environment (http://niche-canada.org), which offers a wide palette of resources, ranging from podcasts, to blogs and events, as well as digital projects. Also, the extremely worthy *Programming Historian* project (http://programminghistorian.org), offering open-access, peer-reviewed, community-sourced tutorials, even if tailored to a broader audience, is tightly linked through personal connections and research interests to environmental history.

Digital tools also help environmental historians tap the public as a source of information—and sometimes even support—for projects that advance the public good. *The Hurricane Digital Memory Bank* (http://hurricanearchive.org), an award-winning digital project of the Roy Rosenzweig Center for History and New Media and the University of New Orleans and others, is the largest free public archive about hurricanes Katrina and Rita (Brennan 2012; Brennan and Kelly n.d.). More than 25,000 digital objects make up the archive, publicly sourced from those affected by the hurricanes. The creators asked contributors to enter a zip code or address, and then geolocated the content in the database using a mashup of Google Maps. Community engagement is essential for ensuring digital projects' sustainability by distributing maintenance efforts. But in some cases the public can be even more directly engaged to crowdfund a project. Writer and environmental historian Jenny Price raised more than $32,000 through Kickstarter to fund production of her app "Our Malibu Beaches," which equips beachgoers with the maps and knowledge necessary to make Malibu's public beaches truly accessible (Coulter 2014).

The digital commons we envision as a way to bring forward the aims of environmental historical scholarship cannot depend exclusively on centralized and institutional efforts. Environmental historians need to be inspired by the example of the open-source code movement if they truly want their data and insights to be active elements in a change in the way in which information is shared. An ecosystemic change can only be realized if many more practitioners and scholars get invested in adopting new ways of sharing or embracing new licensing models for our research products. Just to sit back and wait for more primary sources to become available or our library to subscribe to that specific database or repository will not make documents appear faster on our desks. This is not only an egoistic battle, but also one to free up resources for low-income scholars and countries. The attempts made by W. Caleb McDaniel (2013) to realize what he terms "open notebook history" are exemplary in this regard. As Hardenberg notes, if more environmental historians would adopt an open notebook approach more information would automatically circulate and access to sources would become easier and simpler for all: "we could 'crowdsource' a vast amount of information that would help more and more scholars to perform fruitful research in the environmental humanities" (Hardenberg, in Murphie et al. 2013). Following McDaniel's example, Hardenberg has been attempting to make sources and data available online for a variety of his research projects. A complete list of the sources he has accessed over the years in various European archives is available as Zotero group (www.zotero.org/groups/hardenberg_sources), while a couple of transcriptions from statistical series and GIS data are available on GitHub (https://github.com/wilkohardenberg/data). It is hard to collect more than anecdotal accounts and viable metrics of whether such data find any audience, but it definitely feels good to know that they are not hidden in a drawer (i.e. external hard drive), but free for all to take and reuse. The hope is that this may help to further research and help exchanges among scholars, especially in fields and areas that are underrepresented when it comes to digitalization projects and open-access repositories. This is a very simple way to make your data available to the broader community, which has great potential to be reproduced by anybody. Other options would be to just upload the data on your personal website or, if you have the rights to publish images of your sources, create your own virtual museum with Omeka (http://omeka.org) or Neatline (http://neatline.org). Of course, central to such undertakings is also the need to provide exhaustive metadata, documentation, and references for all datasets (Howard 2013) and to adopt an open license, such as one of the Creative Commons licenses (https://creativecommons.org/licenses), to actually allow people to reuse the data without having to contact the creator for permission first. Also, private, individual initiatives like these, if they became common practice, at a certain stage of development would need some form of institutional support, in the form of new and more advanced forms of rights management by libraries, archives, and museums or—possibly more importantly—searching facilities that make it easier to find relevant datasets and digitized sources (which, as it is, may or may not appear in a standard Google search). To adopt also here the commons metaphor, we would need to think about how to structure what Oakerson calls the "decision-making arrangements" (Oakerson 1992, p. 46) needed to access and manage the common-pool resource, even if we have to create the commons from scratch.

One brilliant example of a community-led attempt to build a real commons around data provided for free by an institutional repository is the Git-Lit project envisioned by Jonathan Reeve (2015), which parses and republishes each digital text in the British Library corpus as version-controlled files on GitHub. In the process, metadata are transformed into human-readable information and collaborative copy-editing is made possible. Another project working along similar lines is GITenberg (http://gitenberg.github.io). This way of treating sources like open-source code is exemplary of how historical research practices could be radically changed in favor of a more collaborative approach by embracing the "spirit of the commons."

Digital environmental commons: Obstacles and opportunities

Digital tools alone have no stake in environmental knowledge. But we do. We want the generation of environmental history knowledge to be deep and broad. We want it to draw on more information than ever before while being able to filter and analyze all this data mindful of tools' powers and limitations. We want its methods and criteria to be challenged by diverse cultural and disciplinary traditions. We want its evaluation to be fair and transparent. We want its distribution not to be limited to a small group of scholars, but to be accessible to policymakers and the public. How can we harness these tools to cultivate fertile conditions for the generation and distribution of environmental history knowledge?

There are obstacles to overcome, mainly based on conservative institutions and profit-driven publishing systems. We need to educate the community on the value of open-access public resources. They should not be compared with subscription-based, profit-driven resources and judged as failing for their inability to present materials hidden behind a paywall. We also need to present new modes in an inclusive and non-threatening manner. Once a project has been piloted, engaging skeptical senior scholars as contributors may eventually win them over as champions. Narrow criteria for promotion and tenure will always be a barrier to engage early-career scholars in digital or experimental work. As publishing, funding, and university structures transform, environmental historians will have to learn to be creative and flexible, and to work not only with one another, but with diverse partners as well.

Digital tools present a wealth of opportunities to cultivate a spirit of the commons in environmental history and related fields. The spirit of the commons is one of transparency and democracy, expanded data, new sources leading to new perspectives and new research questions. Publication and networking opportunities help environmental historians bridge national, disciplinary, and linguistic divides. They provide visibility for the field and its early-career scholars. These gains in knowledge are potentially vast and transformative. And as Giacomo Parrinello (2015) explains, it is not only about knowledge gains; digital tools also offer their own kind of "geeky pleasure." It is this fun that captures the attention and engagement of students and the community, and helps to sustain projects in the long term, even inspiring action. We hope this short essay makes the role of digital tools for the future of environmental history knowledge seem helpful, even if, possibly, in unexpected ways.

References

Brennan, S. (2012, March 26). Collecting and preserving memories of the 2005 Gulf coast hurricanes. Retrieved from www.antspiderbee.net/2012/03/26/hurricane-digital-memory-bank. Accessed 14 September 2015.

Brennan, S., and Kelly, T. M. (n.d.). Why collecting history online is Web 1.5. Retrieved from http://chnm.gmu.edu/essays-on-history-new-media/essays/?essayid=47. Accessed 14 September 2015.

Coulter, K., and Mauch, C. (2011). Introduction. The future of environmental history: Needs and opportunities. *RCC Perspectives* (3). Retrieved from www.carsoncenter.uni-muenchen.de/download/publications/perspectives/2011_perspectives/rcc_issue_6_web_sw.pdf. Accessed 7 June 2015.

Coulter, K. Open access to the beach. Interview with Jenny Price (2014, February 18). Retrieved from www.antspiderbee.net/2014/02/18/open-access-to-the-beach. Accessed 14 September 2015.

Cunningham, S. J., Alqurashi, H., Hinze, A., Vanderschantz, N., Timpany, C., and Heese, R. (2013). Browsing and book selection in the physical library shelves (Working Paper). University of Waikato, Department of Computer Science. Retrieved from http://researchcommons.waikato.ac.nz/handle/10289/7161. Accessed 13 September 2015.

Denbo, S. (2015). AHA council approves guidelines for evaluation of digital projects. Retrieved from www.historians.org/publications-and-directories/perspectives-on-history/september-2015/aha-council-approves-guidelines-for-evaluation-of-digital-proj ects. Accessed 4 September 2015.

Fenner, M. (2013, March 28). Some thoughts on beyond the paper. Retrieved from http://blogs.plos.org/mfenner/2013/03/28/some-thoughts-on-beyond-the-paper. Accessed 14 September 2015.

Grafton, A. (2011, March). Loneliness and freedom. Retrieved from www.historians.org/publications-and-directories/perspectives-on-history/march-2011/loneliness-and-freedom. Accessed 8 September 2015

Hardin, G. (1968). The tragedy of the commons. *Science, 162*(3859), 1243–1248.
Howard, S. (2013, May 18). Unclean, unclean! What historians can do about sharing our messy research data. Retrieved from https://earlymodernnotes.wordpress.com/2013/05/18/unclean-unclean-what-historians-can-do-about-sharing-our-messy-research-data. Accessed 3 September 2015.
König, M. (2015, September 17). Für eine Kultur des Miteinanders in der Wissenschaft. Retrieved from http://rkb.hypotheses.org/1009. Accessed 25 September 2015.
Lloyd, W. F. (1834). *A Lecture on the Notion of Value, as Distinguishable not only from Utility, but also from Value in Exchange Delivered before the University of Oxford in Michaelmas Term 1833*. London: Roake and Varty.
Madison, M. J., Frischmann, B. M., and Strandburg, K. J. (2010). Constructing commons in the cultural environment. *Cornell Law Review*, 95(4), 657–709.
McDaniel, W. C. (2013, May 22). Open notebook history. Retrieved from http://wcm1.web.rice.edu/open-notebook-history.html. Accessed 16 September 2015.
Murphie, A., Jørgensen, D., Gibbs, F., and Hardenberg, W. Graf von. (2013, October 23). Open access and the environmental humanities. Retrieved from http://environmentalhumanities.org/2013/10/23/open-access-and-the-environmental-humanities. Accessed 14 September 2015.
Oakerson, R. J. (1992). Analyzing the commons: A framework. In D. W. Bromley (ed.). *Making the Commons Work. Theory, Practice, and Policy* (41–59). San Francisco, CA: ICS Press.
Parrinello, G. (2015, April 7). Knowledge gains and geeky pleasures: Adding digital to environmental history. Retrieved from www.antspiderbee.net/2015/04/07/knowledge-gains-and-geeky-pleasures-adding-digital-to-environmental-history. Accessed 14 September 2015.
Posthumus, S., and Sinclair, S. (2014). Reading environment(s): Digital humanities meets ecocriticism. *Green Letters*, 18(3), 254–273. Retrieved from http://doi.org/10.1080/14688417.2014.966737. Accessed 14 September 2015.
Priem, J. (2013). Scholarship: Beyond the paper. *Nature*, 495(7442), 437–440. Retrieved from http://doi.org/10.1038/495437a. Accessed 14 September 2015.
Reeve, J. (2015, September 8). Introducing git-lit. Retrieved from http://jonreeve.com/2015/09/introducing-git-lit. Accessed 18 September 2015.
Stockholm Resilience Centre TV. (2009). *Sustainable development and the tragedy of commons*. Retrieved from www.youtube.com/watch?v=ByXM47Ri1Kc. Accessed 28 September 2015.
Stuhl, A. (2015, September 16). On teaching global environmental history with places & events. Retrieved from www.antspiderbee.net/2015/09/16/on-teaching-global-environmental-history-with-places-events. Accessed 16 September 2015.
Svensson, A. (2015). Global plants and digital letters: Epistemological implications of digitising the directors' correspondence at the Royal Botanic Gardens, Kew. *Environmental Humanities*, 6, 73–102.
Underwood, T. (2015, September 12). Twitter post. http://twitter.com/Ted_Underwood/status/642647537220915201. Accessed 12 September 2015.
Unsworth, J. (2000). Scholarly primitives: What methods do humanities researchers have in common, and how might our tools reflect this? Presented at Humanities computing: Formal methods, experimental practice, King's College, London. Retrieved from http://people.brandeis.edu/%7Eunsworth/Kings.5-00/primitives.html. Accessed 12 September 2015.
Wulf, W. A. (1989). The national collaboratory – a white paper. In J. Lederberg, K.

Uncapher, W. A. Wulf (eds), *Towards a National Collaboratory Report of an Invitational Workshop at the Rockefeller University, March 17–18, 1989*. Washington, D.C: National Science Foundation, Directorate for Computer and Information Science Engineering.
Zaagsma, G. (2013). On digital history. *BMGN – Low Countries Historical Review, 128*(4), 3–29.

21 Remote sensing

Digital data at a distance

Sabine Höhler and Nina Wormbs

Introduction: Seeing environments from a distance

To perceive environments from a distance requires highly sensitive instruments. Current remote-sensing devices follow the tradition of instruments that have historically privileged the visual sense: from sonar to radar they claim to 'see' environmental change, translating digital data into visual imagery. This chapter takes two kinds of remote-sensing practices as a starting point for a discussion of techno-scientific approaches to environmental perception beyond the visual sense: sensing of the ocean floor by sonar on ships and sensing polar cryospheres by radar on satellites. Both technologies collect single measurements into synoptic arrangements that can then be viewed as images.

What happens to environmental history if sensory data become the source and digital data management becomes the practice of making sense of the earthly environments? Our exploration will focus, first, on the materiality of the sensing devices as condition and effect of what becomes visible. How can we understand sensory data and their subsequent presentations as visualizations of earthly environments? Second, we focus on the changed view of the archive as a passage to history. In a time of big data and ubiquitous computing, paired with increasing belief in learning machines and massive data analytics, we need to re-pose questions on the matter of historiography, in terms of its objectivity and situatedness. On the one hand, remote-sensing technologies seem merely to scale-up the mediated relation to nature that scientific and technological investigation adopted. Observing and recoding instruments have long outperformed the potentials of direct sensual experiences of seeing, smelling, hearing, tasting and touching a study object. On the other hand, digital data visualizations can evoke unintended or unanticipated affect. Which nature speaks through the digital archive and communication, and how?

The materiality of sensing environments

How did human perception of our environments change with the modern era? Surely humans continued to inhabit their land and travel their countries. But with the rise of the earth sciences in the nineteenth and twentieth centuries emerged also earthly environments of a scale that could not be measured and explored in immediate ways. The perception, for example, of higher atmospheres, deep oceans and outer space relied on complicated sciences and technologies of probing and observing, which provided a situational perspective and a global overview that essentially differed from individual local experiences (Höhler 2015). The historian is challenged not to take these environments as givens but to explore how they gained evidence. As environmental historians with a training and focus on science and technology studies, we study the instruments that made such earthly environments perceivable and recognizable without directly seeing and experiencing them.

The technologies at play in this chapter are of different kinds, but they both build on the possibility of an observer being physically detached from the objects of study. Remote sensing can be understood as sensing objects and phenomena at a distance, without an immediate physical connection. Just as the materiality of the digital itself is challenging, so is the materiality of something which you cannot touch.

Digital data presuppose digital technology. This digital technology is not separate from the production of digital data; rather, as we explore with the cases of ocean floors and icescapes, the instruments are intrinsic to the kinds of things one can investigate and how they can be investigated. The historical perspective demonstrates that there is no 'great divide' between the analog scientific method and the digital scientific method in environmental remote sensing, even though there are tropes suggesting such a divide.[1]

SONAR

Well into the modern era, the world's oceans were largely perceived as transit spaces dividing the continental land masses. Apart from the fishermen harvesting the seas, experienced with local fishing grounds and sea life, ocean knowledge was reserved to navigational skills and coastal mapping projects. In the nineteenth century, hand in hand with national and international endeavors of knowing Earth and with the emerging earth sciences, the prevailing simple techniques of line and lead sounding were refined and organized systematically to give scientific depth to the oceans.

These sounding tools were limited in scope, they were extremely tedious to apply, and they rendered uncertain results. A single measurement with a weight attached to a wire to fathom ocean depths of several kilometers could take many hours, involving the entire ship's crew to operate the winch, hold the ship stable, and determine its position. Acoustic sounding changed the setting drastically. Of interest from the environmental historian's perspective is not just the technological shift from slow and heavy winches to swift and comparatively precise electroacoustic devices available from the early-twentieth century onwards. Remote sensing by acoustic signals put the observation into practice that sound travels much faster in water than in air, echoing the signal off the ocean floor in matter of seconds. The shift in technology was also a media shift: sound waves are material waves; they propagate by using the body of ocean water as their medium (Peters 2015).

The resulting 'datascapes' of ocean depth were mediated in several other ways: measures of time had to be translated into units of distance. Again, the historian needs to be aware that this was no simple and smooth operation. The first echo-sounders involved sound sources like underwater gunshots. The receivers were humans listening to the signals with earphones. And complicated manual conversions of measurements had to be performed *ex post* (Höhler 2002). This sensing technology was analogue until sophisticated sound ranging devices entered the scene in World War II. SONAR, originally the acronym for the US **So**und **N**avigation **a**nd **R**anging technology, became the label for remote ocean sensing devices. Sonar instruments could sound a water body in all directions horizontally and vertically, and they became increasingly available as self-registering devices that could be operated without intricate knowledge of their internal workings, or, to apply Bruno Latour's (1999) notion, sounding technology became 'blackboxed.' Nowadays we find echo-sounding devices inbuilt even in small vessels. While the sound signals remain analogue, senders and receivers are electronic, and data are directly digitalized, ready for further processing.

RADAR

Our second case is concerned with measuring ice in the polar regions from satellites. Contrary to ships, human-made satellites have only been around for about half a century; before their invention, remote sensing from the sky had to use balloons, and later airplanes. In the early 1960s, satellites equipped with sensing instruments started circulating Earth and, in 1972, the first dedicated remote-sensing satellite system was put in place. Remote-sensing satellites often circulate Earth in polar orbits, at a height of ca. 700 km, making consecutive swaths as Earth spins around its axis. The satellites equipped to carry out this task can have several instruments on board. Either they can be passive and just register the emissions from Earth itself, such as heat or reflected sunlight, or the instruments can function as radars (**Ra**dio **D**etection **a**nd **R**anging devices), also a World War II technology of magnificent importance for air warfare. The radar transmits radio waves that bounce on objects and are scattered or absorbed. The reflected signal is received by typically the same system on the satellite and data is sent back to Earth stations on the ground. On Earth, data can be computed and information about the reflected substance can be deduced.

This computational process is core to the outcome of the sensing and is time-consuming and complicated. Sometimes all the data cannot be downloaded and are already managed to some extent by the satellite itself. But most of the work of managing data is done at data-handling centers of different kinds. Algorithms are applied to compute what the sensors have really sensed, transforming data into information. This result is then to be visualized, which is an art in itself. To make images out of the computed sets of data is also a process that is time-consuming and complex. A number of choices have to be made as to brightness and color, stitching together, or leaving different data sets apart. The materiality of these processes is difficult to assess, as is any programming and coding (Wormbs 2013).

However, we argue that the materiality of the sensing devices, sonars and radars, embedded as they are on vessels like ships and satellites, conditions what becomes measurable and visible, and subsequently what is actually measured and displayed. Sonar technologies were a response to the observation that electromagnetic radiation cannot simply 'see' in water; light can penetrate water bodies by only a few meters. In turn, sonar measurements are compromised by solid objects like fish as well as by ocean layers of salinity and temperature that act as reflection or refraction boundaries to sound. Regarding radar, the resolution of a radar instrument decides the final concentration of data on a sensed area. How often the satellite passes over a particular spot affects the possibility of detecting change. The representation of the environment is formed through the means of these technologies; they are environing technologies.

The historian of twentieth and twenty-first century seascapes and icescapes needs to take their construction seriously. Discursively, practically and visually, these earthly environments would not exist without the data and the data performances (Edwards 2010). The experiences of such environments are necessarily mediated, primarily through science and technology. These mediations pose the challenge to study material that is rather unconventional from the historian's perspective. While the historian's sources have been archival material, handwritten or printed documents and photographs, the digital environments presented here ask the historian to use sensory data—analog or digital, data management and data products, the graphs and digital images—as sources. To do environmental history we need to look at new 'archives.'

Data archives

The two cases presented here are examples of new archives that challenge traditional views of the archive as the privileged pathway to environmental history. The traditional archive is characterized by a specific location, by its collections over long time spans, a specific spatial order, a striving for completeness, but also by the authority of selection and sorting. It provides a reference system and a system of accessibility and retrieval (Vismann 2008). The digital archive regulates access to environmental history in very different ways: it organizes time and space differently; it offers new search functions; and, depending on the available computing power, it offers almost unlimited ways of data analysis and combination, including data modeling (Bowker 2006).[2] How will environmental history be affected by this change of tools and perspectives?

To an historian, being in the archive is not merely a procedure or a process of work, but also an existence. Identifying collections, planning the trip, travelling, spending time, being exhilarated and disappointed, and then going home to make sense of it all, have been defining for much historical scholarship. Writing the history of the recent past, and particularly the history of objects and spaces that exist in digital form from their inception, challenges this practice. It should be stressed that sources which are digital from the very beginning are not the same as digitized analogue archives. We are not primarily interested in the possibilities of digitizing documents for the digital humanities. Rather we are interested in the consequences that the organization, storage and circulation of digital material have for the research process itself—the scientists' as much as our own—for the questions asked and the contexts created.

Digital archives as a mechanism for providing information on digital environments can come in various forms, but very often they are simply retrieved through the Internet. This means that, for good or bad, you literally do not need to get out of your chair to do primary research. This also means that your choice of material is affected, since you are not physically bound but can cover scientific research projects in France, Japan and the U.S. in the same afternoon.

One effect of this surgical precision of investigation is that you do not as easily realize the given organizational context, as you might in a traditional archive. In the traditional archive, searching for your particular volumes exposes you to a specific context. For example, you will read about what the same organization was doing at the same time or before and after the particular project that you are interested in. While archival records and catalogs index single items, the folders and boxes may let you find unexpected things in the interstices of these items, allowing access to the wider context almost seamlessly. We do not argue that this is a better or supreme context, but it is a tradition against which we have written history for a long time.

The digital archive does not *give* the same context and thus a specific context needs to be *created*. Contextualization is important, but the conditions are different for projects in the digital era. In the case of sea ice, imagery can be retrieved from various sources and with various sorting procedures. Either images can be collected through specific projects like NASA project ICEsat or European Space Agency (ESA) project CryoSAT. They are often sorted under the web pages of the funding institutions but links or images can also be found in scientific work which analyzes or further processes the primary data. This digital collection resembles the analog in that the funder is the owner and the organizer of the material. However, images can also be retrieved through Internet searches using the tags of the images. This might result in a rather heterogeneous collection, in which several images might be hard to track (true, this could also happen in the analog archive where undated and loose photos could be tucked in a folder). The upside of this is that you have a multitude of images at hand immediately and relatively effortlessly, which you can study and analyze. Their origins, however, might remain unclear.

For the historian it is equally important to consider the way these digital archives are collated and used by the scientists who produce and process the data. Oceanographic research has been a collective and an accumulative endeavor from the beginnings of systematic recording in the early-nineteenth century, and their products—the charts and maps—combine and recombine spatiality and temporality in ways we need to consider. Compared to a satellite survey of the

Earth's surface, ocean-floor sounding is time-consuming. Even though acoustic sounding has reduced measurement times from hours to seconds, any one oceanographic expedition can only cover small oceanic areas. The data and images of larger regions are usually compilations of scientific contributions from different research groups and institutions located in many different places around the world. The archives that the historian browses are not connected to the geographical sites they represent. In addition to the spatial distance, there is also distance in time. Between the different measurement sets lie often not only hours, but weeks or years. The data collections may span large time periods. These temporal layers may become visible in the datasets, as measurements are tagged accordingly, but they are not necessarily visible in the images.

Thus, although they might have been compiled over long stretches of time, digital archives often appear rather flat, temporally. This observation tells us something about the temporalities inscribed into the objects of remote sensing. While in the case of icescape studies the scientists' goal is to capture and record environmental change, sea-floor measurements aim at gaining a picture of a situation that is allegedly static. We also need to take into account the ways that digital data can be shared and processed in versatile ways from different places in the world.

Once data are stored on a server and available digitally through the Internet, there are almost instantaneous possibilities of sharing and duplicating data. There are nearly endless possibilities of calculation and combination and recombination, paving the way for modeling and forecasting efforts. Also, the demands on general data management and data accessibility have been increasing. Much government-funded research has stripped mp 8U data sets from the rigid laws of regulating intellectual property, turning them into the public domain. Accessibility in combination with computing power led to the re-analysis of stored older data sets. Generating new data sets provided new environmental truths. Through their specific formations, digital data and digital archives promote specific epistemological, ultimately ontological, qualities of the environments they represent.

Image and affect

There is a trope that contends that the digital is clean and orderly, objective and efficient, dispassionate and distant. Since 'the digital' is often put in relation to what it is replacing, this trope suggests by implication that 'the analog' is dirty and disarrayed, subjective and slow, personal and bodily. The rhetoric around vinyl recordings suggest that the analog is perceived as being closer to the heart and the real music which it has recorded and is playing. Akin to this reasoning is the idea that things too technological are distant from the affective. We argue differently.

It would be shortsighted to consider remote-sensing technology primarily as scaling-up a process that has been long ongoing, a technological shift from experiencing environments in direct, 'immediate', ways to a heavily mediated environmental experience that inevitably will feel distant and dispassionate. The enormous scales of the environments we have discussed are affective in their own way as they confront us in the form of digital imagery. In this sense, remote-sensing imagery is more material than much of digital research outcomes. This is because images connect immediately to our visual senses and thereby to our bodies (Schneider and Nocke 2014).

From sonar measurements, an elaborate digital data archive of the Earth's ocean floors could now be created. Measurements were perceived as 'continuous' while, technically, the data continued to reflect single measurements. Complete ocean-floor charts were compiled, but more so than in terrestrial mapping, these charts were plot sheets extrapolated across vast data voids. Based on Humboldtian isoline technology, isobaths connected measured points of same depths. The pictures carried the topographical sensation of an emerging underwater space. Scaled and color-coded charts met the eye as coherent pictures. These ocean maps stayed not with the oceanographers but found entrance into geographical textbooks and popular atlases.

Take as an example Bruce Heezen's and Marie Tharp's famous topographical map of the ocean floor before digitization. The map was assembled at the U.S. Lamont-Doherty Geological Observatory from 1950s onward and published in 1977 (Figure 21.1).

The image is a result of several translations, starting from the map, painstakingly plotting available depth soundings manually, to a polished relief painting,

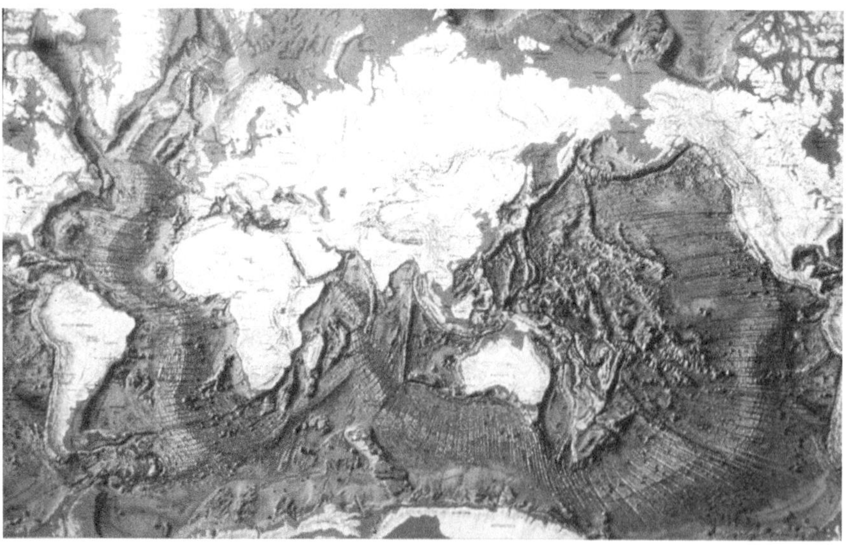

Figure 21.1 Heezen-Tharp map of the world ocean produced by the US Navy 1977

Source: National Oceanic and Atmospheric Administration (NOAA).

and ultimately to the digitized photograph that is accessible online. The image combines conventions of cartography and map design (the plan view, the flat continental presentations, the annotations) with painting and modeling traditions (leaving out the contour lines, highlighting the ocean relief). Data collections and images from other contemporary institutions might have appeared differently, but none could provide such scope, promoting an idea of completeness. The image thus also reflects the US oceanographic institutions' dominance and coverage at the time of the Cold War. From a scientific perspective, the ocean-floor panorama became renowned for its mid-ocean ridges and rifts, the geological features that shifted scientific assumptions about plate tectonics, sea-floor spreading and continental drift. The image has also been described as stunning—as a revelation—because it literally 'discovered' two-thirds of the Earth's surface in an unprecedented way. It unearths numerous previously unseen features. Its awe-inspiring power has been compared to the first views of Earth from space, the Blue Marble.

Similar in optical appearance, but different in construction, is the General Bathymetric Chart of the Oceans (GEBCO) world map of 2014. The shaded relief color map presents the most actual bathymetric knowledge about the world ocean floor in cartographic conventions comparable to the Heezen-Tharp map. The composite picture merges NASA datasets of continental land masses and ocean-sounding data. But the map is also a digital data product, developed from a data set hosted by the British Oceanographic Data Centre. The data set is based on a global grid of international systematic close-spaced ocean soundings, interpolated by satellite altimetry data and interspersed with data sets gathered by other methods. Although it presents another intermediary data product on its way to a stabilized final product, the resulting world map is undoubtedly striking; its beauty lies in the data set and its possibilities beyond the map, both regarding artistic representation and data recombination. The data grids are available for download in different data formats as two-dimensional arrays, together with display software for viewing, accessing and exporting the data, to be run on any PC.[3]

The satellite remote sensing of the sea ice exemplified here is not primarily concerned with further detailing, stabilizing and displaying a physical space by a constantly improved image. On the contrary, the objective is to understand change over time and to compare one situation to another. In the CryoSAT-2 mission, for example, the purpose has been to study changes in the cryosphere, including the thickness of floating sea ice and the elevation of continental ice. This does not prevent scientists from using the same equipment to actually improve the map of the sea floor.[4]

Change can be shown differently. In this particular example of Antarctica (Figure 21.2), change is colored in one single map rather than displayed as a difference between two maps. The map is a compilation of data on the elevation of continental ice collected during three years of sensing (2010–2013) and demonstrates an improvement in detail and coverage in comparison with earlier measurements (and also a loss of Antarctic ice). The area around the South Pole, which due to the inclination of the satellite orbit is hard to sense, is smaller in

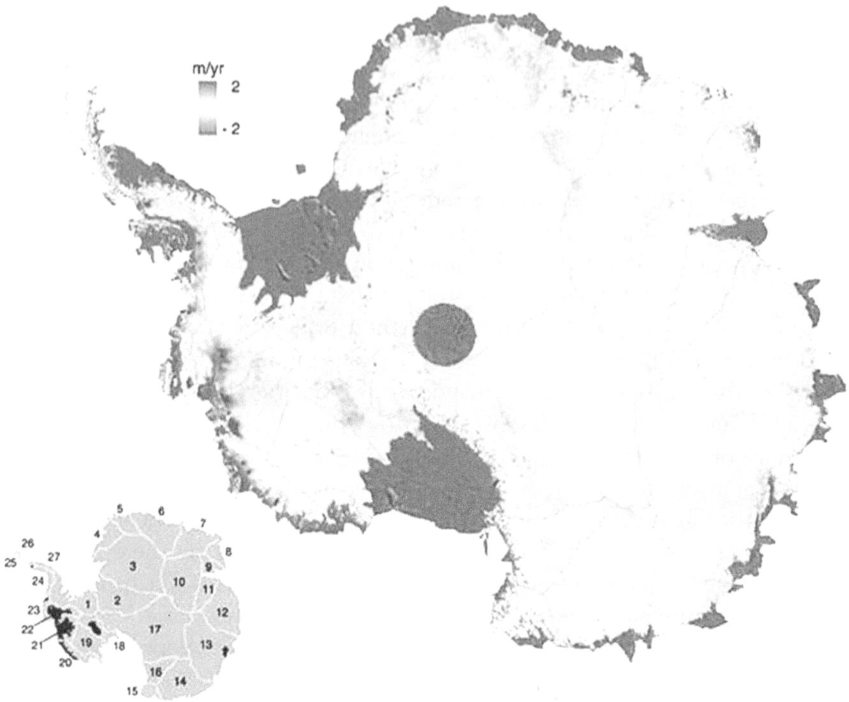

Figure 21.2 Antarctica's ice-loss regions, 2010–2013

Source: ESA European Space Agency.

this mission and hence more information is available. And the problem with stark changes in elevation is also helped through a higher resolution in the instruments. The outcome is a better understanding of the ice-melt in Antarctica, of key importance to climate change science.

What kind of environments are presented through remote-sensing technologies? Sonar creates the ocean visually as a transparent and empty space. This technology cannot 'see' fish and other sea life. The abundant ocean of fish and plant life, of water and salt, of streams and warmth, comes into focus only as 'echoing mistakes' before the data are adjusted. But the transparency brings out the sharp morphological reliefs of the Earth's face that otherwise remain invisible to the human eye. From the transparency, sublimity arises. Satellite remote sensing is similar in many respects. The choice of instruments determines what is visible in the atmosphere, on ground and below ground. Clouds can be invisible or in focus, biomass can be central or peripheral. Sonar and radar are thus clear examples of the materiality of the sensing device as a condition and effect of what becomes measurable and visible. The equipment works as a 'sorting hat', deciding what can go into the picture and what not. Interesting in this respect is how we relate this information to what humans can perceive through their senses.

There are differences between oceans and space, the medium through which the sensing apparatus needs to work. Arguably, the view from space, at a distance that is mind-boggling and awe-inspiring, is still possible to fathom. We have had a good sense of the atmosphere and its different layers for quite some time. Stars and galaxies have long been part of the science and world order of humans. The seas are, on the contrary, late places for synoptic exploration. Continuously moving, these vast amounts of water and biomass have been harder to fathom and more difficult to measure and describe. In this fact also lies the mythical character of the oceans; we simply do not know what is down there and, despite the myriad of ocean measurements, fantasies of strange animals, steep trenches and sunken continents persist. In comparison with space, we might conclude that it actually has to do with the human ability to see through air, but not very well under water. Water is not an element that humans can inhabit or penetrate without technological means.

When sonar serves as our eyes on the water-world, the translation process is different than that of the radar on remote-sensing satellites. Still, they are both techno-scientific processes in which choices have been made. The satellite cannot only see a larger frame than possible for humans, but it can also see things that the human eye cannot, like heat, or the concentration of gases in the sky. In both cases, but in different ways, the resulting images convey an environment otherwise inaccessible to humanity. And in doing so they also create this environment, often in striking and affective ways.

Conclusion

The interpretation of digital data at a distance is a cumbersome, difficult, expensive and time-consuming practice. Contrary to common belief, where the digital is portrayed as instantaneous and clean, we argue that it is in fact slow and messy. The plethora of choices to be made in each and every step of interpretation and understanding is part of creating the digital environment. If we want to avoid interpreting the digital environment naively, we need to understand the methods of data collection. Environmental historians will have to study the complex data assemblages and deal with the difficulties of assessing the origins of the data and the ways they have been processed. The historian's task of making sense of this environment also involves choices and priorities.

We have shown that the digital environments of remote sensing in many respects continue a tradition of analog environmental representation that relied on extensive data collections. But we have also argued that digital collections have opened up new possibilities of environmental perception due to the versatility and scope of digital records and the technologies of swift circulation. Digital data and digital archives have themselves become mediators of our understanding of earthly environments. The research process of studying these emerging environments clearly reveals that while we can provide a multitude of input to substantiate our understanding of Whole Earth, we are definitely not dealing with one simple single digital world.

Acknowledgements

The authors gratefully acknowledge the support for their work by the Swedish Research Council (VR), for the project "Views from a Distance: Remote Sensing Technologies and the Perception of the Earth" (grant nr. 2012–1134) at the Division of History of Science, Technology and Environment, KTH Royal Institute of Technology Stockholm.

Notes

1 Works like Compaine's *The Digital Divide* (2001) discussing the unequal access to information often seem to imply that the perceived gap between haves and have-nots was smaller before the Information Age, when knowledge was available and distributed in analog form only.

2 The question of traditional versus digital archives has been much discussed in connection to the digital humanities as well as for the translations of scientific knowledge about nature into digital archives (e.g. biodiversity databases), but not nearly as much for the digital environmental archives that remote sensing technologies have created.

3 GEBCO world map 2014, www.gebco.net.

4 "CryoSat unveils secrets of the deep", ESA Observing the Earth, October 3rd 2014, visited Dec 1st 2015. (www.esa.int/Our_Activities/Observing_the_Earth/CryoSat/CryoSat_unveils_secrets_of_the_deep). The technology (the methods, apparatus and conversions applied) is very different from sonar technology. Satellite views of ocean depth rely on gravimetry: altimetric measurements of ocean surface height using electromagnetic radiation are converted to measures of ocean altitudes and ultimately to units of depth by taking the gravitational forces on the water bodies into account.

References

Bowker, G. C. (2006). *Memory Practices in the Sciences*. Cambridge, MA: The MIT Press.

Compaine, B. (ed.) (2001). *The Digital Divide: Facing a Crisis or Creating a Myth?* Cambridge, MA: The MIT Press.

Edwards, P. N. (2010). *A Vast Machine: Computer Models, Climate Data, and the Politics of Global Warming*. Cambridge, MA: The MIT Press.

Höhler, S. (2002). Depth records and ocean volumes: Ocean profiling by sounding technology, 1850–1930. *History and Technology*, 18(2), 119–154.

Höhler, S. (2015). Inventorier la terre. In: K. Raj and H. O. Sibum (eds.), *Modernité et globalisation* (Histoire des sciences et des savoirs, ed. Dominique Pestre, Vol. 2) (167–181). Paris: Éditions du Seuil.

Latour, B. (1999). *Pandora's Hope: Essays on the Reality of Science Studies*. Cambridge, MA: Harvard University Press.

Peters, J. D. (2015). *The Marvelous Clouds: Toward a Philosophy of Elemental Media*. Chicago, IL: The University of Chicago Press.

Schneider, B. and Nocke, T. (eds.). (2014). *Image Politics of Climate Change: Visualizations, Imaginations, Documentations*. Bielefeld: Transcript.

Vismann, C. (2008). *Files: Law and Media Technology*. Stanford, CA: Stanford University Press.

Wormbs, N. (2013). Eyes on the ice: Satellite remote sensing and the narratives of visualized data. In M. Christensen, A. E. Nilsson and N. Wormbs (eds). *Media and the Politics of Arctic Climate Change: When the Ice Breaks* (52–69). Basingstoke: Palgrave Macmillan.

22 Walking with GPS

An object lesson

Finn Arne Jørgensen

Many environmental historians have embraced Geographical Information Systems (GIS) as a tool and an analytical framework for their research (e.g. Bonnel and Fortin 2014), but little has been written about its companion technology, the handheld GPS (Global Positioning Systems) device. GPS sensors have become ubiquitous over the last decade, and are now routinely embedded in computers, telephones, tablets, watches, and so on. Yet, the GPS is also functionally blackboxed in discussions about the meaning and experience of place, making it into either an invisible technology or one that has predetermined detrimental effects. This chapter argues that in the dialog between what we see on the screen in our hand and the world out there, the GPS represents a way of seeing and interpreting nature that we need to incorporate in our scholarship. As environmental historians, we should investigate the digital mediation of landscape just as closely as we investigate the landscapes that are being mediated. To do that we need to pay close attention to the workings of technology, challenging what we see on the surface and opening up the black box of technology for cultural analysis (Winner 1993; Latour 2013). In both teaching and research, critical investigation of the GPS allows us to reflect on the influence of technology on the perception of the landscapes surrounding us. It lets us explore the theoretical and methodological implications of writing and teaching environmental history in a world where people increasingly experience nature through or with the assistance of a diverse range of technologies.

This chapter proposes that *object lessons* can be productively applied to uncover the methodological implications of thinking about the GPS as a technology for engaging with place and space. I am here not thinking about the most common contemporary usage of the phrase "object lesson," typically used to "demonstrate some inviolable but self-evident principle—and often offered as a warning to those who might stray from it" (Young 2013, p. 6). Instead, I propose the older history of object lessons, developed as a pedagogical method of teaching attention to and awareness of the world through objects. In doing so, this chapter serves a dual methodological purpose. First, it asks how we as environmental historians can interrogate material objects, both made and found, in our research and teaching. Second, it examines in detail one particular object—the handheld GPS unit—and places it in its larger societal and infrastructural context. The chapter ends with a reflection on whether space and place has come to mean something else in the digital age than before. How does the sensory and epistemological experience of a landscape change with the tools and technologies we have at hand to experience it through? The attention-building structure of the object lesson will serve to shift our attention from the object itself towards the relationships that the GPS enables, between the mind, the body, and nature.

I will begin by introducing object lessons and their history more thoroughly, before I turn our attention to the GPS unit, loosely following the structure set up by Elizabeth Mayo in her 1831 book *Lessons on Objects*. I will argue that a GPS device can hold an object lesson for environmental historians and as such has a place in our teaching and research. It is an extraordinarily complex object that links age-old navigational practices, everyday landscape experience, and Cold War space technology. The GPS unit begs us to engage with a set of questions about how we know and make sense of the world around us, rather than simply being passive users of a technology that tells us where we are. The GPS becomes an object lesson which uncovers the entanglement of the digital and the material, of media and environment. It offers us a way to think about the presence of data in and about nature.

Object lessons

The Swiss education reformer Johann Heinrich Pestalozzi (1746–1827) introduced the idea of material object lessons as a hands-on pedagogical approach at the end of the eighteenth century. Pestolazzi's objects lessons can be seen as a radical pedagogy linked to Chartist educational philosophy, aiming to create "a closer connection of words and things" (Lovett and Collins 1840, p. 91). "Observation is the absolute basis of all knowledge," Pestolazzi stated, arguing that the first goal of education must be "to lead a child to observe with accuracy; the second, to express with correctness the result of his [sic] observations" (quoted in Calkins 1870, p. iii).

A number of educators across Europe sought to implement and refine his pedagogical approach for use in schools, including Elizabeth Mayo (1793–1865) who published *Lessons on Objects*, a book that contained a method for teaching school children to examine objects based on Pestalozzi's philosophy (Mayo 1831). Mayo's lesson plans contained lists of objects that children should experience and the qualities they should learn from them, though they were more evocative than exhaustive. In Mayo's view, students could only write and talk about an object after "the sensory and material groundwork" enabled by the progression of the object lesson was in place (Young 2013, p. 16).

What all these object lessons in books from the mid-nineteenth century have in common is a deep concern for the training of the senses, and this is where we as environmental historians should take note. "The importance of a proper cultivation of the senses by means of home training can not be over-estimated. The child has access to the material world only through its senses," emphasized

Norman A. Calkins (1871, p. 24). Bloomsbury's recent *Object Lessons* book series, edited by Christopher Schaberg and Ian Bogost, which aims to tell stories about the hidden lives of ordinary things, echoes this approach, though arguably with more inspiration from Barthes' *Mythologies* (2013 [1957]) than from Pestalozzi's disciples.

How can object lessons function as an experiment with methods in nature-culture-history research? While many things can be said about what method is and should be, I will here simply treat the scientific method of the humanities as the art of observing with attention, and this attention can be structured or guided through various procedures. The object lessons method is based on the premise that when observing objects directly, and then expanding the sphere of observation to larger phenomena, we can learn more about the world around us. I will here apply this approach to a study of the GPS as a technology, loosely following five steps defined by Mayo.

Step 1

> To lead children to observe with attention the objects which surround them, and then to describe with accuracy the impressions they convey, appears to be the first step in the business of education.
>
> (Mayo 1831, p. 1)

As a first step, Mayo instructs us to focus on naming objects and their components. Let us then start by examining the physical GPS device. We typically encounter it as specialized device, produced by manufacturers such as Garmin or Magellan, or as a feature of smartphones or tablets of all kinds, enabled by a computer chip that can determine the geographical location of the device. (I will note that how the machine works is not evident from the initial observation of the device: the GPS chip is assisted by other environmental sensors, such as electronic compasses and altimeters, as well as triangulation off cell phone towers and Wi-Fi networks for smartphones.) Almost all GPS devices are designed to be portable; some are for use in cars, others can be handheld for trekking, hunting, and fieldwork. Many dedicated GPS devices such as the Garmin Oregon 650 are small, black, and somewhat chubby gadgets made of plastic and glass, weighing approximately 200 grams, and easily pocketable. Devices are typically rugged and waterproof with a small screen that can be used with gloves.

Mayo encourages us to ask questions about the object's function and dependencies. People obviously use a GPS device to find out where they are, and how to get where they want to go. Many users find that the GPS unit provides security when traveling in new locations. Both the GPS chip and the map display requires electricity, generally provided by a rechargeable battery so that the device can be mobile. The key hardware components—GPS chip, screen, battery—must be linked to the software that interprets the signals from the chip and displays the result on a digital map, and as users we can interface with this software as long as the device has power. Each component can be the subject of its own extended inquiry into its making and dependencies, as well as the upstream and downstream chains of materials involved in the production and disposal process.

Step 2

> ... the children should be much exercised upon the qualities already remarked, but these should now be presented to them in other objects ... enabling them to form the abstract idea of the quality. Having had all their senses brought into action, they may be led to determine the sense by the exercise of which any particular property was observed.
>
> (Mayo 1831, p. 24)

In this stage, Mayo introduces more abstract concepts which can be observed, such as elasticity, transparency, opacity, and a great many others, to the children. Here would be a good place to ask what it is we sense through the GPS and how the GPS is able to provide this information. Let us start with considering a walk in the forest, which engages all the senses, as well as the whole body. Obviously, we encounter topography, geography, flora, and fauna through our senses, all the basic components of landscape. Looking deeper, we can see traces of human activity, such as paths, cut-down trees, and perhaps some litter. We can also see traces of animal activity, such as tracks, excrement, the occasional skeleton, and so on. Every now and then we come across power lines and the clear-cut corridors that surround them. Looking up, we can occasionally see condensation trails from planes flying above, maybe even hear the sound of planes or traffic. If it gets dark, we can often see light pollution from buildings, roads, and cities. All these are just obvious signs that there are no places undisturbed by humans in the Anthropocene. They are signs of history and that nature is thoroughly entangled with human and non-human activity, constantly changing. We can also observe another, and often less obvious, layer of signs or inscriptions on nature. Particular sites in nature have gained a place in human memory and culture—they have been given names and have been used for particular activities, and so on, over time. Often these names reflect activities that are traditionally undertaken here. For instance, in the forest near my hometown in Northern Norway, there is a place called "Appelsinhålla" ("Orange Hollow"), since this is where people would stop to eat an orange while skiing in the winter. This is a layer of cultural knowledge which does not reside in the physical place itself, but in memory and media (you can now check in to Appelsinhålla on Facebook).

But what happens if we bring the GPS with us when we go for our walk in nature—to turn it on, enter a destination, and follow the instructions for how to get there? We still have access to all the sensory information discussed above, but can also observe spatial data on the screen. Our current physical location is displayed as a point on the digital map, which can be combined with directions for how to get to another location. One thing we would notice is that there is a big difference between GPS units designed for use in cars, which are designed to give directions along roads, and GPS units designed for handheld use, which typically emphasize topography more than roads. When traversing a landscape on foot, the lay of the land matters, and here the GPS can be a useful tool in planning routes. More advanced GPS units can also switch base maps, including 2D, 3D, and satellite imagery, allowing for different representations of landscape. Other information layers are also possible, such as the locations of gas stations, restaurants, lodging, or geocaching locations. The spatial information presented by the GPS is thus not all that different from that provided by traditional maps, with the added feature of automatically placing the user's location on the map. In some cases, we have additionally delegated the wayfinding to algorithms that have calculated the best way to the destination.

A GPS is linked to the world in two fundamental ways. The first is through the digital maps their location data is displayed on. Maps are not neutral representations of landscape, but interpretations that are laden with history, culture, and power (Crampton 2001). The GPS should prompt us to ask whose view of the world we are seeing on the digital map, because it is never just a visualization of raw spatial data (Gitelman 2013). The second way is through the network of GPS satellites orbiting Earth. The device we hold in our hands is only an access point to what Geoff Manaugh (2014) has called:

> … a 50,000-kilometer wide super-device barreling through 'cosmic kinks' in spacetime, […] an artificial constellation [and] an immersive, semi-weaponized, three-dimensional spacetime instrument, sloshing back and forth with 170-second-long tides of darkness, the black ropes of spacetime being strummed by the edges of a 32-point star.

It is only when tethered to this network of satellites that GPS devices allow users to quickly and accurately determine their coordinates. While full coverage requires a minimum of 24 satellites, the US Department of Defense-owned GPS network currently has 32 operational satellites in orbit (there are other GPS networks, such as the Russian GLONASS, which many GPS units can also access for better accuracy, as well as European, Chinese, and Indian networks under development). A GPS unit depends on having at least four satellites in sight to work, so the satellite orbits are carefully divided among six orbital planes (Easton and Frazier 2013, p. 80). Through trilateration and synchronized atomic clocks, the GPS unit determines its location based on the reception of high-frequency radio waves from the satellites, which can be used to calculate the distance to each satellite.

A key point in this object lesson is that we do not stand at a distance from these infrastructural technologies (Kurgan 2013, p. 14). We are embedded in them. Adrian MacKenzie's (2010) concept of *wirelessness*, which "designates an experience trending toward entanglements with things, objects, gadgets, infrastructures, and services, and imbued with indistinct sensations and practices of network-associated change," (p. 5) is a useful term for expressing this particular quality of the GPS-user-nature relationship. Many go to nature to cut the wires of modern technology, but wirelessness, "being connected to objects and infrastructures without knowing exactly how or where," (ibid. back cover) follows us there. What the GPS does is to expand our idea of what *environment* means to also include the digital, which is an important lesson for environmental historians.

Step 3

> In this series the children may be led to the observation of qualities which cannot be discerned merely by the senses. Thus by showing them at the same time wool and woolen cloth, and questioning them as to the difference of the two, they will readily conceive the ideas of natural and artificial. In this manner they may be led to remark the distinction between foreign and native; exotic and indigenous; animal, vegetable, mineral, &c.
>
> (Mayo 1831, p. 39)

This third step in Mayo's object lessons extends the previous steps through contrast and comparison, and calls for students to begin making explanations and justifications. Here we can turn the larger contexts within which we place GPS devices, as well as their relationships to other forms of navigation. In contemporary media it is not hard to find many objections to the GPS as being detrimentally different to more traditional wayfinding, in ways that are harmful to the spatial cognitive capabilities of people (e.g. Milner 2016). In my Zotero library, I have a growing collection of newspaper articles about GPS navigation going wrong as people blindly follow the instructions on the screen rather than listening to common sense, such as a recent viral story about the American tourist in Iceland who typed one wrong letter in a street address and was sent to a small town six hours away when trying to get to his Reykjavik hotel from Keflavik airport (Bilefsky 2016). These stories are typically presented in a way that blames the GPS for making people stupid, as opposed to writing that people occasionally use GPS devices in stupid ways.

Scholars have conceptualized GPS wayfinding similarly. For instance, in John Edward Huth's *The Lost Art of Finding Our Way* (2013), navigation by GPS is swiftly dismissed. Since the user is simply following instructions from a computer, it does not count as proper wayfinding. In such concerns, we find clear hints of what the geographer Alastair Bonnett (2014, p. 2) calls a geographical anxiety about the disenchantment of landscapes. This disenchantment process where places become generic and lose their unique and diverse characters is a complex process with many influencing factors, but the GPS is certainly one of the components intertwined with such narratives (Kingsnorth 2009; Augé 2009; Casey 2013). Much of the resistance we find stems from an idea that the GPS's directed navigation makes serendipity impossible – it takes away our ability to observe and to pay attention, and does not reflect the complex network of things in space.

These narratives possess a clear preference for a direct and unmediated experience of place, landscape, and nature. But does landscape really have a privileged position that we can access directly, without mediation and translation of various kinds, what Bruno Latour (2005) calls "opaque layers of translations, transmissions, betrayals, without any complicated machinery of assembly, delegation, proof, argumentation, negotiation, and conclusion" (Latour 2005, p. 33)? If we pay close attention to the workings of the GPS and how it links us to the world, the story becomes more complicated. When combined with a digital map, the GPS bridges the gap between the user's body and the pixels on the screen as a mediating technology. With the GPS in hand, the user seeks a place in space and time. As such, it is part of the experience of the places one encounter on the way to the final destination. Walking brings our attention to the technologies through which we experience nature. It is a sensory experience, though not unaided.

This form of mediation is not optional: it is how we interface with the world (Kember and Zylinska 2012). Technologies like the GPS are not the only ones that shape our relationship to our surroundings, but they exemplify the pervasiveness of media by enabling and helping formulate a sense of place through *tuning* it, a subtle process of calibration (Coyne 2010). We should here note that historians of technology have long demonstrated that technology is never neutral, including its mediation (Kranzberg 1986). As I have argued elsewhere, the particular constellations of technologies and users in use at any given point in time "shapes our understanding of both the mediated phenomenon and ourselves" (Jørgensen 2014b). The GPS encourages us to pay attention to the nature of wirelessness, one that is not experienced through the body, but distributed through data, media, and infrastructures. Human interpretation and experience is still relevant for making sense of this nature, but we need to understand how it is filtered through machines, technologies, models, and database structures.

Step 4

> The chief aim proposed in this series is, to exercise the children in arranging and classifying objects; thus developing a higher faculty than that of simply observing their qualities. The complex operation of connecting things by their points of resemblance, and at the same time distinguishing them individually by their points of dissimilarity, prepares for one of the highest exercises of our reason ...
>
> (Mayo 1831, p. 63)

This step refines and combines the previous steps through deeper reflection on the objects studied. This is where we should revisit our walk with the GPS and question what it really means to walk. The human body, in movement, with a GPS, brings together all the components of object lessons, and is thus a powerful analytical engine. Many scholars have turned their attention to walking, finding that walking is more than just a way of getting from one place to another —it is also a mode of attention. For instance, the philosopher Frédéric Gros argues that walking exposes you to nature where "everything talks to you, greets you, demands your attention" (2014, p. 54). Others have pointed to the knowledge aspect of walking. In *Wanderlust*, Rebecca Solnit writes how "walking shares with making and working that crucial element of the body and the mind with the world, of knowing the world through the body and the body through the world" (2000, p. 29). Tim Ingold and Jo Vergunst argue that "the movement of walking is itself a way of knowing" (2008, p. 5). Similarly, Tim Edensor calls walking a reflexive practice, "a strategy for capturing particularly treasured views unobtainable by other modes of transport" (2000, pp. 85–86). Finally, Tim Ingold (2008) argues that "inhabitants … make their way through a world-in-information rather than across its preformed surface" (2008, p. 1802). This is just a selection of the large literature that treats walking as a way of developing a particular set of knowledge about a place, experienced through the body.

This opens up the possibility of walking as a deliberate method, a knowledge practice that can be deployed systematically. Scholars, both practitioners and theorists, in a number of disciplines have developed a number of methodological approaches to systematic walking. We can call these protocols, scripts, or algorithms for walking—instructions that encourage serendipity and reflection. Karen O'Rourke (2013) writes about deliberate scripting in art projects (O'Rourke 2013, p. 48), but also the landscape itself has scripts, such as paths, hedges, ditches, and so on. The walker can choose to follow scripts or to do something else, but often at a cost. The most well-known of these is *psychogeography* as a tool for reading a landscape. As Guy Debord (2008 [1955]) describes it, psychogeography should examine "the specific effects of the geographical environment … on the emotions and behavior of individuals" (Debord 2008, p. 23). As with the object lessons, the work of psychogeography is to stimulate a particular form of attention to one's surroundings, often aiming to make visible hidden and unexpected connections.

Could one not say something similar about the potential roles of the GPS in nature-culture-history research? When combined with other ways of sensing and experiencing a landscape as a place composed of physical things, media, and information, the GPS adds another layer of meaning.

Step 5

> The following lessons may be advantageously used as a first exercise in composition. The object should be presented to the children, and they should continue, as before, to make their own observations upon it. Questions should then be addressed to them, calculated to elicit their knowledge of its natural history, and further particulars should afterward be communicated by the teacher, to render their information more complete. After having re-arranged and repeated the matter so obtained, the teacher should examine the class, and require a written account. … It stimulates their attention, furnishes a test of their having well understood the lesson, and leads them to arrange and express their ideas with clearness and facility.
>
> (Mayo 1831, p. 97)

In this final step, Mayo presents object lessons as active and comparative, paying attention to history, context, and materiality, in a processual manner. We have mastered the basics of observation and can now turn our attention to composition. So far, I have talked about nature and landscape in a quite generic way, never actually digging into the tangled histories of actual places. This can change now, at this final stage, where I will tell a story, or rather retell and reinterpret one I have told before (Jørgensen 2014a). This particular case starts during a drive with a GPS unit, which we can read as an example of how the digital plays a critical role in the making of place, and in how we encounter wirelessness in nature.

While driving through the Northern Swedish countryside one summer day a few years ago, a place caught my attention. It appeared suddenly and unexpected on my phone's screen as "The little island in the middle of the lake" in the smartphone app Gowalla, listing nearby points of interest. The app had few such points along this section of the European Route E10 road, in a region with a very low population density and mostly flat stretches of industrial forest plantations. The road connects the industrial city Luleå in Sweden (home of Facebook's massive new data center, cooled by the northern climate and powered by cheap local hydropower) and the picturesque fishing village Å at the very tip of the Lofoten Islands in Norway. On the way there, the road passes through mountains and Arctic tundra, boreal forests and lake-dotted plains.

As we passed through this landscape by car, we were mostly struck by the monotony of it all. Our car was filled to the brim with children and dogs and bags. Halfway through the two-day trek to see family in Norway, my wife was at the wheel. I was fiddling around with my brand new iPhone, which truly seemed a magical product after years with "dumb" Nokia and Sony Ericsson phones. The large touchscreen, the integrated GPS, and the 3G data connection made the phone a window to the world where data and the physical finally coexisted and interacted in a way that felt right to me. It was also a window into the behaviors of other people accessing this world.

Someone—I do not know who or why—had used Gowalla to tag this little island in the middle of the lake as a point of interest, and so the island became interesting to me, mostly because it was so incongruous. The chance meeting with the strange, digital place this island was, existing to me only as a data point, has remained in my mind for years. What did the existence of the island, not as a physical place in the middle of a lake in Northern Sweden, but as a digital place discovered on my phone, really mean?

From the road, places like the little island in the middle of the lake were invisible and undiscoverable. The countryside could seem mostly empty for outsiders and people, like us, passing through on the way to somewhere else. Yet, landscapes are storied, inhabited for centuries by people with shifting and sometimes competing ways of using them. There are traces everywhere. Signs by the road alert drivers to officially designated points of interest such as national parks, war memorials, the massive World Heritage Site, Laponia. The fact that the road signs are also written in Sami reminds the people passing through that we are actually in Sápmi, the traditional lands of the Sami people. Travelers interested in what historian of technology David Nye (1996) calls the technological sublime can observe giant mines, hydropower plants, and the occasional wind turbine—all signs of significant human interventions in the landscape. Rickety moose-hunting towers, animal-crossing signs, and black plastic sacks tied to sticks along the road to indicate the presence of herded reindeer hint at humans and animals co-existing and interacting, no matter how empty the road may seem. Look closely enough, and the land is suddenly very crowded with layers upon layers of history.

Visitors and travelers have been able to learn about these and other points of interest through media such as maps and travel guides for a long time. Throughout the nineteenth century, urban travelers turned the Scandinavian countryside into a tourist destination by exploring, mapping, and writing extensively about what they considered areas worth seeing. Published since 1886, the Swedish Tourist Association's annual book (Svenska turistföreningen 1886) demonstrates how its authors directed visitors to particular places and vistas. For instance, the first volume provided directions to the most accessible glaciers in Lapland. Descending from the Walli mountain toward the Njåtso valley, the book advised, would give you the most spectacular views of massive glaciers. This is one example of how media shapes and directs the attention of travelers to specific sights and locations.

More modern travel guides such as *Lonely Planet* and *Rough Guides* belong in this tradition, but have recently come under strong pressure from digital media. GPS-enabled geosocial networking apps are, in other words, only the latest in a long history of making places shareable and findable through media (Jørgensen 2014b). They should, however, be considered platforms rather than stand-alone products, dependent on crowdsourced information for all but the most iconic points of interest. The GPS-enabled smartphone gives us a dynamic window into this world. While certainly not the only possible view of space and its meanings, Gowalla and other geosocial networking apps opened up the landscape and covered it in data generated by people who had passed by, lived here, experienced these places, and then chosen to share them.

While stories of the GPS and its influence on the experience of landscape generally focus on people being lost or not lost, my encounter with lost little island tells another story of place and location in the era of the GPS. When Gowalla launched in 2009, a little over a year before my road trip, it was one of the first location-based social networks. Designed for GPS-enabled smartphones, Gowalla experimented with game-like features, allowing users to check into "Spots" that could again be organized into "Trips" in different categories. More importantly, users could add locations themselves, so Gowalla's world map was populated as the service gained users around the world. Through Gowalla and competing services such as Foursquare, Yelp, Brightkite, and Findery (just to mention a few) people tagged, ranked, and annotated the world around them, creating and interacting with complex shared representations of the places that mattered to people. These networks function as platforms for sharing and discovery of place and location, processes that generate this meaning.

While beautifully designed and loved by many of its users, Gowalla lost ground against its main competitor (Foursquare) before being purchased by Facebook in 2011. A promised data-export option never materialized, so all the Spots and Trips registered by Gowalla's users were lost (Raymond 2012). All the places that people had found and shared—the places they had made discoverable —were gone. I can never again find the little island in the middle of the lake. The lost little island tells us something about the ephemeral nature of places in the digital world. We know that space becomes place as people imbue it with meaning over time. But places can be unmade, too, with the disappearance of the services that first made us aware of them.

Conclusion: mediation and the digital environmental humanities

When walking with a GPS, we are reminded that the subjects of our studies in nature-culture-history research also experience nature through technology and that they have done so for a long time. Their understanding of the environment is as mediated by technology as ours, though we can never take for granted exactly what this mediation implies. As environmental historians we need to incorporate this technology into our sites of investigation if we are to meaningfully talk about the use and experience of nature in history. Furthermore, if we are to use the GPS as a way of tracing the spread of new media through nature, it should also prompt us to consider old media, and the relationships to nature that formed when old media was new.

While landscapes envisioned, experienced, contested, and constructed, abound in environmental history scholarship, environmental historians have written surprisingly little about actually walking around nature in their research process, with or without GPS units. In his classic article "Doing Environmental History" Donald Worster admonished environmental historians to "get out of parliamentary chambers, out of birthing rooms and factories, get out of doors altogether, and ramble into fields, woods, and the open air. It is time we bought a good set of walking shoes, and we cannot avoid getting some mud on them" (1988, p. 289). We get the impression that this is (or should be) what separates the environmental historian from other types of historians, where putting our bodies in direct contact with nature through walking provides us with some form of privileged knowledge or insights that cannot be acquired from traditional archival studies. Walking might also be a metaphor, a heuristic image, for the kind of work we must do and attention we must exercise if we are to tell stories of nature, history, and what Worster calls "their working intricately together" (ibid. p. 307).

GPS, just like other forms of navigation, is a way of locating ourselves in the world. This means that it is a way of making sense of the world. Both mapping and wayfinding are age-old practices of organizing and imposing order on landscapes. When navigating with or without a GPS, we are interacting with visualizations of data. The ground you walk on is not only grass, stone, soil, asphalt. It has a history and a presence in media beyond the physical.

Environments are simultaneously material and digital, as digital ethnographers Sarah Pink and Larissa Hiorth (2014) argue: "These digital and physical contexts are conjoined by human movement, of persons who weave their ways through both environments within the same sequence or flow of practical ability".

Using a GPS requires technical skills, for sure, and they can go from the very basic to the seriously advanced, all depending on what you want to do and how much time and effort you can devote to it. Like so many other forms of digital humanities research, collaboration with technically skilled people in other disciplines may be a good solution where you, as a humanities scholar, suggests a research question that you then in partnership with others refine and explore through the use of technology.

But it is equally important that we bring our analytical skills to bear on the technologies both we and our historical subjects use. The use and impact of GPS technologies are not something to be taken for granted. We have long traditions in particularly science and technology studies, including history of technology, for critically engaging with technology in this way, though mostly at a theoretical basis. New developments in so-called *critical making* allows for "material forms of engagement with technologies to supplement and extend critical reflection" (Ratto and Ree 2012). What we can label *digital environmental humanities*, the simultaneous use of digital tools and methods for analyzing historical and contemporary natures on the one hand, and the study of how the digital has become part of nature on the other hand, is still a nascent research area. If it is to become a successful field, it needs to embrace this dual nature of the digital, where objects like the GPS become part of nature in the process of telling us about nature.

References

Augé, M. (2009). *Non-Places: An Introduction to Supermodernity*. London: Verso.

Barthes, R. (2013 [1957]) *Mythologies: The Complete Edition, in a New Translation*. New York: Hill and Wang.

Bilefsky, D. (2016, February 4). GPS mix-up brings wrong turn, and celebrity, to an American in Iceland. *The New York Times*. Retrieved from www.nytimes.com/2016/02/05/world/europe/iceland-american-tourist-gps.html.

Bonnel, J. and Fortin, M. (eds) (2014). *Historical GIS Research in Canada*. Calgary: University of Calgary Press.

Bonnett, A. (2014). *Off the Map: Lost Spaces, Invisible Cities, Forgotten Islands, Feral Places and What They Tell Us About the World*. London: Aurum Press.

Calkins, N.A. (1871). *Primary Object Lessons, for Training the Senses and Developing the Faculties of Children*. New York: Harper & Brothers.

Casey, E. (2013). *The Fate of Place: A Philosophical History*. Oakland, CA: University of California Press.

Coyne, R. (2010). *The Tuning of Place: Sociable Spaces and Pervasive Digital Media*. Cambridge, MA: The MIT Press.

Crampton, J. W. (2001). Maps as social constructions: Power, communication and visualization. *Progress in Human Geography*, 25(2), 235–252

Debord, G. (2008 [1955]). Introduction to a critique of urban geography. In Bauder, H. and Mauro, S. E-D. (eds), *Critical Geographies: A Collection of Readings*. Kelowna, BC: Praxis (e) Press.

Easton, R.D. and Frazier, E. F. (2013). *GPS Declassified: From Smart Bombs to Smartphones*. Dulles, VA: Potomac Books.

Edensor, T. (2000). Walking in the British countryside: Reflexivity, embodied practices and ways to escape. *Body & Society*, 6(3–4), 81–106.

Gitelman, L. (ed.) (2013). *"Raw Data" is an Oxymoron*. Cambridge, MA: The MIT Press.

Gros, F. (2014). *A Philosophy of Walking*. London: Verso.

Huth, J.E. (2013). *The Lost Art of Finding Our Way*. Cambridge, MA: Harvard University Press.

Ingold, T. (2008). Bindings against boundaries: Entanglements of life in an open world. *Environment and Planning A*, 40(8), 1796–1810.

Ingold, T. and Vergunst, J. (2008). *Ways of Walking: Ethnography and Practice on Foot*. London: Ashgate.

Jørgensen, F.A. (2014a, June 24). What happens to a place when the data about it is lost? *The Atlantic* online. Retrieved from www.theatlantic.com/technology/archive/2014/06/discontinued-gps-networks-and-the-mystery-islands-that-disappear-with-them/373234.

Jørgensen, F.A. (2014b). The armchair traveler's guide to digital environmental humanities. *Environmental Humanities*, 4, 95–112.

Kember,.S and Zylinska, J. (2012). *Life after New Media: Mediation as a Vital Process*. Cambridge, MA: The MIT Press.

Kingsnorth, P. (2009). *Real England: The Battle Against the Bland*. London: Portobello Books.

Kranzberg, M. (1986). Technology and history: "Kranzberg's Laws". *Technology and Culture*, *27*(3), 544–560.

Latour, B. (2005). From realpolitik to dingpolitik, or how to make things public. In Latour, B. and Weibel, P. (eds) *Making Things Public: Atmospheres of Democracy*. Cambridge, MA: The MIT Press.

Latour, B. (2013). *An Inquiry into Modes of Existence: An Anthropology of the Moderns*. Cambridge, MA: Harvard University Press.

Lovett, W. and Collins, J. (1840). *Chartism; A New Organization of the People, Embracing a Plan for the Education and Improvement of the People Politically and Socially*. London: Watson, Hetherington, and Cleave, 91.

Manaugh, G. (2014, December 27). Glitches in spacetime, frozen into the built environment. *BLDGBLOG*. Retrieved from www.bldgblog.com/2014/12/glitches-in-spacetime-frozen-into-the-built-environment.

Mayo, E. (1831). *Lessons on Objects*. London: R.B. Seeley and W. Burnside.

Milner, G. (2016, February 11). Ignore the GPS. That ocean is not a road. *The New York Times*. Retrieved from www.nytimes.com/2016/02/14/opinion/sunday/ignore-the-gps-that-ocean-is-not-a-road.html.

Nye, D. (1996). *American Technological Sublime*. Cambridge, MA: The MIT Press.

Pink, S. and Hjorth, L. (2014). The digital wayfarer: Reconceptualising camera phone practices in an age of locative media. In Goggin, G. and Hjorth, L. (eds) *The Routledge Companion to Mobile Media*. London: Routledge, 488–498.

Ratto, M., and Ree, R. (2012). Materializing information: 3D printing and social change. *First Monday*, 17(7). Retrieved from http://firstmonday.org/ojs/index.php/fm/article/view/3968/3273.

Raymond, S. [@sco]. (2012, October 7). @nertzy @dwfrank I'm afraid that there aren't any

plans to resurrect it, unfortunately. the servers have all been spun down. sorry. :-/ [tweet] Retrieved from https://twitter.com/sco/status/255014236982505472.
Solnit, R. (2000). *Wanderlust: A History of Walking*. New York: Penguin Books.
Svenska turistföreningen. (1886). *Svenska turistföreningens årsskrift 1886*. Stockholm: R. Almqvist and J. Wiksell's Boktryckeri. Retrieved from http://runeberg.org/stf/1886.
Winner, L. (1993). Upon opening the black box and finding it empty: Social constructivism and the philosophy of technology. *Science, Technology, and Human Values*, *18*(3), 362–378.
Worster, D. (1988). Appendix: Doing environmental history. In Worster, D. (ed.) *The Ends of the Earth: Perspectives on Modern Environmental History*. Cambridge: Cambridge University Press.
Young, A. (2013). Material wisdom: Learning from object lessons. *Cabinet*, 50, 16–19.

23 "But where am I?"

Reflections on digital activism promoting Indigenous People's presence in a Canadian heritage village

L. Anders Sandberg, Martha Stiegman and Jesse Thistle

"But where am I?" This is the question one of us, Jesse Thistle, asked himself when stepping into Black Creek Pioneer Village, a heritage village that celebrates and documents life in a mid-nineteenth century farming community in Toronto, Ontario. Jesse is Métis-Cree and traces his heritage to the Métis Rebellion in 1869, a resistance that was triggered when the newly formed Canadian government, in taking over the Indigenous lands previously granted by the British government to the Hudson's Bay Company, threatened the culture and rights of the local Métis residents.

Jesse, a recent graduate of York University and scholar of Indigenous history in the Greater Toronto Area, is a resident of the Black Creek area that borders the village. The visit prompted Jesse to write, as an assignment for an undergraduate course at York, an open letter to the management of the Village telling how, as a young elementary school child, he had visited and been infatuated by the experience but how he had then come back as a mature student to see the village in a very different light. Most notably, he noticed that Indigenous peoples were not represented in the village. Specifically, he critiqued the Village's British-only foundation narrative of Upper Canada (the colonial name of the province of Ontario) and the traumatic effect that it had on him and on Indigenous people more generally. The letter, however, was not presented to the village administration at the time.

"But where are they?" This is the question that another one of us, Anders Sandberg, a Professor in Environmental Studies at York University, asked as he also noted the absence of Indigenous peoples in Black Creek Pioneer Village. Anders and collaborators have since spoken about this absence on campus walks that they conduct regularly for various groups on and off-campus. These tours are organized under the name the Alternative Campus Tour, a tour that encourages participants to think critically about their everyday surroundings, that gives voice to different and marginalized voices on campus, and that offers students the opportunity to venture out of the classroom (Bardekjian et al. 2013; Sandberg 2013, 2015). Besides a walking component, the tour has a digital form in the shape of a website with text and image-based stories about different sites on the campus. New material is added on a continuous basis to the website (Alternative Campus Tour 2015).

In 2014, Anders heard Jesse read his open letter at a presentation at the university. This was the beginning of a collaboration. Anders felt that Jesse's letter could become a valuable web-based contribution to the Alternative Campus Tour. And they both felt the letter should be presented to the village administration, but were not sure about its exact form or the timing. They were more certain about the basic premise of the project. They wanted to set out to directly challenge Canadian history as it was represented by Black Creek Pioneer Village and to ask why Indigenous people have been left out, what that means, and how it affects Canadians and Indigenous people alike.

They also hoped that the project would contribute to the reassertion of an Indigenous presence at York University, in Toronto, Upper Canada, and Canadian history as a whole, and in doing work to make places like Black Creek Pioneer Village include, rightfully so they felt, Indigenous peoples in the formation narrative of Canada. They also felt that strengthening Indigenous people's inclusion in the metanarrative of Canada would make the public aware of deeper issues like treaties and cohabitation, and the rights and responsibilities Indigenous peoples and Canadians have on Turtle Island (the Indigenous name for North America) in regard to both. It might also open a discussion about the settler responsibilities that derive from the treaty relationships in this part of the country, that bind Indigenous and settler peoples to each other, and the lands we now share. The project then would be part education and part active resistance.

They also resolved that the project should be web-based and have a biographical focus on Jesse's life, as is a customary Indigenous storytelling strategy; they decided it best to work through his experiences as a focus, given the potency of biography. This is where Martha Stiegman, also a professor in the Faculty of Environmental Studies, joined the project. Martha is a videographer with an extensive background working with Indigenous community members. Anders and Jesse asked her to co-develop the script with Jesse and to shoot the video. She would thus support the creation of a site story for the Alternative Campus Tour with not only a written and photographic content but also a video and audio component.

On a broader methodological scale, their project speaks to a number of themes of this book, including the de-colonizing of research, and the affective and digital turns in nature-culture and environmental history research. They were especially interested in using digital methods to convey an affective and artistic message to a broader public and to political and administrative decision-makers rather than confine a discussion to the academy. Their aim in employing a digital-affective method was to highlight "something that moves, that triggers predominantly personal reactions … [and where] [t]he senses, the personal and the body are thus at the forefront …" (Berberich et al. 2015, p. 1). This would in turn aim to defy the dominant language, tropes and narratives of settler society "to enact new forms of expression, production and reception that try, in differing ways, to uncover 'the story-telling of the people to come' and through various 'art practices' intervene in and challenge the 'general distribution of the sensible' and help 're-define the 'social'" (Berberich et al. 2015, p. 40).

Together, the three of us set out to do a three-part project on Black Creek Pioneer Village, all of the three parts calling attention to the erasure of Indigenous history in the village and with the aim of changing this situation. One part is the current chapter, which provides a background, context, and record of the progress of the project as it unfolded and is unfolding. The actual methodology and process of conducting research is thus ad hoc and contingent, where we learn and change our procedure and tactics as we go along and learn from our interactions. Part of this methodology is to consider and reflect on the ethical and strategic issues surrounding our dealings with the Black Creek Pioneer village administration. The other part consists of the two digital activist interventions. One is composed of a short written introduction and several photos of Black Creek Pioneer Village as well as a reproduction of Jesse's open letter. The other is a video that challenges the Village's erasure of Indigenous peoples in its constructed pastoral imaginary by drawing on Jesse's biography as Métis-Cree visiting the village as a young boy and a university student. These items are posted on the Alternative Campus Tour website.

Black Creek Pioneer Village

The mission of Black Creek Pioneer Village, as stated on its website, is to help

> people apply the history of life in the Toronto region to build a better future in their own communities. We invite everyone to join in this exploration. We care for and show a historical collection of value to all, create experience of benefit to all, and tell meaningful stories for all. Our programs and exhibits focus on experiential learning and encourage people to draw connections between the mid-1800s and contemporary life. Our services enhance these connections and encourage diverse usage of the site (www.blackcreek.ca/v2/museum/mission.dot).

Such a mission statement is not unusual for heritage sites. They typically have a nostalgic and moral message that tell us more about the present than the history they purport to describe. In this case, the present denies the place of Indigenous nations, justifies Canadian claims of sovereignty, bolsters the myth of terra nullius, and refuses a space for Indigenous claims for recognition of lands and resources. The distortions and erasures in "pioneer villages" are well described and evidenced in the academic literature. Tivy, for example, says it well: "Museum brochure pictures show happy, healthy people engaged in domestic, agricultural or social activities. These are people the tourist would want to meet, activities we can join" (1993, p. 42) (Figure 23.1). She goes on to state:

> Few museums have challenged this narrative with stories of local contention, desolation or social struggle. Social issues such as industrial

Figure 23.1 The focus in Black Creek Pioneer Village is on the "village folk," the white settlers who formed the core of the community. Indigenous peoples and the knowledge and practices they would have shared with early settlers are absent in these depictions.

Source: Photo courtesy of L. Anders Sandberg.

> labour, cultural or racial conflict, suffrage, poverty, crime and so on are minimal in exhibits in these museums. Instead, one sees, even in local history museums with chronological gallery exhibits, mainly local culture in the form of three dominant groups: pioneers, Victorians, and Indians.
>
> (ibid. p. 38)

Black Creek Pioneer Village falls into this category, though without the "Indians." When exploring the village, we only found a few signs of Indigenous history. In the gift shop, there is the standard assortment of items, such as dream catchers, miniature canoes, jewelry, and books, depicting renditions of "Plains Indian" life and other stereotypical ideas of Indigenous life completely detached from the local context. In the village itself, we found an Inuit doll in one of the displays and a canoe made according to Indigenous peoples' conventions, but with no attribution (see Figure 23.2). At the village's website, there is one page dedicated to the First Nations' presence in southern Ontario, but the information is highly general and is not specific to the Black Creek area. The text, accompanied by a photo of a group of Ojibwa in front of a tipi, points out that Aboriginal peoples have occupied the Great Lakes region for at least 12,000 years, naming the Ojibwa, the Iroquois, the Ottawa, the Huron (Wendat), the Fox and the Mississauga. It also mentions that Aboriginal peoples survived on hunting, gathering, and fishing and that they taught the early settlers about medicinal herbs, sources of food (such as maple sugar and corn), the use of animal hide for clothing, and how to travel on snowshoes or in birch bark canoes

Figure 23.2 There are very few Indigenous artifacts displayed in Black Creek Pioneer Village. This canoe is an exception, built according to Indigenous peoples' principles, but with no attribution

Source: Photo courtesy of L. Anders Sandberg.

(Black Creek Pioneer Village 2015). A more detailed history could, among other things, acknowledge the presence of several pre-eighteenth century Wendat villages a few hundred metres from the Black Creek Pioneer Village's site, the Toronto Carrying Place (a major transportation route across the southern Ontario peninsula) only about three kilometres west of the village, the land claims against the area by both the Mississaugas of the New Credit and the Six Nations of the Grand River, the past occupancy of other First Nations groups, and the growing Indigenous presence among institutions, faculty, staff and students on the York University Campus in the present (Sandberg, 2013, 2015).

Overall, the Village resembles what Jean Baudrillard has called a simulacrum, a mirror image of something that never existed. Journalist John Bentley Mays alludes to this situation in the following:

> Black Creek is annoying, because of everything from the cloying sweetness and phony rusticity of the "village" itself to the vast, junk-crammed boutique and fries-and-burger outlet masked by all that sentimental, schmoozy exaltation. But it is also rich in delights for the connoisseur of wicked ironies. The buildings so diligently hauled to Toronto and carefully set up here to evoke nostalgia for a "harder, simpler" time, for instance, were abandoned precisely because the hard lifestyle to which they belonged was being made obsolete by Ontario's swift modernization and consumer revolution in late Victorian times.
>
> (Mays 1992)

In academic and private conversations, village staff and administrators readily admit to the selective and romanticized rendition of village life but rationalize it on the basis of maintaining its revenue stream and serving a public demand. The visitors, the majority of whom are school children, families with young children, and pensioners, so the argument goes, favour a sanitized and partial view of history.

On the basis of the candor of the Village staff, who admitted to the partial and distorted stories told about mid-nineteenth century rural life in Ontario, we decided to first explore whether there was any aspiration within the administration to expand its mandate to explore in more detail the presence of the original inhabitants in the area. We felt that such support might help if we were to put pressure on the institution to change its mandate, and perhaps also help us shape how that pressure might be applied. We were inspired by such a tactic by President Franklin Roosevelt who allegedly responded to one lobby group: "Ok, you have convinced me, now go out and put pressure on me to change." However, we found very little sympathy for such change. The staff and administration's standard and rehearsed answer is that the village's mandate is not to cover the original inhabitants and, besides, so their argument continues, there were no Indians in the area at the time of European settlement.

There are several problems with such an argument. To begin with, it is typically based on written sources, such as the census and legal documents, where Indigenous peoples are notably absent. But that does not provide conclusive evidence that Indigenous peoples did not travel, camp, extract sustenance, or work in the area seasonally or temporarily as has been documented in other places. Their presumed absence, in other words, is a question for closer investigation not only in the oral record of Indigenous peoples themselves but perhaps also in the archival record, such as newspapers, settlers' diaries, the correspondence of the Department of Indian Affairs, and other forms of daily records.

The bias towards acknowledging a permanent rather than temporary presence of "settlers" is by no means only a characteristic of Indigenous peoples in Black Creek Pioneer Village. It is also clearly shown in a small exhibit on Black settlers' history in the area that is displayed in the village. Once again, the census constitutes the basis for establishing Black people's presence. In addition, the narration biases the recording of "settlers" to those who fit into the "settlers' mold", as illustrated, for example, by the statement: "Like people from other ethnic groups, people of African descent worked in a variety of jobs. They were farmers, labourers, artisans, home makers, and professionals." In such schemes that recognize presence only in censuses, property records, and permanent residency, there is no allowance for documenting Black people who occupied the area temporarily or informally. In a similar way, Indigenous peoples who moved seasonally through the area to work, sell products of their own making, or to hunt, fish, or gather, are erased. We hypothesize that if equal resources had been put into researching the Indigenous presence in the area as had been afforded to the settler presence (as, for example, the history of one single barn (TRCA, 2015)), the Black Creek Pioneer Village of the 1860s may have looked very different.

It is also, we feel, wrong to rigidly hold onto the mandate of the village to only depict life in the 1860s. The artifacts and displays in the village in fact stray from this rigid mandate on many occasions, including the inclusion of buildings from other periods and showing long-range histories of other ethnic groups in the area. By contrast, there are few references to First Peoples, including any mention of the nearby remnants of the pre-eighteenth century Wendat communities, the series of land and resource claims to the area that Mississauga and Haudenasaunee nations have asserted and that Jesse mentions in his appearance in our video, and the recognition of a growing Indigenous people's presence in Toronto today (Sandberg 2013, 2015).[1]

But more broadly and fundamentally, the Black Creek Pioneer Village narrative speaks against persistent efforts to make connections between historic treaties and other agreements between settler society and Indigenous peoples. These agreements have relevance for how we live together in the present: how we embody the spirit and intent of those agreements (despite their being undermined by colonial government interpretation and application), and how we use them as a framework to work towards reconciliation. One of the things that struck us in reading the village mandate—to "encourage people to draw connections between the mid-1800s and contemporary life"—is that the version of history presented at the village encourages people do the very opposite. This is particularly disturbing at a time when there are at least some efforts to address the historical injustices inflicted on Indigenous Peoples in Canada (see, for example, Truth and Reconciliation Commission of Canada 2015).

Pursuing the project

Given the seemingly entrenched view on the exclusion of First Nations histories and present realities in the village administration, and the negative reaction our challenge might pose, we decided that our very first task should be to seek official permission to film *Thistle in the village*, a process we started on 16 July 2015. This proved more difficult than we anticipated. In spite of the fact that our project lacked a budget, was non-obtrusive (our plan was to have Martha filming Jesse for a couple of hours, a plan that changed only slightly at the time of filming), and was non-commercial, the village administration insisted we go through the formal application process that is standard for the dealings with large-scale commercial productions of the Hollywood variety. This involved providing evidence of insurance and workers' compensation coverage for the principals and having the Village administration review the film script in advance of filming. Other provisions in the contract stipulated that we were not allowed to interview the public or village staff. We also had to provide a $300 security deposit and pay a $150 hourly fee for filming, a rate that was half the regular rate, a concession made by the village administration in recognition of the small scale and low budget of our project.

We recite this process not because we think that the village administration wanted to scuttle the project, but to show the disciplining that typically takes place through formal rules and regulations within an institutional setting that make difficult any challenges to its everyday routine activities. On 19 November

2015 our project was approved and on 20 November 2015 we filmed inside the village. Besides featuring Jesse, the video also contains one jingle dancer and one grass dancer, both in full regalia, whose performance disrupted the dominant narrative told in the village (Figure 23.3).

The video challenges the version of history presented at the village, and its erasure of Indigenous peoples. We open with images of the village itself: a woman passes across the screen dressed in a 1850s bonnet, shawl and full skirt. Sheep bleat, fenced in their pasture. A young native boy, roughly ten years old, watches in amazement as a tinsmith crafts a small tool. Jesse's voice comes in, remembering his first visit to the site as a schoolboy, and his fascination with the craftsmanship on display. "The leather workers, blacksmith, print shop workers—they all mesmerized me," he recalls. "My brothers and I played pioneers for weeks after that visit".

Jesse links the spark that trip ignited in him, to his now-love of history and his recent position as a historian at York University where he is tracing the road of his people, the Cree/Métis Road Allowance community of Park Valley Saskatchewan. In the video we move from images of York campus and a defiant Jesse, facing the camera in front of York's Vari Hall, to a series of images from his Métis community. "I trace my road back to Louis Riel and the Battle at Batoche" he asserts, over a series of archival photos of his ancestors who fought as part of the Riel Resistance: Jesse's great uncles, the Arcane brothers who fought there as child soldiers, and his great grandmother, Marianne LeDoux, who would spend the rest of her life traumatized, hiding in the bush.

Figure 23.3 Anishinabeg-Metis jingle dancer Wenzdae Brewster-Dimaline disrupts the settler narrative in Black Creek Pioneer Village

Source: Photo courtesy of Martha Stiegman.

Jesse's current research documents the effects of inter-generational trauma, and links the trauma his family would have faced to the years he spent homeless on the streets of Toronto. He is researching the impacts on his great grandfather, St. Pierre Arcand, who fought at Batoche as a child soldier, the impacts on his people of being hunted as military enemies by the Royal Canadian Mounted Police, and the breakdown of *wahkootiwin* (Cree for kinship) bonds of trust and family that held Métis-Cree society together prior to 1885. With the Riel Resistance, his people went from a lucrative status as middlemen in the fur trade to a landless people, spending generations in exile. Forced off their traditional territory and dispossessed of land from the so-called scrip process instituted by the new Canadian government, the only place they were able to establish somewhat permanent settlement were on the ten meters reserved on the side of public roads and railways for maintenance, and in between square block settlements, thus the name "Roadside Allowance". In the video, Jesse remembers the parting words of his grandfather, or Mushum, when he left at age three with his father and brothers for Toronto: "Don't tell them you're Métis or they'll hang you."

Jesse spent years homeless on the streets of Toronto, an experience he links, in part, to shame he felt about his native heritage—a shame to which the history presented at Black Creek Pioneer Village, echoed by a thousand daily erasures of Indigenous presence, contributed. If this version of history that glorifies pioneer triumph over terra nullius is justified as a "simplified" or "sanitized" account, Jesse's experience demands viewers consider for whom that narrative is experienced as sanitized, and for whom this account is in fact quite bloody. The story presented at the village is a story with very real, and violent, consequences.

An urban metropolis like Toronto might be viewed as an endgame of settler colonialism. It is a space where the land itself has been erased and buried under concrete, a space where all traces of Indigenous history and presence seems to have been removed—a canvas for the postmodern multicultural Canadian project. It is at the heart of this concrete jungle our video asserts Indigenous history and presence. "This land is Anishnabee territory. It's Huron-Wendat territory. Haudenosaunee territory" Jesse reminds us, over images of spaces where this urban triumph feels the most complete. And yet these are places that have never ceased to be of deep significance for the original peoples of this land (Freeman 2010; Johnson 2013, 2015; Woodsworth 2012). The four-storey-high billboards and video screens at Dundas Square lie over Yonge Street, a road that was built by the early citizens of colonial York and the Northwest Company (a fur trading enterprise based in Montreal) to replace the north-south axis of the Toronto Carrying Place along the Humber River, which is why Yonge Street goes from Lake Ontario to the city of Barrie and beyond, into the Canadian Northwest—roughly the same path the Toronto Carrying Place took (Robinson 1933, 1943). The crowds of pedestrians on Spadina Road cross an ancient pathway Ishpadina, an Ansihnabee word referring to the rise in the land, sloping up from the shoreline of Lake Ontario. The CN Tower framed by the Toronto skyline, viewed from the Toronto islands, is a place of deep spiritual significance for Mississauga people, and contested in terms of the treaty of the Mississauga, the First Nations inhabitants of the Toronto region at the time of British Loyalist immigrations in the 1760s to the 1780s. The Mississauga have contended for over 200 years that they did not relinquish the island in the Toronto "Purchases" of 1787, 1788, or 1805, but only the territory stretching inland, back from the shore line.

While listing the Indigenous and historical relevance of these sites in the film, Jesse says, "That's the version of history I'm here to assert." Indeed, the history Jesse is "revising", the deeper foundational Indigenous history of Toronto, has been erased by concrete and time, but more so by the settler narrative Canada uses to justify the often-hostile takeover of Indigenous lands and territory.

Conclusion

The digital revolution provides endless opportunities for academics conducting nature-culture and environmental history research to communicate their messages in different ways to different audiences. At York University, we have for many years conducted an Alternative Campus Tour for students, faculty, staff, community neighbours and the general public. The tour is advertised and available digitally on the World Wide Web. One of us has argued elsewhere that the tour can be a critical and subversive pedagogical tool for students as well as a means to establish closer ties with surrounding communities. It is also a way of conducting an affective and embodied learning of the everyday, learning not only from a text and a voice, but from walking, experiencing, enjoying and enduring a place with your full body (Sandberg 2015). In this context, we argue that a digital record can effectively illustrate the lack of representation of First Nations both in the past and present on and near the campus. The account draws in particular on a personal narrative of one of us, using an individual experience and associated emotions, everyday scenes, and an interview and dance to disrupt the dominant narrative of a settler pioneer village that neighbours the university campus. Our ultimate aim is to rally support that challenges and convinces local politicians and the village administration to include a fair representation of First Peoples in the region.

Note

1 Some examples of wider Indigenous presence in Toronto around the mid-1800s include: Indigenous warriors playing a prominent role in the defense of Fort York against Americans in 1813; the Mississaugas living at the Credit River reserve from 1826–1847; Indigenous peoples selling crafts and other goods at Canadian National Exhibition and St. Lawrence Market in the mid-1800s; the Mississaugas protesting to the Duke of Newcastle in 1860 that their council grounds at the site of Centre for Addiction and Mental Health were still owned by them; Oronhyatekha (Peter Martin) finishing his medical degree at the University of Toronto in 1864; Peter Jones (Kahkewaquonaby) writing an article in *Toronto's Daily Mail* in 1875 to defend Indigenous peoples against negative portrayals as 'savages'; Buffalo Bill's Wild West Show (featuring many Indigenous leaders and performers) being held at Woodbine Park in 1885; E. Pauline Johnson performing in Toronto towards the end of nineteenth century; and at least one Indigenous family having a camp on the Rouge River until about 1915. We are indebted to Jon Johnson for this list.

References

Alternative Campus Tour. (2015). Accessed 1 October 2015 at http://alternativecampustour.info.yorku.ca.

Black Creek Pioneer Village. (2015). Aboriginal people. Accessed 15 December 2015 at www.blackcreek.ca/for-kids/homework-help/aboriginal-people.dot.

Bentley Mays, J. (1992). Toronto's 'pioneer village' is no village at all, but a series of non-nostalgic bits and pieces put together in a nostalgic way, *Globe & Mail*, 16 December, p. C1.

Berberich, C., Campbell, N. and Hudson, R. eds. (2015). Introduction: Affective landscapes. In Berberich, C., Campbell, N. and Hudson, R., eds. *Affective Landscapes in Literature, Art and Everyday Life: Memory, Place and Senses* (1–10). Aldershot: Ashgate.

Bardekjian, A., Classens, M. and Sandberg, L.A. (2013). Reading the urban landscape: The case of a campus tour at York University, Toronto, Ontario, Canada. *Journal of Environmental Studies and Sciences*, 2(3), 249–56.

Freeman, V. (2010). 'Toronto has no history!' Indigeneity, settler colonialism and historical memory in Canada's largest city. PhD dissertation, University of Toronto, Canada.

Johnson, J. (2013). The Indigenous environmental history of downtown Toronto. In Sandberg, L. A., Bocking, S., Coates, C. and Cruickshank, K., eds. *Urban Explorations: Environmental Histories of the Toronto Region* (58–71). Hamilton, ON: L.R. Wilson Centre for Canadian Studies.

Johnson, J. (2015). Pathways to the eighth fire: Indigenous knowledge and storytelling in Toronto. PhD dissertation, York University, Toronto, Canada.

Robinson, P. J. (1933). *Toronto During the French Regime: A History of the Toronto Region from Brule to Simcoe 1615–1793*. Toronto: University of Toronto Press..

Robinson, P. J. (1943). Yonge Street and the Northwest Company. *Canadian Historical Review*. XXIV, 253–265.

Sandberg, L. A. (2013). On stockades and bridges: From York University to the Jane and Finch neighbourhood. In Sandberg, L.A., Coates, C. and Cruikshank, K., eds, *Urban Explorations: Environmental Histories of the Toronto Region* (97–120). Hamilton, ON: L.R. Wilson Centre for Canadian History, McMaster University.

Sandberg, L.A. (2015). Subverting the enterprise university: The case of the alternative campus tour at York University, Toronto, Canada. *International Studies in Widening Participation*, 2(2), 12–19.

TRCA. (2015). The Dalziel Barn. Accessed 25 September 2015 at www.dalzielbarn.com/home.html.

Truth and Reconciliation Commission of Canada (2015). Honouring the truth, reconciling the future. Summary of the Final Report of the Truth and Reconciliation Commission of Canada. Accessed 16 December 2015 at www.myrobust.com/websites/trcinstitution/File/Reports/Executive_Summary_English_Web.pdf.

Tivy, M. (1993). Museums, visitors and the reconstruction of the past in Ontario. *Material Culture Review*, 37(1), 35–51.

Woodworth, B. Raweno:kwas. (2012). Urban place as an expression of the ancestors. In Ingrid Leman Stefanovic and Stephen Bede Scharper, eds, *The Natural City: Re-Envisioning the Built Environment* (223–230). Toronto: University of Toronto Press.

Index

For Product Safety Concerns and Information please contact our EU representative GPSR@taylorandfrancis.com Taylor & Francis Verlag GmbH, Kaufingerstraße 24, 80331 München, Germany